INDUSTRY 5.0

Innovation and Adoption for Sustainability

Electronic Commerce Management for Business

INDUSTRY 5.0

Innovation and Adoption for Sustainability

Tilottama Singh, PhD
Pradeep Suri, PhD
Zulkarin Jahangir, PhD
Editors

APPLE ACADEMIC PRESS

First edition published 2027

Apple Academic Press Inc.
1265 Goldenrod Circle, NE,
Palm Bay, FL 32905 USA

760 Laurentian Drive, Unit 19,
Burlington, ON L7N 0A4, Canada

CRC Press
2385 NW Executive Center Drive,
Suite 320, Boca Raton FL 33431

4 Park Square, Milton Park,
Abingdon, Oxon, OX14 4RN UK

Library and Archives Canada Cataloguing in Publication

Title: Industry 5.0 : innovation and adoption for sustainability / Tilottama Singh, PhD, Pradeep Suri, PhD, Zulkarin Jahangir, PhD, editors.
Other titles: Industry 5.0 (Palm Bay. Fla.)
Names: Singh, Tilottama, editor. | Suri, Pradeep, editor. | Jahangir, Zulkarin, editor.
Description: First edition. | Series statement: Electronic commerce management for business | Includes bibliographical references and index.
Identifiers: Canadiana (print) 20260143464 | Canadiana (ebook) 20260143537 | ISBN 9781779641861 (hardcover) | ISBN 9781779641878 (ebook)
Subjects: LCSH: Industry 5.0—Social aspects. | LCSH: Social responsibility of business. | LCSH: Sustainable development. | LCSH: Technological innovations—Environmental aspects.
Classification: LCC T14.5 .I53 2026 | DDC 303.48/3—dc23

Library of Congress Cataloging-in-Publication Data

Names: Singh, Tilottama editor | Suri, Pradeep editor | Jahangir, Zulkarin editor

Title: Industry 5.0 : innovation and adoption for sustainability / Tilottama Singh, PhD, Pradeep Suri, PhD, Zulkarin Jahangir, PhD, editors.

Description: First edition. | Palm bay, FL, USA : Apple Academic Press, Inc. ; Boca Raton, FL : CRC Press, 2026. | Series: Electronic commerce management for business | Includes bibliographical references and index. | Summary: "Industry 5.0 is not merely a continuation of its predecessor, Industry 4.0, but rather a transformative vision that intertwines human-centric innovation with environmental viability. Unlike Industry 4.0, which predominantly focused on automation and digitization, Industry 5.0 emphasizes a synergistic relationship between technology, ecology, and societal well-being. It envisions a future where progress is not measured solely by productivity but by the collaborative incorporation of human intelligence, artificial intelligence, and sustainable practices. This new book, Industry 5.0: Innovation and Adoption for Sustainability, explores these concepts, offering an in-depth discussion of how Industry 5.0 changes industries and communities while addressing global concerns including resource depletion, social injustice, and climate change. It emphasizes how essential Industry 5.0 is to balancing humanity's responsibility to the environment with its demand for innovative practices. Through detailed discourse, the book covers a broad range of topics, ranging from corporate social responsibility as a foundation for a better future to the profound impact of blockchain, artificial intelligence (AI), and Lean Six Sigma methodologies in driving efficiency. The book's emphasis on the sociological effects of Industry 5.0 is one of its most distinctive features. Today, we need to address the practices of how communities could be empowered by this worldview, change financial ecosystems, and provide comprehensive setup that support the Sustainable Development Goals (SDGs) of the UN. The chapters cover progressive initiatives like digital entrepreneurship, which makes accessible opportunities while addressing ecological challenges, and sustainable tourism, which reflects shifting traveller preferences toward eco-friendly experiences. The amalgamation of trends in the banking, agricultural, and hospitality sectors illustrates the versatility of Industry 5.0 in driving sustainable development across diverse domains. Strengthened with case studies, research analyses, and proactive conceptual models, this volume offers a valuable repository of actionable insights for academics, business leaders, and policymakers. From redefining operational strategies to enhancing decision-making processes, the book indicates how technological enablers can be utilized to create a balanced and durable global economy. The diverse topics in the book include sustainability and corporate social responsibility, the ecological effects of digital entrepreneurship, a Lean Six Sigma approach toward industry 5.0, finance with focus on sustainability, the integration of artificial intelligence in human resource management, enhanced digital banking services and environmental sustainability, artificial intelligence in the restaurant industry, workplace spirituality and organizational performance, and more. Examining the social impacts of e-commerce in the contemporary business environment in aspects of sustainability, technology, finance, and socio-economic consequences, and emphasizing the coordination of modernization with environmental management, ethical governance, and human-centric design, this book invites readers to anticipate a future where progress does not come at the expense of global well-being"-- Provided by publisher.

Identifiers: LCCN 2026011864 (print) | LCCN 2026011865 (ebook) | ISBN 9781779641861 hbk | ISBN 9781779641878 ebk

Subjects: LCSH: Industry 5.0--Social aspects | Industry 5.0--Economic aspects | Sustainability | Artificial intelligence--Industrial applications | Sustainable Development Goals (Project)

Classification: LCC T59.6 .I483 2026 (print) | LCC T59.6 (ebook)

LC record available at https://lccn.loc.gov/2026011864

LC ebook record available at https://lccn.loc.gov/2026011865

ISBN: 978-1-77964-186-1 (hbk)
ISBN: 978-1-77964-187-8 (ebk)

DOI: 10.1201/9781779641878

About the Book Series Electronic Commerce Management for Business

Editors-in-Chief:

Dr. Vikas Garg
Associate Professor, Symbiosis Institute of Business Management, Noida, Symbiosis International (Deemed University), Pune, India

Dr. Richa Goel
Symbiosis Centre for Management Studies, Noida, India
Symbiosis International (Deemed University), Pune, India

Electronic trading, typically called e-commerce, involves procurement and distribution of electronic systems and other digital networks of goods or services. E-commerce stands for electronic commerce and concerns the electronic medium trade in goods and services. As the Internet grew, the volume of online trade has increased significantly. B2B, B2C, C2C, and related resources help to build online technology for future customer needs and markets. E-commerce has revolutionized industry, disrupted trade with the Internet, and created an e-commerce market for customers and enterprises through the electronic communications system.

The advances in internet and web-based technology have steadily eroded boundaries between traditional economies and the new online economy, including the amount of corporate money. The growth in e-commerce is enormous. Another explanation for this is the low cost of the PC and the rising use of the Internet. The corporate community is rapidly conscious of the possibilities provided by e-commerce.

This series mainly discusses the role of e-commerce in today's business. It further highlights some critical issues in e-commerce, suggestions, and future strategies for e-commerce in years to come.

Topics of Books in the Series:

The primary endeavor of this series is to introduce and explore contemporary research developments in a variety of rapidly growing research areas. The volumes deal with the following topics but not limited to:

- E-Business and Mobile Commerce
- Big Data Analytics
- Business and Data Analytics
- Implications of Technological Disruption
- Technology and the War for Talent
- Impact of New Technologies on Business
- Artificial Intelligence and Its Impact on Business
- Managing Global Competitiveness Using Technology
- Impact of New Technology on Global Labor Markets
- Blockchain in International Business
- IoT and Smart Cities
- IoT Strategies and Business models
- Industrial Internet of Things (IoT)
- AR (Augmented Reality) in Product Design and Manufacturing
- AR (Augmented Reality) and Training
- Use of AR (Augmented Reality) in Sales and Marketing
- Smart Manufacturing
- Cryptocurrency
- Strategies of Emerging Countries in the Era of New Technology
- The Roadmap to Thriving in the New Skills Economy
- Future of Mobile Apps
- Social Media—A Powerful Tool
- Big Data Powers Organizations
- Integration of Technology in HR Practices
- Skillset for Industry 4.0
- Managing Virtual Teams and Environments for Innovation
- Innovation in Business
- Effects of Shifts in Globalization on Emerging Economies
- Smart Global Supply Chain Management
- Corporate Governance in the New Global Technology-Driven Economy
- Business Intelligence Tools & Applications
- Cyber Defense & Information Technology
- Cloud Computing
- Data & Intelligent System Modeling
- Digital Marketing & Internet Marketing
- Smart Connected Products
- Impact of Technology on Financial Decisions
- Finance and Risk Management for Using New Technology

- Digital Transformation and Financial Services
- Banking services in the digital era
- Virtual Reality
- Robotics and Its Use in Business
- Industry 4.0 and Labor Issues

The series serves as a valuable resource for researchers, academicians, and industry professionals and undergraduate, postgraduate, doctoral students in electronic commerce, Internet marketing, business innovation, digitization, technology in product design, management, and more.

Features of the volumes under this series explore recent trends, model extensions, developments, solutions to real-time problems, and case studies.

BOOKS IN THE SERIES

Applications of Artificial Intelligence in Business and Finance 5.0
Editors: Richa Goel, PhD, Vikas Garg, PhD, and Michela Floris, PhD

Blockchain Technology: Unlocking the Value of 3Ts in Business World
Editors: Richa Chauhan, PhD, and Vidhi Kaul, PhD

Entrepreneurship and Sustainable Business Development
Editors: Ruchi Khandelwal, PhD, Deepa Kapoor, PhD, Nidhi Gupta, PhD, and Neha Gupta, PhD

The Role of Smart Technology and Industry 4.0 in Shaping Smart Cities
Editors: Vinay Kandpal, PhD, Manoj Kumar Nallapaneni, PhD, and Deep Chandra, PhD

Fintech and Blockchain Technology in Financial Markets
Editors: Neha Puri, PhD, Anjali Munde, PhD, and Emilia Alayerdov, PhD

E-Financial Strategies for Sustainable Development: A Handbook for Promoting Financial Inclusion and Ending Poverty (SDG 1)
Editors: Ernesto D. R. Santibanez Gonzalez, PhD, Rajeev Sijaria, PhD, and Vikas Garg, PhD

Green Finance and Sustainable Development Goals
Editors: Vikas Garg, PhD, Krishnendu Ghosh, PhD, and Ronita Ram, PhD

Industry 5.0: Innovation and Adoption for Sustainability
Editors: Tilottama Singh, PhD, Pradeep Suri , PhD, and
Zulkarin Jahangir, PhD

Global Sustainable Development: Integration of Digital and Human-Centric Pathways
Editors: Kavita Indapurkar, PhD, Neeru Sidana, PhD, and
Iqbal Thonse Hawaldar, PhD

About the Editors

Dr. Tilottama Singh, PhD

Tilottama Singh, PhD, is a certified HR analyst from the Indian Institutes of Management (IIM) and a proficient academic, researcher, and trainer with more than 15 years of experience in the field of emotions, spirituality, and anthropology. Presently working as Head of the Department in the School of Management at the Uttaranchal Institute of Management, Uttaranchal University, Dehradun, India, she has been named Teacher of the Year Award 2023 by the government for her exemplary contribution in the field of building innovative learning pedagogy and education welfare initiatives in the state. Her areas of expertise are human behavior, emotions, and psychological testing. She started her career in the hotel industry with Leela Kempinski in the corporate head office of human resources and training. Later, she joined the education sector with the topmost universities, such as IMS Unison Group and Amity University, the leading university in Pan Asia, with her versatile international exposure in industry and academia. Having an enriched research portfolio with Scopus and international refereed journals, book chapters, and conference presentations, she has received best presentation awards at international conferences like IEOM, MIDAS, and many more. She also serves as a member of AIMA and acts to lead and liaise between the student community and the industry delegates, with a keen interest in training and building sustainable models for business and society. A prolific EDP trainer, she has organized many international conferences and workshops. Having more than seven books of international repute and acting as the series editor for top publishers like Emerald and Taylor & Francis, her work involves a deep dive into sustainable development goals (SDGs) and policy suggestions for circular economies.

Dr. Pradeep Suri, PhD

Prof. (Dr.) Pradeep Suri, PhD, is well-known in the management faculty fraternity. He is presently Dean UIM (Management Department) of Uttaranchal University, Dehradun (NAAC –A+), India. He has qualified for the UGC NET examination in Management and holds a PhD in Commerce and Management. In addition, he has MBA and PGDBM degrees in Finance, and he has also been qualified by the Institute of Cost Accountants of India (ICWAI). He has a total work experience of 29 years, including four years in industry. He has worked at various institutions of national and international repute. He has written many research papers, including has published 18 papers in Scopus-indexed journals and several books. He is actively engaged in consulting assignments, and to date, he has done consulting for five different organizations under the flagship of Uttaranchal University. He has already done some projects of national repute in full, and some on the verge of completion. He has successfully completed many faculty development programs (FDPs), management development programs, and skills development programs. Last year he had completed an international FDP for the delegates of the Republic of Tanzania on "Price Capping." In addition to the above, he is very active in the field of portfolio management and also is managing portfolios for his clients in his personal capacity. He has written five study materials on the subjects of finance and accounting for the University of Dehradun. He has two books to his credit on the core area of finance. He is a currently guiding seven students for the award of PhD, and five have been successfully awarded with PhD degree under his able guidance. He has organized and attended many conferences, seminars, workshops, and symposiums at both national and international levels.

Dr. Zulkarin Jahangir, PhD

Zulkarin Jahangir, PhD, is a specialized research professional with experience of over 20 years with universities, international NGOs, and social enterprises. Presently, he is associated with United Nations Development Programme (UNDP) projects. He is a faculty member in the School of Management in North South University in Bangladesh. He has extensive experience in collaborating research with global organizations such as the United Nations High Commissioner for Refugees (UNHCR), United States Agency for International Development (USAID), and World Food Programme to develop policy and advocacy plans. He also has significant experience as an academic and expert in the emerging subdiscipline of technology in international development. Having sound research and conference organizing experience, he is an expert on practical strategies to address the UN's Sustainable Development Goals at the global level.

Contents

Contributors

Shipra Agarwal
Department of Commerce, Graphic Era Deemed to be University Dehradun, Uttarakhand, India

Megha Aggarwal
Uttaranchal Institute of Management, Uttaranchal University, Dehradun, India

Vasim Ahmad
Uttaranchal Institute of Management, Uttaranchal University, Dehradun, India

Mohammed Arif Hussain
GITAM School of Business, GITAM University, Hyderabad, Telangana, India

Manish Badoni
Uttaranchal School of Hotel & Hospitality Management, Uttaranchal University, Dehradun, India

Sangeeta Bagga
Department of Commerce, Graphic Era Hill University, Dehradun, Uttarakhand, India

Abhilasha Chauhan
Assistant Professor, Jaypee Business School, Jaypee University of Information Technology, Waknaghat – Solan, Himachal Pradesh, India

Priyanshu Chauhan
Graphic Era Deemed to be University, Dehradun, Uttarakhand, India

Mamta Chawla
Amity Global Business School, Amity University, Noida, Uttar Pradesh, India

Rituparna Choudhury
School of Business, University of Petroleum and Energy Studies (UPES), Dehradun, Uttarakhand, India

Sayonee Das
Uttaranchal University, Dehradun, Uttarakhand, India

Himani Grewal
Department of Management, Moradabad Institute of Technology, Moradabad, Uttar Pradesh, India

Sumeet Gupta
School of Business, University of Petroleum and Energy Studies (UPES), Dehradun, Uttarakhand, India

Vijay Kumar Jain
Department of Management Studies, DIT University, Dehradun, Uttarakhand, India

Ruchitra Joshi Kaparwan
Department of Commerce, Graphic Era Hill University, Dehradun, Uttarakhand, India

Vanshika Kakkar
Operational Risk & Analytics Analyst, BNY, India

Bipin Chandra Kandpal
Uttaranchal Institute of Management, Uttaranchal University, Dehradun, Uttarakhand, India

Rakesh Kumar
Department of Management, Uttaranchal Institute of Management, Dehradun, Uttarakhand, India

Rakesh Kumar
Uttaranchal University, Premnagar, Dehradun, Uttarakhand, India

Pankaj Kumar
Department of Management Studies, DIT University, Dehradun, Uttarakhand, India

Sayali Santosh Mahadik
Department of Computer Engineering, ABMSP's Anantrao Pawar College of Engineering and Research, Pune, Maharashtra, India

Rupa Khanna Malhotra
Department of Commerce, Graphic Era Deemed to be University, Dehradun, Uttarakhand, India

Satish Chandra Mishra
Department of Management Studies, DIT University, Dehradun, Uttarakhand, India

Jitendra Musale
Department of Computer Engineering, ABMSP's Anantrao Pawar College of Engineering and Research, Pune, India

Arvind Sanjay Musale
Department of Computer Engineering, ABMSP's Anantrao Pawar College of Engineering and Research, Pune, Maharashtra, India

Sambhaji Nawale
Department of Computer Engineering, ABMSP's Anantrao Pawar College of Engineering and Research, Pune, Maharashtra, India

Divya Negi Ghai
School of Management and Commerce Studies, Shri Guru Ram Rai University, Dehradun, Uttarakhand, India

Babita Rawat
Uttaranchal Institute of Management, Uttaranchal University, Dehradun, Uttarakhand, India

Deepak Sahni
School of Management and Commerce Studies, Shri Guru Ram Rai University, Dehradun, Uttarakhand, India

Priti Sharma
Department of Commerce, Graphic Era Deemed to be University, Dehradun, Uttarakhand, India

Rajeshwari Sharma
Associate Professor, IQ City United World School of Business, Kolkata, West Bengal, India

Himanshi Sharma
Assistant Professor, Graphic Era deemed to be University, Uttarakhand, India

Preeti Sharma
Department of Management Studies, DIT University, Dehradun, Uttarakhand, India

Asmita Shivaji Deore
Department of Computer Engineering, ABMSP's Anantrao Pawar College of Engineering and Research, Pune, Maharashtra, India

Shrish Singh
School of Leadership and Management (SLM), Manav Rachna International Institute of Research and Studies, Faridabad, India

Aditi Singhal
Department of Management Studies, DIT University, Dehradun, Uttarakhand, India

Ekta Sodha
Uttaranchal Institute of Management, Uttaranchal University, Dehradun, Uttarakhand, India

Satyam Prakash Srivastava
Graphic Era Deemed to be University, Dehradun, Uttarakhand, India

Riktesh Srivastav
City University, Ajman, United Arab Emirates

Vaishnavi Sandeep Tilekar
Department of Computer Engineering, ABMSP's Anantrao Pawar College of Engineering and Research, Pune, Maharashtra, India

Tanuja Tomer
Assistant Professor, School of Business Studies, Jigyasa University, Dehradun, India

Himani Upreti
Department of Commerce, Graphic Era Deemed to be University Dehradun, Uttarakhand, India

Sudha Vemaraju
GITAM School of Business, GITAM University, Hyderabad, Telangana, India

Abbreviations

AARs	Average Abnormal Returns
AI	artificial intelligence
ANOVA	analysis of variance
API	application programming interface
AVE	average variance extracted
CAARs	Cumulative Average Abnormal Returns
CL	customer loyalty
COBOT	Co-operative Robot
CPPS	Cyber-Physical Production System
CR	composite reliability
CS	customer satisfaction
CSR	corporate social responsibility
DMAIC	Define-Measure-Analyze-Improve-Control
EE	employee engagement
EFA	exploratory factor analysis
EPR	extended producer responsibility
ESG	environmental, society, and governance
FLCs	financial literacy centers
GAFLT	Grameena Abhyudaya Financial Literacy Trust
GNPAs	gross nonperforming assets
HRM	human resource management
IBC	Insolvency and Bankruptcy Code
IBI	Imperial Bank of India
ICT	information and communication technologies
IFSC	International Financial Services Centre
IIoT	industrial Internet of Things
IMF	International Monetary Fund
IoT	Internet of Things
ISM	interpretive structural modeling
JIT	just-in-time

KMO	Kaiser–Meyer–Olkin
KPIs	key performance indicators
LSS	Lean Six Sigma
MPS	market price per share
MSMEs	micro-, small-, and medium-enterprises
NGO	nongovernmental organizations
NNPAs	net nonperforming assets
NPAs	nonperforming assets
NPCI	National Payments Corporation of India
OCB	organizational citizenship behavior
OECD	Organisation for Economic Co-operation and Development
ONGC	Oil and Natural Gas Corporation
OP	organizational performance
OT	organizational trust
PB	perceived benefit
PCA	principal component analysis
PEU	perceived ease of use
PLCs	programmable logic controllers
PNB	Punjab National Bank
PR	perceived risk
PSBs	public sector banks
PT	perceived trust
PU	perceived usefulness
RBI	Reserve Bank of India
RFID	radiofrequency identification
RIL	Reliance Industries Limited
RMR	root mean square residual
RNW	return on net worth
RoHS	restriction of hazardous substances
ROS	robot operating system
RPA	robotic process automation
RQ	research question
RSETIs	rural self-employment training institutes
SAIL	Steel Authority of India Limited
SBI	State Bank of India

SDGs	sustainable development goals
SEM	structural equation modeling
SHGs	self-help groups
SIF	sustainable innovation framework
SPSS	statistical analysis software
SSA	sub-Saharan Africa
SSIM	structural self-similarity matrix
SVMs	support vector machines
TAM	technology acceptance model
TBL	triple bottom line
TCS	Tata Consultancy Services
USP	unique selling proposition
UTAUT	Unified Theory of Acceptance and Use of Technology
VSM	value stream mapping
WEEE	waste electrical and electronic equipment
WPS	workplace spirituality

Acknowledgments

First and foremost, we would like to express our deepest gratitude to our All-Powerful God. Our success in finishing this book has been largely attributed to your persistent faith in us.

We express our sincere gratitude to everyone who helped make this book a reality. With immense pleasure and gratitude, we would like to acknowledge the people and institutions that were essential in making this initiative a reality. The editors would like to express their gratitude to all authors, reviewers, and collaborators for their tremendous help in seeing this project through to completion. Your substantial contributions have improved the content offered in the chapters in terms of quality, coherence, and organization. Your work has greatly influenced how this book turned out in the end. We owe an appreciation to the referees who offered in-depth and perceptive criticism of many chapters. Your criticism forced us to rephrase our thoughts, examine certain areas in greater detail, and defend particular recommendations, which improved the content and guaranteed its accuracy.

We also extend our deepest appreciation to the individuals, our families, our colleagues, mentors, and the organizations whose contributions were crucial in making this effort a reality. Your support, whether direct or indirect, has been invaluable. Your guidance and insights have profoundly shaped our understanding and application of the principles discussed in this book.

We will conclude by thanking our readers for their participation and attention. We really hope that this book will offer insightful analysis and make a significant contribution to the current conversation around Industry 5.0 and sustainable innovation.

Foreword 1

It is with great pride that we present *Industry 5.0: Innovation and Adoption for Sustainability*. This book offers testimony of how creativity, sustainability, and technology advancements are working together to redefine industries and change our shared future.

In an era where ecological preservation must coexist with a steady progression in technology-driven approaches, this book discusses the necessity of sustainable practices as integral to the next industrial revolution. It emphasizes how essential Industry 5.0 is to balancing humanity's responsibility to the environment with its demand for innovative practices. Through detailed discourse, the book covers a broad range of topics, ranging from corporate social responsibility as a foundation for a better future to the profound impact of blockchain, artificial intelligence (AI), and Lean Six Sigma methodologies in driving efficiency.

The book's emphasis on the sociological effects of Industry 5.0 is one of its most distinctive features. Today, we need to address the practices of how communities could be empowered by this worldview, change financial ecosystems, and provide comprehensive setup that support the Sustainable Development Goals (SDGs) of the UN. The chapters cover progressive initiatives like digital entrepreneurship, which makes accessible opportunities while addressing ecological challenges, and sustainable tourism, which reflects shifting traveler preferences toward eco-friendly experiences. The amalgamation of trends in the banking, agricultural, and hospitality sectors illustrates the versatility of Industry 5.0 in driving sustainable development across diverse domains.

Strengthened with case studies, research analyses, and proactive conceptual models, this volume offers a value repository of actionable insights for academics, business leaders, and policymakers. From redefining operational strategies to enhancing decision-making processes, the book indicates how technological enablers can be utilized to create a balanced and durable global economy.

The contributors to this book bring a wealth of expertise, explaining the future challenges and opportunities of our time. By emphasizing the coordination of modernization with environmental management, ethical governance, and human-centric design, this book invites readers to anticipate a future where progress does not come at the expense of global well-being.

We cordially encourage you to join us in this engaging discussion as we discuss the various angles of Industry 5.0 and its potential for transforming both industries and societies. Together, these contributions inspire a collective movement toward an industrial blueprint that values sustainable practices, inclusivity, and the stable balance between progress and preservation.

—Prof. Vijayan Immanuel
Vice Chancellor, Presidency University, Bengaluru, India

Foreword 2

We are pleased to present the first edition of *Industry 5.0: Innovation and Adoption for Sustainability*. This book explores resilient business models within the framework of Industry 5.0, highlighting a cutting-edge technological platform that offers vast economic opportunities for digital experts, creatives, and business executives. It serves as a comprehensive guide for Industry 5.0 strategy, sustainable business practices, and corporate governance policies.

In an era defined by unprecedented challenges and limited resources, the call for adopting sustainable practices in business and e-commerce has grown louder. This volume delves into the transformative intersection of sustainable practices and Industry 5.0, tracing the evolution of modern business from early industrialization to the cutting-edge landscape of today and beyond. It emphasizes the importance of environmental concerns, social responsibility, and ethical business practices, providing a comprehensive guide for businesses aiming to thrive in a sustainable world.

Through rigorous research and real-world case studies, this book presents practical insights and strategies for adopting sustainable practices, reducing environmental footprints, and enhancing societal impact. Topics include circular economy principles, green supply chains, renewable energy integration, and the role of technology in driving sustainable innovation. The book demonstrates that sustainable practices can drive growth while fostering a responsible society, and resilient business models exert pressure on organizations to manage resources responsibly. It argues for an integrated approach, showing that responsibility is a collaborative effort between businesses and policymakers.

We invite readers from diverse backgrounds business leaders, policymakers, academics, and anyone interested in the future of industry to explore the insights and strategies presented in this book. Together, we can harness the power of Industry 5.0 to create a more sustainable and prosperous world.

Join us on this journey, with innovation as our guiding principle and sustainability as our goal. This aims to be the essential resource for understanding and implementing the technologies and practices that will shape the future of industry and lead us toward a greener, more resilient world.

—Shri Jitender Joshi
Founder and Chancellor, Uttaranchal University, India

Preface

The concept of Industry 5.0 represents a persistent shift toward a human-centric, sustainable industrial framework. This book, *Industry 5.0: Innovation and Adoption for Sustainability*, is an outcome of the international conference dedicated to clarifying the transformative potential of this reference in addressing global challenges.

The chapters together explore the amalgamation of technology, environment, and social welfare, presenting various perspectives on topics such as lean innovation frameworks, blockchain integration, green supply chains, and the ecological effects of digital entrepreneurship. By bridging academic research and industry practices, the book offers a roadmap for adopting innovative strategies that match with the Sustainable Development Goals (SDGs).

We aim to provide a resource that not only identifies challenges but also offers viable solutions. Whether it is observing the satisfaction of travelers in sustainable tourism or assessing the role of AI in banking and HR management, this book emphasizes practical insights that businesses and policymakers can use to encourage sustainable growth.

We extend our sincere appreciation to the contributors whose expertise has motivated us to write this book. We also thank the readers for joining us on this journey toward understanding the opportunities and responsibilities inherent in Industry 5.0. Let us work together to support creative mindset while prioritizing eco-friendly practices for generations to come.

Introduction to the Book

Industry 5.0 is not merely a continuation of its predecessor, Industry 4.0, but rather a transformative vision that intertwines human-centric innovation with environmental viability. Unlike Industry 4.0, which predominantly focused on automation and digitization, Industry 5.0 emphasizes a synergistic relationship between technology, ecology, and societal well-being. It envisions a future where progress is not measured solely by productivity but by the collaborative incorporation of human intelligence, artificial intelligence, and sustainable practices. This book focuses thoroughly on these concepts, offering an in-depth discussion of how Industry 5.0 changes industries and communities while addressing global concerns including resource depletion, social injustice, and climate change.

The text builds its foundation on sustainability and corporate social responsibility (CSR), two integral components for building a better future. These principles act as guiding frameworks for businesses seeking to balance profitability with accountability. Organizations are now incorporating sustainable practices into their core strategies, recognizing that it is not merely a moral imperative but a vital component of long-term growth. By aligning with frameworks like the United Nations' Sustainable Development Goals (SDGs), businesses are making strides toward reducing their environmental footprint while creating shared value for stakeholders. This book discusses how various industries integrate these principles, emphasizing the need for collective efforts to build a future where economic development coexists with environmental stewardship.

A key theme of this book is the role of technological enablers in driving Industry 5.0's capabilities. Advanced technologies such as artificial intelligence (AI), blockchain, the Internet of Things (IoT), and robotics are not only transforming operations but also preparing the grounds for sustainable practices. These technologies are thoroughly investigated, displaying how they might improve productivity, openness, and flexibility in a variety of industries. For instance, AI enables data-driven decision-making, encouraging businesses to optimize resource allocation. Blockchain ensures greater accountability in supply chains, addressing ethical sourcing and waste reduction. This book discusses how these innovations, when applied carefully, contribute to creating systems that are both state-of-the-art and environmentally conscious.

Another interesting aspect discussed in this book is the ecological impact of digital entrepreneurship. The growth of digital platforms and tools has equalized access to global markets, supporting innovation. However, these improvements come with ecological concerns, such as increased energy consumption and electronic waste. The book provides a detailed analysis of how entrepreneurs can incorporate sustainability into their business models, promoting principles like the circular economy. This discussion focuses on the dual role of technology as both a driver of innovation and a tool for addressing environmental challenges.

The hospitality and tourism sectors are also reviewed through the viewpoint of Industry 5.0. In a post-pandemic era, travellers increasingly prioritize eco-friendly accommodations and authentic cultural experiences. The book also focuses on how sustainable practices in tourism—such as reducing carbon footprints, conserving resources, and supporting local communities—align with changing consumer values. It also analyses how technology enables energy-efficient operations and customized visitor experiences, transforming sustainable tourism into a motivator for both social change and economic recovery.

Innovation framework such as Lean Six Sigma is presented as an important tool for operational excellence in a sustainable future. This book discusses how these frameworks emphasize waste management, resource optimization, and continuous improvement. By combining these methodologies with developing technologies, industries can adopt systematic approaches to sustainable innovation. This perspective supports the book's pervasive theme: the need for inclusive thinking in transforming industrial practices.

Financial ecosystems are another focus, with the book assessing their role in fostering sustainable economies. Discussions such as green finance initiatives and the use of blockchain and AI in financial services demonstrate how the sector supports sustainable development. The study addresses the concept of incorporation in finance by looking at how these technologies promote equity and inspire overlooked groups. This arrangement between economic progress and ecological responsibility is an ongoing theme throughout the book.

Community support, particularly through CSR efforts, is highlighted as a significant factor in driving meaningful social change. The book describes how financial organizations can go beyond conventional profit-driven models to generate societal value through looking at the CSR initiatives of public sector banks in India. These examples display the capacity of CSR to reengineer education, infrastructure development, and skill enhancement, organizing financial institutions with the broader goals of Industry 5.0.

The unification of blockchain technology into supply chain management and quality assurance is another convincing discussion point. This book describes how combining blockchain with Six Sigma methodologies can lead to operational quality and greater dependability. These approaches are presented as part of the broader perspective of cross-disciplinary collaboration, which is essential for addressing complex challenges in the modern industrial landscape.

Artificial intelligence's transformative role across various industries is another focal area. The book discusses its applications in human resource management, digital banking, and even the culinary industry. In HR, AI-driven analytics enable organizations to identify talent gaps and improve employee engagement. In banking, AI enhances customer experiences by personalizing services. In the culinary world, AI revolutionizes operations by predicting consumer preferences. These discussions underline AI's potential not only to enhance efficiency but also to contribute to sustainable practices.

The transition from Industry 4.0 to Industry 5.0 indicates a shift in how progress and innovation are conceptualized. The book emphasizes the transition from automation-focused strategies to human-centric development. It provides a framework for reviewing traditional business models, urging industries to integrate ethical practices and ecological balance into their operations.

This book represents a comprehensive approach to addressing the challenges of the 21st century. By combining themes technological innovation, inclusivity, and long-term preservation of resources, it draws a comprehensive picture of Industry 5.0's potential to redefine industrial and societal practices. Whether talking about digital business, blockchain, AI, or sustainable tourism, the book gives readers a thorough understanding of how these topics come together to create a future that puts both development and the health of the planet first.

CHAPTER 1

Sustainability and Corporate Social Responsibility: Integral Components for Building a Better Future

HIMANI UPRETI,[1*] RUPA KHANNA MALHOTRA,[1] and RAKESH KUMAR[2]

[1]*Department of Commerce, Graphic Era Deemed to be University, Dehradun, Uttarakhand, India*

[2]*Department of Management, Uttaranchal Institute of Management, Dehradun, Uttarakhand, India*

*Corresponding author

ABSTRACT

In the face of global challenges, such as climate change, resource depletion, and social inequality, there is an urgent need for businesses to adopt sustainable practices. Thus, the current study is conducted to investigate the crucial role of corporate social responsibility (CSR) in making a more promising future for both businesses and society on a broader scale. The study seeks to investigate on how organizations can adopt sustainability practices as their business strategies through various CSR initiatives.

1.1 INTRODUCTION

Conceptually, corporate social responsibility (CSR) and sustainable development are closely intertwined. Sustainable development is defined to be the approach toward the economic development that meets the current

Industry 5.0: Innovation and Adoption for Sustainability.
Tilottama Singh, Pradeep Suri, and Zulkarin Jahangir (Eds.)

DOI: 10.1201/9781779641878-1

needs without compromising the fulfillment of the needs of future generation. Alike CSR is the combination of social and environmental concerns into enterprise operations. It represents the contributions made by using the groups to the society via such activities and investments, which are not always a part of their vital commercial enterprise concerns. The idea of CSR is connected with the precept of "sustainability," which means that commercial enterprise should be worried now not simply about the profits or dividends but also about the long-term social and environmental results of their business activities. In essence, CSR is the compensatory responsibility of an enterprise entity for any detrimental impact of its activities on the society and environment. The interdependence of sustainable development and CSR is often conceptualized within a theoretical framework that recognizes the mutual benefits and dependencies between these two concepts. One such framework is the "triple bottom line" (TBL) or "three pillars of sustainability" model. This model emphasizes the interconnectedness of economic, social, and environmental dimensions in achieving sustainable development.

Economic dimension of TBL represents the traditional bottom line, which focuses on the financial performance and profit maximization aspect of a business. In the context of sustainable development, economic sustainability involves generating profits in a way that contributes to long-term economic health without compromising social and environmental well-being. Social pillar of sustainability considers the impact of business activities on people and communities. CSR is a key component of the social dimension, emphasizing ethical business practices, social responsibility, and positive contributions to society. This involves considerations such as fair labor practices, community development, and stakeholder engagement. The environmental pillar highlights the importance of minimizing negative environmental impacts and promoting ecological sustainability. Sustainable development recognizes the finite nature of natural resources and aims to protect ecosystems. CSR initiatives often include environmentally responsible practices, such as energy efficiency, waste reduction, and the use of renewable resources.

1.2 CORPORATE SOCIAL RESPONSIBILITY AND TRIPLE BOTTOM LINE: INTERCONNECTEDNESS AND SYNERGY

Attention to people involves the obligation of companies toward labor, the people within the network, and the area in which or in near proximity of the company where it operates and conducts its operations. The human beings-orientated principle within the CSR is that the corporate entity needs to be touchy and supportive to the wishes of the society just because the society's popularity and assist is wanted for the sustainable operation of the corporate entity. "Planet" points to the corporate responsibility to defend the environment in which it operates through sustainable environmental practices. This thing acknowledges the causal courting between corporate and the planet Earth and demand from the agency various measures required to be undertaken by it such as desisting from polluting the soil, water, and air with the aid of manner of harmful and unfavorable products or by-products such as toxic chemicals and radioenergetic or different dangerous metals. Profit, defined as the added cost created by the business corporation at some point of its operation, is the logical primary subject and the primary objective of a commercial enterprise business enterprise. Business agencies try to maximize income through numerous control and operational strategies, together with the simplification of the approaches involved in manufacturing, disposing of inefficiency in the operations, saving of time, number of substances via adoption of latest technology, and many others. The TBL framework suggests that these three dimensions such as (1) people, (2) planet, and (3) profit are interdependent and mutually reinforcing.

1.2.1 SYNERGY

By integrating CSR practices into their business, they create a synergy between economic, social, and environmental factors. For example, investing in environmentally friendly technologies not only reduces ecological impact (environmental dimension) but can also lead to cost savings and operational efficiency (economic dimension).

1.2.2 LONG-TERM VALUE CREATION

Active engagement with the social and environmental concerns is leading toward long-term value creation and facilitating their brand reputation, cultivating robust relationships with stakeholders, and securing the enduring success of their operations. The generation of long-term value is facilitated by sustainable development and CSR.

1.2.3 RISK MITIGATION

It is a well-known fact that risk mitigation contributes to economic stability. Companies that prioritize sustainable practices are often better positioned to mitigate risks associated with environmental regulations, social unrest, and changing consumer preferences.

Through their CSR initiatives, companies are committed toward the TBL approach of profit, people, and planet and, thus, operating ethically toward social and environmental concerns.

1.3 REVIEW OF LITERATURE

Khanna and Gupta (2011) conducted a study on CSR practices adopted by some selected companies such as Reliance Industries, Mahindra & Mahindra, Oil and Natural Gas Corporation (ONGC), Steel Authority of India Limited (SAIL), Bharat Heavy Electricals Limited (BHEL), Indian Airlines, and GlaxoSmithKline were examined to understand the status of CSR, and its need and relevance in Indian context and challenges in the implementation of CSR were highlighted. Khandelwal and Bakshi (2014) conducted a study that explored the prospective impact of new CSR regulations on the future of CSR. The researchers examined this through selected case studies and concluded that India, in recent years, has actively revised its corporate laws and legal framework. These revisions aimed to enhance transparency, bolster accountability, and bring the nation in line with international business standards. The study emphasized that the introduction of the new CSR regulation positioned India to become a benchmark in CSR practices, marking it as one of the pioneering nations in Asia to implement such regulatory measures. Dhanesh and Ganga (2014) conducted a study through 19 semistructured interviews in 16 companies to explore drivers for conducting CSR activities and found that there is coexistence of two drivers the moral imperative and the four economic imperative, which motivates the companies under toward social responsibility. Tilt (2016) conducted the study to examine influence of factors that impact CSR and its reporting in developing nations with special reference to the influence of political ideology, the sociocultural understandings, and stage of economic development on CSR reporting, and the study concluded that examining CSR practices and CSR reporting could be of no use without proper contextualization, which should be done on the basis of different political, social, cultural, economic, and environmental factors. Dey and Das (2018) concentrated on different exploration papers on corporate social obligation concerning Indian organizations. The analyst saw that none of the investigations have thought about all realities of corporate social obligation to partners. A study concentrated on the responsiveness of chosen financial area organizations toward corporate social obligation. It also discovered that education and health are the two areas in which each company has performed exceptionally well. The scientist presumed that larger part of the organizations offered not exactly the endorsed sum in that frame of mind after the alteration in Organizations act; hence, government and other lawful specialists should make a move for the prosperity of the general public. Krishnan (2018) compared India's manufacturing and service industries' CSR spending. The scientist found that assembling organizations offer more toward the ecological issues instead of administration area. It is likewise seen that in FY 2015, 2016, and 2017, the assembling organizations have spent more. The downside of this exploration paper was exclusively to cover a little example of Indian organizations. Saxena (2016) breaks down the investigation of CSR in private and public area banks. In the wake of doing all the examinations, he reasoned that every one of the organizations are doing perfect yet there are a few regions, which need additional consideration. There are a few regions where organization needs to take areas of strength to make CSR a major achievement. He also said that more and more businesses are realizing that CSR is more than just a collection of individual practices; rather, it is a comprehensive set of policies and practices that should be incorporated into the activities and operations of the business. Chopra and Arju (2017) under the study comparison among public and private sector banks was conducted with respect to the CSR contribution made by them. It has been revealed in the study that the CSR spent made by the banks is not equivalent to the prescribed limit of 2%. The study also revealed both the private banks and the public sectors banks are contributing equally toward CSR expenditure, and no distinction in the patterns has been found in their expenditures with regard to the amount spent. As per their examination, public sector bank such as National Bank of India is a best entertainer and UCO Bank is least entertainer with respect to CSR initiatives and spending. Kakade (2019) dissected the relative investigation of CSR rehearses in India. The researcher comes to the conclusion that

CSR varies according to a company's turnover and size. This is because a small company with a low profit margin cannot engage in CSR to a greater extent. A comparative study of Indian and multinational companies through annual reports was explored, revealing divergent focuses on CSR communication. Multinational companies tended to emphasize softer, quality-driven aspects, while Indian companies provided more detailed information about financial benefits and price advantages for human resources and customers. Kaur and Tandon (2017) concentrated on "the job of corporate social obligation in India." They found that many large organizations had understood the significance of CSR for the purpose of altruism and notoriety and expanding business intensity. In addition, they discovered that public awareness is required for CSR to be successful. Gupta and Agarwal (2015) concentrated on the CSR report of 10 banks in India. They reasoned that none of the banks distribute separate CSR report and, surprisingly, a portion of these does not actually distribute business obligation report moreover. They likewise observed that fixation on friendly issues is more than ecological issues. Grewal and Dutt (2019) compared SBI and Axis Bank's CSR policies and investments. They saw that as both the attempting to serve the general public in their most ideal way. However, they presumed that as far as speculation, SBI is putting more in CSR in examination with Pivot Bank. From 2013–2014 to 2017–2018, SBI contributed Rs. 6314.5 million in CSR, while Pivot Bank contributed Rs. 5919.6 million. However, assuming we see area wise or action wise, both the banks have put most elevated sum in CSR spending plan. They reasoned that there is no significant distinction in CSR of SBI and Hub Bank. Dhingra and Mittal (2014) concentrated on the CSR rehearses in Indian financial area. They observed that there are a couple of banks, which report their CSR exercises in triple primary concern rehearses. They additionally found that among the banks which they decide for their examination, some are putting forth bogus signals with regard to their attempts for socionatural worries. The vast majority of the banks use CSR rehearses as a showcasing device, and many are just putting forth token attempts such as gifts to magnanimous trusts, nongovernmental organizations (NGOs), and so on. There are not many banks who have shown their reasonable CSR theory. Singh and Singh (2015), in their investigation on CSR in broad daylight and confidential areas of banks, observed that Indian banks are putting forth great attempt yet there are still a few regions where they need to concentrate more. As per their exploration, public sector banks such as SBI and PNB are giving more commitment instead of HDFC and ICICI. Despite the RBI's financial literacy guidelines, banks are not taking the necessary steps. Kewlani and Bhatt (2019) analyzed CSR in selected banks both private and public and they revealed the changing trends in CSR expenditure for public banks and a static upward trend for private banks. They suggested a positive impact of CSR initiatives on the share prices of private banks but a negative impact on public sector banks.

1.4 RESEARCH OBJECTIVE

CSR initiatives mentioned under the Seventh Schedule of the Indian Companies Act, 2013, Section 135 has been introduced with an aim to seek positive effect of CSR activities on ensuring and promoting environmental and safety management. Thus, the current study is conducted with the primary objective to investigate case of Tata Consultancy Services (TCS) and Reliance Industries Limited (RIL), the top two CSR discharging companies of India, with respect to the CSR initiatives undertaken by them and finding an insight toward how businesses can flourish while simultaneously contributing to environmental preservation, community well-being, and ethical governance. Additionally, the research examines potential challenges and obstacles that organizations may face in their pursuit of sustainability. By identifying these challenges, the study aims to propose practical strategies and solutions for businesses to surmount obstacles on their sustainability journey through CSR.

1.5 RESEARCH METHODOLOGY

Research methodology adopted is conceptual in nature, which is based on secondary data and review of literature. Data is extracted from web-based sources related to CSR initiatives undertaken by top two companies undertaking CSR initiatives, namely (1) TCS and (2) RIL, for a period of 2014–2015 to 2020–2021. In order to maintain reliability of work and fulfill the research objective, authentic websites of companies, government, and other national and international bodies were accessed for collecting the secondary data.

1.6 DATA ANALYSIS AND DISCUSSION

The Seventh Schedule of the Indian Companies Act, 2013, outlines list of multiple activities that can be considered as part of their CSR initiatives. The list is inclusive of multiple activities from eradication of poverty and hunger to rural development, from Clean Ganga Kosh to PM CARES Fund (csrxchange.gov.in). As per Section 135 of the Indian Companies Act, 2013, certain qualifying companies are required to spend a percentage of their profits on CSR activities. The activities listed in Schedule VII cover a broad spectrum of social and environmental initiatives. The above figure shows the major development causes covered under the purview of CSR, which are segregated into different sectors ranging from disaster management, education, eradication of poverty and hunger, Clean Ganga, encouraging sports, and gender equality to rural and slum area development. Companies subject to CSR requirements need to identify and undertake projects falling within the scope of Schedule VII, and the expenditures on such activities are disclosed in their annual reports. The specifics of the CSR activities undertaken depend on the company's size, profitability, and other criteria as outlined in the Indian Companies Act, 2013. For conducting the current study, top two companies (TCS and RIL) discharging their CSR obligations have been investigated highlighting instances of companies successfully integrating sustainable practices into their operations.

1.7 CORPORATE SOCIAL RESPONSIBILITY INITIATIVES DEVELOPMENT SECTOR WISE

1.7.1 CASE I: TATA CONSULTANCY SERVICES

TCS has been involved in various skill development programs aimed at enhancing employability and creating opportunities for underprivileged youth. Through its "Tata STRIVE" initiative, TCS has provided vocational training in areas such as information technology, health care, and agriculture. TCS has been recognized for its commitment to sustainable practices in the IT industry. A continuous progression can be observed in total CSR spent made by the company since financial year 2014–2015 till 2020–2021. It can be culminated from Table 1.1 underneath that the major chunk of CSR spent has been allocated to the development sector under the head "other central government fund," which includes contribution toward TCS foundation and overheads for various CSR initiatives almost in all the seven years followed by health care except in the year 2020–2021 where second largest contribution has been made toward COVID-19 support fund, which was PM CARES. A significant contribution toward education has also been made by the company for the period under study. For digital empowerment, TCS has undertaken projects to promote digital literacy. The company has supported initiatives that provide digital education to marginalized communities, contributing to social inclusion and bridging the digital divide. Several initiatives have been implemented by the company to reduce its environmental footprint and to act as a frontrunner for environment safeguard. This includes energy-efficient practices in data centers, waste reduction measures, and the promotion of green technologies; beside, this company is also spending its CSR funds toward "Jal Jeevan Mission," which includes restoration and rejuvenation of Siruseri Lake, desilting, repairing, and maintenance of Chinnappanahalli Lake.

1.7.2 CASE II: RELIANCE INDUSTRIES LIMITED

Reliance Industries has focused on initiatives promoting the circular economy, especially in the plastics sector. Through its "Recron Green Gold" initiative, the company has developed sustainable and recyclable materials for various applications, contributing to a reduction in environmental impact. Reliance has also invested significantly in renewable energy projects. The company aims to be a net carbon-zero company by 2035. This includes the development of solar and wind power projects to meet a substantial portion of its energy requirements sustainably. Reliance Foundation, the CSR arm of Reliance Industries, has been actively involved in community development programs. These initiatives include healthcare services, education programs, and rural development projects, with a focus on improving the quality of life in local communities. Further, the company has also implemented water conservation and management initiatives across its operations. This includes the adoption of water-efficient technologies and practices to reduce water consumption and promote responsible water management. A continuous progression can be observed in total CSR spent made by the company since financial year 2014–2015 till 2020–2021. It can be culminated from Table 1.2 underneath that the

TABLE 1.1 Development Sector-wise Contribution by TCS (2014–2015 to 2020–2021)

S. No.	Year (FY) Development Sector	2014–2015 Amount Spent (in Cr)	2015–2016 Amount Spent (in Cr)	2016–2017 Amount Spent (in Cr)	2017–2018 Amount Spent (in Cr)	2018–2019 Amount Spent (in Cr)	2019–2020 Amount Spent (in Cr)	2020–2021 Amount Spent (in Cr)
1	Art and culture	1.68	0.3	0	0	0	0	1
2	Environmental sustainability	9.78	0.06	1	1	3	6	0
3	Education	32.43	71.16	88	86	36	114	16
4	Gender equality	0.26	1.09	0	0	0	0	0
5	Poverty, eradicating hunger, malnutrition, sanitation, and health care	18.85	71.3	86	131	296	176	28
6	Other Central Government funds	153.94	131.87	205	182	99	306	373
7	Prime Minister's National Relief Fund	1.49	4.44	0	0	0	0	256
	Total	218.43	280.22	380.00	400.00	434.00	602.00	674.00

Source: https://www.csr.gov.in/

major chunk of CSR spent has been allocated to the development sector under the head education and rural development and health care. A significant contribution can be observed toward training in sports specifically for the previous 4 years (Table 1.2).

A through insight toward the CSR spent patterns of both the companies highlights that the common areas where these top-notch companies are contributing significantly are health care and education, while gender equality, environmental sustainability, and promotion of art and culture are some of the significant areas of concern, which are still under played or lack focus.

1.8 OBSTACLES IN THE PURSUIT OF SUSTAINABILITY THROUGH CORPORATE SOCIAL RESPONSIBILITY INITIATIVES

Organizations often face various challenges and obstacles while carrying their journey of sustainability through CSR initiatives. Some of these challenges include:

- Resource allocation and budget constraints: Implementing sustainability initiatives may require significant financial investments. Organizations might face challenges in allocating resources and managing budgets for CSR projects, especially if they are operating in industries with thin profit margins.
- Short-term versus long-term goals: One of the major obstacles faced by the corporates while carrying their journey of sustainability through CSR is creation of a balance among the short-term financial goals of the company, which are essential for survival with the long-term sustainability objectives, which are essential for future growth and development. Usually, companies may prioritize and focus on immediate profits over long-term sustainability, which can create tension between economic interests and CSR initiatives.
- Lack of awareness and understanding: Lacking the awareness or understanding among the employees and stakeholders with respect to sustainability can be challenging for getting support these stakeholders for CSR initiatives. This obstacle can be addressed through effective communication and education.

TABLE 1.2 Development Sector-Wise Contribution by RIL (2014–2015 to 2020–2021)

S. No.	Year (FY)	2014–2015	2015–2016	2016–2017	2017–2018	**2018–2019**	**2019–2020**	**2020–2021**
	Development Sector	**Amount Spent (in Cr)**	**Amount Spent (in Cr)**	**Amount Spent (in Cr)**	**Amount Spent (in Cr)**	**Amount Spent (in Cr)**	**Amount Spent (in Cr)**	**Amount Spent (in Cr)**
1	Art and culture	0	0.28	27.9	0.86	0.84	0.8	452
2	Environmental sustainability	0.11	0.85	2.91	0.34	0.37	0.37	2
3	Education	21.34	216.74	219.35	374.87	527.34	253.82	0
5	Poverty, eradicating hunger, malnutrition, sanitation, and health care	613.6	394.04	256.59	147.89	113.4	34.7	309
6	Other Central Government funds	0	4.9	0	0	0	0	0
7	Other initiatives (not mentioned)	4.09	0	0	0	0	0	0
8	Rural development	121.44	14.25	138.58	181.9	158.72	577.37	110
9	Training to promote sports	0	8.64	0	43.11	48.65	41.65	49
	Total	760.58	639.70	645.33	748.97	849.32	908.71	922.0

Source: https://www.csr.gov.in/

- Global supply chain complexity: Applying sustainability practices throughout the supply chain can be challenging. This can include issues related to monitoring, compliance, and accountability.
- Regulatory compliance: Emerging diverse regulations with regard to sustainability practices can pose a challenge for the organizations, which can become challenging in a global business environment. Thus, organizations must stay abreast of changing laws and regulations to ensure compliance.
- Measuring and reporting impact: Establishing clear metrics and measurement frameworks for sustainability performance is crucial but challenging, as it often involves both qualitative and quantitative aspects. Quantifying the impact of CSR initiatives can be difficult and, thus, challenging.
- Resistance to change: Resistance by the employees and management to adopt sustainable practices can hinder progress of sustainability. Employees or management may resist changes, which are resulting into disruption in the established routines.
- Green washing concerns: Exaggeration or misrepresentation by companies of their commitment to sustainability may raise a risk of organizations engaging in green washing. This can erode trust among consumers and stakeholders and harm the reputation of the organization.
- Engaging and aligning the stakeholders: As discussed, CSR is the combination of social and environmental concerns into enterprise operations, which revolves around the multiple stakeholders associated with the company from employees to local community. One of the major challenge the companies come across is aligning these stakeholders. Further, engaging various stakeholders is again a challenge faced by the companies while carrying on CSR initiatives. Building of consensus among stakeholders who are geographically scattered and have different needs and expectations and ensuring their diverse perspectives requires significant effort.
- Infrastructural and technological limitations: Implementing sustainable technologies and infrastructure while carrying on CSR initiatives may be hindered by technological limitations or the lack of appropriate infrastructure. Upgrading systems to more sustainable alternatives can be costly and may face resistance.

- Economic uncertainty: Existence of downturns and economic uncertainties can make the organizations to focus more on the immediate financial stability for survival and underplay sustainability initiatives. As it is hard form the companies to maintain long-term commitment toward sustainability during the time of recession and downturns. This acts as a significant obstacle in the pursuit of sustainable development.

1.9 ADDRESSING THE CHALLENGES: PRACTICAL SOLUTIONS

Rigorous strategic planning and a commitment to integrating sustainable practices into the core of business operations can facilitate in addressing the obstacles which companies come across while carrying the journey of sustainability through CSR. Some of the practical strategies and solutions have been provided through the study for the businesses to surmount these challenges given as follows:

- Integrate sustainability into business strategy: Sustainability should be integrated to the core business strategy since inception. Thus, commitment toward integration of sustainability into decision-making process is of utmost importance so as to bring in balance between short-term and long-term goals of the organization.
- Employee education and engagement: Fostering a culture of sustainability through employee's engagement in the decision-making process and encouraging their active participation in CSR initiatives can help in meeting out the challenge of resistance and unawareness in the part of employees. Further training and awareness can be conducted for educating employees to enhance their understanding toward relevance of sustainable practices.
- Clear communication and transparency: Transparent system of reporting should be adopted to formulate clear articulation of the impact of CSR initiatives on both environmental and social fronts. This will result into building of trust among stakeholders. Thus, sustainability goal, their benefits, and achievement should be transparently reported.
- Developing key performance indicators (KPIs) for sustainability: For quantifying the impact of CSR initiatives, there is a need to

establish measurable KPIs, which can track and report on sustainability performance. This helps in demonstrating the organization's commitment to achieving sustainability goals.

- Stakeholder collaboration: Engagement of stakeholders such as local communities, customers, and suppliers while deciding CSR initiatives can make it easier for the companies to understand their expectations and concerns, which make the journey of sustainable practices more comprehendible and attainable. Thus, stakeholder collaboration may play a pivotal role in understanding their expectations from corporate and, thus, working in sync with them.
- Innovative technologies and practices: Technological advancement and application of tools such as artificial intelligence (AI) and Internet of Things (IoT) can also act as a facilitator in enhancing the impact of CSR initiatives toward achievement of sustainability. Adopting and investing in innovative technologies such as energy-efficient systems, waste reduction measures, and environmentally friendly production processes while carrying on CSR activities can make the companies to reach out to the achievement of sustainable goals.
- Regulatory compliance: Complying with the various provisions laid down by the regulatory bodies can assist in addressing the challenges encountered in the pathway of sustainability. Thus, staying informed about evolving regulations related to sustainability and ensuring compliance of such policies through various CSR initiatives can assist in promoting sustainable practices at a broader level.
- Third-party certification and audits: Conduction of regular audits by the external auditors in order to enhance the transparency with respect to the company's commitment toward sustainability can also encourage the companies to have an approach toward sustainability through CSR initiatives. Further, third-party certification can also act as a validation for sustainability practices adopted by companies.
- Incentivize sustainable behavior: Giving recognition and incentivizing the sustainable practices can encourage ongoing commitment of all involved in the initiatives. Thus, incentive programs for employees and suppliers can be implemented who

are substantially contributing toward sustainability goals. This can act as a motivator for them.

- Adaptability and continuous improvement: Regularly checkout the sustainability strategies and embrace a culture of continuous improvement. Regularly reassess sustainability strategies, learn from experiences, and make adjustments to enhance effectiveness.

These practical strategies can not only overcome obstacles on their sustainability journey but also position themselves as responsible corporate citizens who are committed toward posing a positive impact on the environment and the society while carrying on their economic commitments.

1.10 CONCLUSION

Through the above study, it can be culminated that in the modern business landscape, the success of any business house is not solely defined by financial metrics but also by a commitment toward the environmental stewardship, ethical conduct, and social equity. The corporate has to simultaneously work upon TBL people, planet, and profit. Further, the paper underscores the importance of aligning business practices with sustainable and socially responsible principles through the synthesis of diverse perspectives and the examination of real-world instances. The actionable recommendations provided in this paper are intended as a roadmap for organizations to navigate the complexities of integrating sustainability and CSR into their core strategies. Recommendations given in the paper are designed to empower businesses not only to mitigate negative impacts but to actively contribute to positive societal and environmental change, thus leading to achievement of sustainability goals. Finally, the study emphasizes that businesses should transcend traditional profit-centric paradigms and actively participates in building a world that is more sustainable, resilient, and equitable. It is hoped that by accommodating the mentioned recommendations, companies can not only enhance their own long-term viability but also play a pivotal role in fostering positive global change. In doing so, we move closer to a future where business excellence is synonymous with a genuine commitment to the well-being of our planet and its inhabitants.

1.11 LIMITATION OF THE STUDY

Dependence on secondary data is the major limitation of the current study; more accurate and detailed inferences could be made by conducting primary study by way of collecting information related to various CSR initiatives undertaken by the companies through conduction of interviews of CSR managers.

KEYWORDS

- **corporate social responsibility**
- **Schedule VII**
- **sustainable development**
- **sustainability**

REFERENCES

Chopra, I. P., Arju. (2017). A comparative study of corporate social performance in public and private banks in India. *Asian Journal of Research in Banking and Finance, 7*, 52.

Dey, S. K., Dash, A. P. (2018). CSR practices by Indian Companies: A review of literature. *International Journal for Research in Engineering Application & Management*, *04*(02), 2018.

Dhanesh, S., Ganga, 3. (2014). Why corporate social responsibility? An analysis of drivers of CSR in India. *Sage Publication, 29*, 114–129. https://doi.org/10.1177/0893318914545496

Dhingra, D., Mittal, R. (2014). CSR practices in Indian banking sector. *Global Journal of Finance and Management*, *6*(9), 853–862, ISSN 0975-6477.

Grewal, H., Dutt, R. (2019). A comparative study of impact of CSR awareness on employee satisfaction with special reference to select Indian public and private sector banks. *Research Review International Journal of Multidisciplinary*, *O4*(06), 1485–1489.

Gupta, R., Agarwal, G. (2015). CSR: A check on Indian banks for responsible investment. In: *AIMS International Conference on Management*, ISBN: 978-81-924713-8-9. https://www.csr.gov.in

Kakade, Dr. (2019). A comparative study of corporate social responsibility practices in India. *Research Review International Journal of Multidisciplinary*, *04*(06), 2019, 10.13140/RG.2.2.28202.47047.

Kaur, S., Tandon, N. (2017). The role of corporate social responsibility in India. *Research Journal of Commerce & Behavioral Science-RJCBS*, *6*, 29–34.

Kewlani, H., Bhatt, K. (2019). A comparative study on CSR practices in selected public and private banks in India. *GAP Interdisciplinarities*, *02*(01).

Khandelwal, R., Bakshi, S. (2014). The new CSR regulation in India: The way forward. *Procedia Economics and Finance, 11*, 60–67. https://doi.org/10.1016/S2212-5671(14)00176-2

Khanna, P., Gupta, G. (2011). Status of corporate social responsibility: In Indian. *APJRBM, 2*(1), ISSN 2229-4104.

Krishnan, A. (2018). Comparative analysis study on CSR expenditure in India: The case of manufacturing and service industries. *International Journal of Pure and Applied Mathematics*, *118*, 421–443.

Saxena, S. (2016). Comparative study of CSR of public and private sector banks in India. *Worldwide Journal of Multidisciplinary Research and Development WWJMRD*, *2*(1), 21–23

Singh, I., Singh T. D. (2015). Comparative study on CSR in public and private sector banks of India. *International Journal of Research in Management, Economics & Commerce*, *05*(03), 2015, ISSN 2250-057X.

Tilt, C. A. (2016). Corporate social responsibility research: The importance of context. *International Journal of Corporate Social Responsibility, 1*, 2. https://doi.org/10.1186/s40991-016-0003-7

CHAPTER 2

A Comprehensive Exploration of Technology Enablers in Industry 5.0 for Sustainable Development Goals

RAJESHWARI SHARMA[1] and HIMANSHI SHARMA[2,*]

[1]*IQ City United World School of Business, Kolkata, West Bengal, India*

[2]*Graphic Era deemed to be University, Uttarakhand, India*

*Corresponding author

ABSTRACT

Industry 5.0 represents the latest paradigm shift in industrial evolution, marking a significant departure from its predecessors by emphasizing the harmonious collaboration between humans and advanced technologies. Unlike the predominantly automated processes of Industry 4.0, Industry 5.0 envisions a future where humans and machines work together synergistically, leveraging each other's strengths to enhance productivity, creativity, and innovation. This new era prioritizes the integration of human skills, creativity, and emotional intelligence alongside cutting-edge technologies such as artificial intelligence, cloud computing, and the Internet of Things. As a concept that transcends traditional industry boundaries, Industry 5.0 sets the stage for a holistic and collaborative future, reshaping the landscape of production and fostering a new era of sustainable and socially responsible innovation. Industry 5.0 holds significant potential in advancing the achievement of the Sustainable Development Goals (SDGs) by fostering a human-centric, collaborative, and technologically integrated approach to industrial practices. This research aims to review the role of

Industry 5.0: Innovation and Adoption for Sustainability.
Tilottama Singh, Pradeep Suri, and Zulkarin Jahangir (Eds.)

DOI: 10.1201/9781779641878-2

technology enablers in Industry 5.0 in contributing to the achievement of the SDGs. The study focuses on identifying the specific technologies that can drive sustainable development across various industries.

2.1 INTRODUCTION

The onset of the First Industrial Revolution 1.0 marked a shift from manual to mechanized production powered by steam or water. Subsequently, the Second Industrial Revolution 2.0 introduced modern production lines, fostering increased productivity and substantial economic growth. The Third Industrial Revolution 3.0 incorporated field-level computers such as programmable logic controllers and communication technologies that led to automated production processes. In the era of Industry 4.0, production systems, represented by cyber–physical production systems, can intelligently decide through real-time communication and collaboration among interconnected manufacturing components. Industry 4.0 initiated in Germany and has gained global recognition over the past decade. At the onset of the 10-year milestone of Industry 4.0's, the European Commission introduced Industry 5.0, a term coined by Rada (2018).

The concept of Industry 5.0 represents a paradigm shift in industrial evolution, emphasizing the fusion of advanced technologies with a commitment to sustainable development (Tiwari et al., 2022). Industry 5.0 is envisioned as a framework that not only drives economic growth but also aligns with the principles of environmental sustainability and social responsibility (Mourtzis et al., 2022). The ongoing industrial evolution has ushered in an era marked by the integration of cutting-edge technologies that promise to reshape the landscape of industry. The proliferation of technology enablers such as Internet of Things (IoT), artificial intelligence (AI), and blockchain has emerged as a catalytic force, driving innovation across diverse sectors. These technologies enable autonomous industry, efficient operations, and secure systems, with applications ranging from location tracking to smart devices and security (Ivanov, 2023). Amid these advancements, a critical consideration arises concerning the potential synergies between Industry 5.0 and technology enablers and their transformative implications for addressing global challenges, particularly in the context of the Sustainable Development Goals (SDGs). Industry 5.0 focuses on personalized manufacturing and empowering humans in the manufacturing process (Alves, 2023). Collaborative robots play a key role, mitigating the risks by perceiving, understanding, and interacting with human operators, adapting to their goals and expectations. These robots observe and learn from human tasks, help operators in their task performance. Additionally, Industry 5.0 involves the integration of technologies into human life to enhance human capabilities. Supply chain management, AI, big data, digital transformation, machine learning, and the IoT are prominent enablers recognized by researchers to drive Industry 5.0 (Tallat et al., 2023). The industry actively employs these technology enablers to benefit and facilitate human workers. Industry 5.0 recognizes the industry's potential to achieve social objectives that strive to become a sustainable source of development (Ghobakhloo et al., 2023). It emphasizes production considerations aligned with the health of employees. As a reliable system for individuals seeking a fulfilling and healthy career, Industry 5.0 contributes to the necessary technological advancements toward sustainable development in health care. Industry 5.0 prioritizes worker well-being, leveraging new technology to generate wealth while respecting the constraints of our planet and empowering workers to adapt to changing skill and training requirements (Saikia, 2023). The future's disruptive technology has the potential to have significant repercussions for enterprises, consumers, and employees. Leaders and innovators would need to reassess their entrepreneurial and corporate strategies for the future as these technologies pose a challenge. The market is anticipated to witness the growth and proliferation of many advanced technologies in the year 2020 and beyond.

The adoption of the "2030 Agenda for Sustainable Development" occurred under the title "Transforming our World: the 2030 Agenda for Sustainable Development" in 2015 by the United Nations and 193 countries. The 2030 Agenda establishes a globally enforceable structure that comprehensively considers all aspects of sustainable development, thereby guaranteeing a high quality of life for everyone. The SDGs are a set of 17 interconnected objectives aimed at addressing global challenges such as poverty, inequality, and climate change. The idea is to have a future that is smart, safe, and good for everyone because of technology. These goals provide a framework for collaboration between different sectors, including public and private, to achieve sustainable development (Rashed and Shah, 2021). However, progress toward these goals varies, with some, such as economic targets, showing promising advancements, while others, such as education and climate action, require accelerated efforts (Halkos and Gkampoura, 2021).

The integration of advanced technologies in Industry 5.0 can significantly contribute to the achievement of SDGs. For instance, the integration of AI and IoT in smart manufacturing processes has the potential to significantly improve resource efficiency, reduce waste, and lower carbon emissions, thereby supporting SDGs 12 and 13 (Parmentola et al., 2022; Matsunaga et al., 2022). This aligns with the principles of Industry 5.0, emphasizing sustainable industrialization, innovation, and responsible resource consumption, which are directly linked to SDG 9 and SDG 12 (Kasinathan et al., 2022). Various methodologies and tactics can facilitate the attainment of the SDGs, but the most efficient approach would expedite the achievement of the SDGs within a shorter time frame. A comprehensive examination of strategies to accomplish the SDGs is crucial for the SDG community. Upon examining the several types of ways to achieve SDGs, it becomes evident that a technology-driven strategy is the least ambiguous when compared to approaches centered around human development, economics, and the environment (Elavarasan et al., 2022). Thus, technology can be seen as the instrument to accelerate the advancement of SDGs. Disruptive technologies, among the different technological advancements, are the most groundbreaking and possess significant future potential to propel humanity toward sustainable development. In such circumstances, it is important to analyze the impact and role of disruptive technologies on SDGs and how they can aid in reaching SDGs in a shorter period. This study aims to investigate the research question and provide a qualitative analysis of how disruptive technologies impact the SDGs. Furthermore, the SDGs' interaction caused by the disruptive technologies is also identified and mapped. An extensive analysis has been conducted to examine the effects of disruptive technologies on every SDG.

The use of digital technologies, including AI and IoT, can also promote circular economy models, which are crucial for sustainable production and consumption. Furthermore, the application of these technologies can optimize energy efficiency in smart manufacturing, leading to significant improvements in energy consumption and costs (Matsunaga et al., 2022). The global market size for AI is projected to increase by USD 75.54 billion from 2019 to 2023. According to a report issued by McKinsey, the influence of AI on the worldwide economy is projected to reach a staggering 13 trillion USD by 2030 (Gonzalo et al., 2020). AI is projected to provide an annual GDP growth rate of up to 1.2% until the year 2030. By the year 2022, it is projected that 30% of cyberattacks on AI will focus on exploiting training-data manipulation, AI model pilfering, or the use of adversarial samples to target AI-powered systems (Cearley, 2020).

The IoT, in particular, has been identified as a key technology for sustainable development in the manufacturing sector, facilitating real-time data sharing and decision-making (Singh, 2023). With extensive applications across diverse industries, IoT plays a crucial role in farming, water management, and recycling and holds immense potential in shaping smart cities for enhanced community management and stability. The IoT market is dominated by major firms, including Facebook, Hewlett Packard Enterprise, Microsoft, Dell Technologies, Apple, Intel Corporation, Google, and Cisco. The global IoT market had a market size of 190 billion USD in 2018. It is projected to reach 1111.3 billion USD by 2026, with a compound annual growth rate of 24.7% (Stephens et al., 2023). The IoT refers to the semantic relationship between the terms "internet" and "things." The term "internet" refers to the interconnected networks that enable the connectedness of many entities, including individuals, physical objects, and logical agents. The IoT can be stated simply as "IoT allows people and things to be connected anytime, anyplace, with anything and anyone, ideally using any path/network and any service" (Furstenau et al., 2023). While the concepts of IoT have been around for decades, the introduction of industrial Internet of Things (IIoT) aims to utilize and implement IoT within the framework of the Industry 4.0 Revolution.

Concurrently, blockchain technology has garnered recognition for its ability to contribute to various SDGs by bolstering transparency, traceability, and accountability in various sectors (Fregonese, 2023). As an emerging technology facilitating decentralized and transactional data sharing among untrusted participants, blockchain introduces new paradigms in distributed software architectures. Components can reach consensus on shared states without relying on a central integration point or specific participating entities, showcasing promising advancements across industries (Masood et al., 2024). Blockchain functions as a decentralized database, similar to a ledger that records all transactions (Howell et al., 2023). Blockchain's decentralized nature, absence of central trust, and unchangeable and cryptographically secure features have made it increasingly relevant in the realms of government decision-making, financial institutions, and major corporations (Merrell, 2022; Shari and Malip, 2023). Blockchain has transitioned from theoretical discourse to tangible implementation in various domains, such as digital currency, payment settlement (Papadis and Tassiulas, 2020), credit financing (Zheng et al., 2022), financial transactions (Saqib et al., 2023), securities (Wenhua et al., 2023), insurance (Trivedi, 2023), and others. Furthermore, the utilization of blockchain technology has the potential to significantly contribute to the attainment of carbon neutrality (Qin et al., 2023; Patil et al., 2023). Carbon neutrality refers to the practice of balancing carbon dioxide or greenhouse gas emissions by activities such as afforestation, energy saving, and emissions reduction (Yang et al., 2023; Chen et al., 2022). Blockchain technology has far-reaching implications across various industries, offering solutions for secure and transparent recordkeeping, decentralized applications, and novel approaches to conducting transactions and managing data.

S*ection 2.2—Literature Review*, through a comprehensive review of existing work, describes the evolution of Industry 5.0, the technology key enablers, and their applications in the industry along with the SDG. *Section 2.3—Discussion* discusses the application areas in Industry 5.0 of the various key enablers and the contribution of the technology enablers for achieving various SDG's. *Section 2.4—Conclusion* discusses the main findings and contributions of the study in the existing knowledge of research in Industry 5.0.

RQ1. How can the strategic integration of technology enablers within the framework of Industry 5.0 contribute to the advancement of SDGs, considering their transformative potential? RQ2. What are the key insights derived from the convergence of cutting-edge technologies such as AI, blockchain, and the IoT?

2.2 LITERATURE REVIEW

The notion of Industry 5.0 revolves around the collaboration of humans and robots as partners rather than competitors (Nahavandi, 2019). The transformation associated with Industry 5.0 involves addressing the taxonomy for integrating blockchain and machine learning in an IoT environment (Miglani and Kumar, 2021). This includes expanding technological infrastructure, providing budgetary support based on sustainable business models, standardizing and synchronizing protocols, and enhancing stakeholder engagement and involvement (Mbunge et al., 2021). The Industry 5.0 transformation also encompasses taxonomy analysis to facilitate the implementation of methods and algorithms for various IoT applications

(Chegini et al., 2021). It involves identifying techniques to enhance the security and efficiency of data transmission between IoT devices (Vithanage et al., 2021), exploring the utilization of deep learning and AI for monitoring purposes (Sujith and Sajja, 2021), and utilizing Amazon Web Services for equipment monitoring through IoT and cyberphysical systems (Dineva and Atanasova, 2021). Additionally, Industry 5.0 addresses challenges and impediments related to IoT (Omolara et al., 2022 and Miraz et al., 2018), and it explores energy efficiency and assessment models using big data and AI (Anthopoulos and Kazantzi, 2022).

Sachs's (1999) explored sustainable development in fundamental concepts and offered a ranking of diverse developmental trajectories. His emphasis lies in underlining the significance of integrating economic, social, and environmental considerations to achieve authentic sustainability. Concurrently, Sachs (2001) underscored the indispensable role of technology in advancing sustainable development. His examination scrutinized how technological progress can effectively confront global challenges, contributing to enduring economic, social, and environmental well-being. Nambiar and Sridharan's work (2016) is anticipated to revolve around the interplay among sustainability, innovation, and technology transfer. Their discussion centered on the mechanisms through which technological innovations are transferred and applied to foster sustainable development. Ilmola-Sheppard and Hukkinen (1997) contributed to the discourse surrounding the role of technology in sustainable development. Their exploration touched upon the perception of technology, examines whether it is regarded as a myth or a concrete reality in the pursuit of sustainability. Heeks and Ospina (2016) focused on the role of Information and Communication Technologies (ICT) in sustainable development. They may discuss how ICT can contribute to addressing global development challenges and propose a research agenda for further exploration. In their study, Heeks and Ospina (2016) concentrated on the role of ICT in sustainable development. Their focus extends to the ways in which ICT can play a crucial role in addressing global developmental challenges. Additionally, they proposed a research agenda to further investigate the potential contributions of ICT to sustainable development. This collective literature review provides insights into different dimensions of the relationship between technology and sustainable development as explored by these notable authors.

2.2.1 SYNERGIES OF INDUSTRY 4.0 AND 5.0, ARTIFICIAL INTELLIGENCE, AND SUSTAINABILITY

Schwab (2017) underscored the confluence of Industry 4.0 and AI for a transformative era in manufacturing and industry, whereas Brown's work (2018) accentuated AI's substantial role in sustainability, optimizing resource usage and reducing environmental impact through advanced analytics. In the context of smart manufacturing within Industry 4.0, Wang et al. (2016) emphasized AI's ability to enhance sustainability by streamlining the production processes and minimizing waste. Furthermore, UN Global Pulse (2019) highlighted the AI's application in addressing SDGs that aligned with the technological advancements and global sustainability objectives. However, Floridi et al. (2018) cautioned that AI can contribute to sustainability, but it brings challenges, including energy consumption and ethical considerations, and emphasized the need for responsible AI practices. This review elucidates the multifaceted relationship between Industry 4.0, AI, and sustainability, spanning from operational enhancements to global sustainability. The study by Ghobakhloo et.al. (2022) provided insights into the effective realization of human-centricity, socioenvironmental sustainability, and resilience within the context of Industry 5.0. Their research concluded by saying that Industry 5.0 has the potential to enhance sustainable development through 16 interdependent functions. These functions exhibited considerable diversity in their aims and characteristics, encompassing microtechnical aspects such as real-time communication to macro-operational elements such as supply chain adaptability.

2.2.2 SYNERGIES OF INDUSTRY 5.0, INTERNET OF THINGS, AND SUSTAINABILITY

Salam (2019) stated that the IoT is one of the enabling technologies of sustainable digital transformation and environmental protection for industry and society. The IoT is rapidly advancing as a potent facilitator in various fields, including water and health, energy administration, environmental surveillance, smart cities, intelligent industry, and supply chain management (Rosca et al., 2021). IoT devices find applications in various domains such as appliance remote monitoring (Blanco-Novoa et al., 2017; Suárez-Albela et al., 2016) and precision agriculture (Pérez-Expósito et al., 2017). In industrial settings, the application of IoT principles, known as Industrial IoT, allows for the deployment of numerous sensors, actuators, and smart machinery that can be remotely monitored. The continuous evolution of IoT devices, characterized by increased power, affordability, and energy efficiency (Mulay, 2015), is anticipated to contribute significantly to the global economy, with projections indicating USD 14 trillion in economic value by 2030, particularly in industrial applications (Sustainability Guidelines, 2018). Over the past two decades, the convergence of the traditional Internet of People concept and the Internet Protocol has given rise to the Internet of Everything, integrating individuals, objects, procedures, and information to enhance the quality of people's lives (Miranda et al., 2015). In the context of Industry 5.0, the integration of IoT is deemed crucial for achieving SDGs as it has the potential to drive sustainability and efficiency across various industrial processes (Fraga-Lamas et al., 2021; Fatima et al., 2022; El Jaouhari et al., 2022). This digitalization of operations through IoT in Industry 5.0 enhances productivity, communication, and performance in production plants and warehouses (Fatima et al., 2022). Moreover, IoT contributes significantly to sustainable supply chain management, reshaping the future of supply chains and promoting sustainability (El Jaouhari et al., 2022). The synergies between IoT and Industry 5.0 hold promise in advancing the industrial sector toward greater sustainability and efficiency, aligning with SDGs. The alignment of Industry 5.0 with the SDGs can be seen in various ways. Nastic et al. (2015) introduced SDG-Pro, a programming framework for software-defined IoT cloud systems, which can contribute to SDG 9 (industry, innovation, and infrastructure) by promoting the development of IoT cloud applications. El Jaouhari et al. (2022) highlighted the role of IoT in sustainable supply chain management, linking it to SDG 12 (responsible, consumption, and production) and SDG 13 (climate action). Romero-Gázquez (2018) presented a software architecture solution for an IIoT scenario, which can support SDG 9 by enhancing the efficiency and productivity of smart factories. Last, Liu and Mishra (2022) discussed the challenges of green IoT and its potential contribution to SDG 7 (affordable and clean energy) and SDG 13. These studies collectively demonstrate the potential of IoT in Industry 5.0 to advance various SDGs.

2.2.3 SYNERGIES OF INDUSTRY 5.0, BLOCKCHAIN, AND SUSTAINABILITY

Blockchain technology has the potential to significantly contribute to the achievement of SDGs in Industry 5.0. Verma et al. (2022) and Javaid (2021) both highlighted the role of blockchain in enhancing security, privacy, and data transparency in Industry 5.0, which are crucial for achieving SDG 9 (industry, innovation, and infrastructure) and SDG 12 (responsible, consumption, and production). Li (2022) further emphasized the importance of blockchain in Industry 4.0, which can be extended to Industry 5.0, in ensuring data security and trust in the manufacturing sector, contributing to SDG 9. Aysan et al. (2021) underscored the potential of blockchain-enabled solutions as SDG accelerators, providing a framework for their development and implementation. Therefore, the use of blockchain technology in Industry 5.0 can contribute to the achievement of various SDGs, particularly SDG 9 and SDG 12.

In the context of Industry 4.0, Javaid (2021) identified blockchain as a key enabler in various spheres such as supply chain management, smart contracts, and data security. The combination of blockchain and the IoT presents both opportunities and challenges, particularly in the context of Industry 4.0 and Society 5.0 (Tyagi et al., 2023). The role of blockchain in Industry 4.0 is further emphasized, with its potential to revolutionize the manufacturing sector and enable secure data transfer. Blockchain technology can be used to achieve the SDGs in Industry 5.0 by providing transparency, traceability, and security in supply chains. It can help improve the sustainability of the food supply chain by enabling transparent traceability and reducing environmental impact (Jiewu et al., 2023). Blockchain can also contribute to the sustainability, resilience, and human-centricity pillars of Industry 5.0 by providing trust and security in complex ecosystems such as industrial supply chains. Additionally, blockchain can enhance the efficiency, security, and transparency of large-scale industry networks, leading to increased levels of security, transparency, and efficiency (Chandan et al.,2023). By using blockchain technology, the manufacturing process can achieve resilience under disturbances and disruptions, enhancing the decentralized autonomous intelligence of the system (Oscar et al., 2023). Overall, blockchain technology has the potential to address issues related to sustainability and contribute to the achievement of the SDGs in Industry 5.0.

2.3 DISCUSSION

In Figure 2.1, Industry 5.0 represents the next phase of industrial evolution, emphasizing human–machine collaboration and the integration of advanced technologies. The convergence of Industry 5.0 with AI, blockchain, and the IoT can amplify their impact on SDGs in several ways.

FIGURE 2.1 Industry 5.0.

2.3.1 ENHANCED DATA CONNECTIVITY

IoT integration: Industry 5.0's focus on connectivity can enhance the integration of IoT devices across industrial processes. This interconnectedness enables real-time data collection and analysis, supporting informed decision-making for sustainable practices.

2.3.2 SMART MANUFACTURING AND ARTIFICIAL INTELLIGENCE

AI-driven optimization: Industry 5.0's smart manufacturing principles can be further enhanced by AI applications. Machine learning algorithms can optimize production processes, reduce waste, and enhance energy efficiency, aligning with SDG 9 (Industry, innovation, and infrastructure) and SDG 12 (responsible, consumption, and production).

2.3.3 BLOCKCHAIN FOR TRANSPARENT SUPPLY CHAINS

Decentralized and transparent transactions: Industry 5.0 can leverage blockchain to establish transparent and traceable supply chains. This is critical for responsible sourcing and ethical production, contributing to SDG 12.

2.3.4 HUMAN–MACHINE COLLABORATION

AI augmentation: Industry 5.0 promotes collaboration between humans and machines. AI algorithms can assist workers in decision-making, optimize resource utilization, and improve overall efficiency, supporting SDG 8 (decent work and economic growth).

2.3.5 PREDICTIVE MAINTENANCE AND SUSTAINABILITY

IoT sensors and predictive analytics: Industry 5.0''s adoption of IoT devices for predictive maintenance aligns with SDG 7 (affordable and clean energy). Predictive analytics powered by AI can reduce downtime, extend equipment lifespan, and optimize energy usage.

2.3.6 MAPPING KEY TECHNOLOGY IN INDUSTRY 5.0

Figure 2.2 represents key technology enablers and its application areas in Industry 5.0, which addresses the RQ1 of the present study.

Artificial Intelligence (AI)	Internet of Things (IoT)	Blockchain
Energy efficiency and Assessment models Improve resource efficiency Reduce waste Lower carbon emissions Sustainable industrialization & innovation Recycle Responsible resource consumption Improvements in energy consumption and costs	Water and energy management Environmental monitoring Health Smart cities Smart industry Farming Equipment monitoring Improve resource efficiency Improvements in energy consumption & management	Data Security Data Privacy and Transparency Smart Contracts Data Traceability Enhancing the decentralized autonomous intelligence of the system

FIGURE 2.2 Application areas of technology enablers.

AI plays a pivotal role in addressing critical environmental challenges by promoting energy efficiency and sustainability. Through advanced assessment models, AI enables industries to analyze and optimize their energy consumption, leading to improvements in efficiency and reduced operational costs. This not only benefits businesses but also contributes to the broader goal of lowering carbon emissions. AI-driven systems can assess resource usage patterns, identify inefficiencies, and propose strategies for improved resource efficiency. By streamlining processes and minimizing waste, AI aids in sustainable industrialization and fosters innovation that aligns with environmental conservation. The integration of AI in recycling processes further enhances the circular economy by identifying opportunities for reusing materials and reducing the overall environmental impact. Responsible resource consumption is promoted through AI algorithms that guide decision-makers toward sustainable practices. Overall, AI emerges as a powerful tool in creating a more environmentally conscious future, driving positive change in energy consumption, waste reduction, and sustainable industrial practices.

The IoT transforms water and energy management, enhancing efficiency across diverse sectors. In environmental monitoring, IoT devices enable real-time data collection for informed conservation. In health care, IoT ensures personalized patient care. Smart cities optimize resource usage and sustainability, while industrial IoT improves efficiency and reduces environmental impact. In farming, IoT aids precision agriculture for increased crop yield. Equipment monitoring prolongs machinery lifespan, reducing waste. IoT's data-driven insights identify inefficiencies, leading to resource conservation and cost savings. Notably, IoT advances energy management by monitoring usage patterns and suggesting energy-saving measures. Overall, IoT fosters a more sustainable and technologically advanced future across water and energy management, environmental monitoring, health care, smart cities, industries, farming, and equipment monitoring.

Blockchain technology revolutionizes data security through decentralization, ensuring privacy, transparency, and traceability. Smart contracts automate agreements, reducing fraud and errors. The immutability of blockchain records enhances data traceability, while decentralized autonomous intelligence prevents single points of failure, reinforcing system security. Overall, blockchain provides a robust solution for secure and trustworthy data management.

2.3.7 MAPPING OF SUSTAINABLE DEVELOPMENT GOALS WITH KEY TECHNOLOGY ENABLERS

Figure 2.3 highlights the main SDGs with the key technology enablers, which address the RQ2 of the present study.

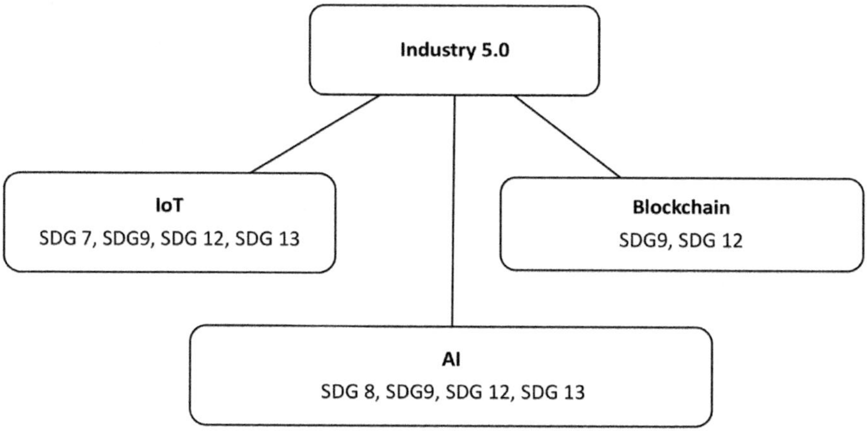

FIGURE 2.3 SDG alignment with key technology enablers.

IoT is aligned mostly with SDG 7 (affordable clean energy), SDG 9 (industry, innovation, and infrastructure), SDG 12 (responsible, consumption, and production), and SDG 13 (climate action).

AI aligns with SDG 8 (decent work and economic growth), SDG 9 (industry, innovation, and infrastructure), SDG 12 (responsible, consumption, and production), and SDG 13 (climate action).

Blockchain mainly mapped with SDG 9 (industry, innovation, and infrastructure) and SDG 12 (responsible, consumption, and production).

The SDG 9, SDG 12, and SDG 13 were the most important for AI, IoT, and blockchain enabling technologies in Industry 5.0.

2.4 CONCLUSION

The collaborative impact of blockchain, AI, and IoT forms a robust toolkit that effectively addresses diverse facets of sustainable development, encompassing economic, social, and environmental objectives. This amalgamation of technologies not only encourages innovation and inclusivity but also plays a pivotal role in advancing a more sustainable and equitable future. As Industry 5.0 integrates seamlessly with AI, blockchain, and IoT, a synergistic effect is achieved, nurturing sustainable practices across different sectors. The collective capabilities of these technologies significantly contribute to attaining various sustainability development goals, emphasizing the promotion of efficiency, transparency, and responsible resource management. In conclusion, the harmonious integration of these advanced technologies emerges as a key driver in steering societies toward a more sustainable and well-balanced future. By systematically delving into the pivotal role of technology enablers and categorizing them within the framework of Industry 5.0, this research provides actionable insights. These insights are instrumental for strategically aligning these elements to effectively contribute to the realization of SDGs. This comprehensive research work serves as a resourceful guide for those keen on navigating the intricate landscape of Industry 5.0 and its intersection with technology to drive sustainable development initiatives. Technology acts as a powerful enabler in achieving the SDGs, providing innovative solutions and tools to address global challenges. However, it is crucial to ensure that these technologies are deployed ethically, inclusively, and with consideration for potential negative impacts. This research is valuable for policymakers and industry leaders seeking to understand the interplay between Industry 5.0 and technology enablers in the context of sustainable development. By systematically examining the pivotal role of technology enablers and categorizing them within the framework of Industry 5.0, the research provides actionable insights for strategically aligning these elements to support the attainment of SDGs.

KEYWORDS

- **Industry 5.0**
- **technology**
- **sustainability**
- **sustainability goals**
- **sustainable development**
- **artificial intelligence**
- **cloud computing**
- **Internet of Things**

REFERENCES

Alves, J., Lima, T. M., Gaspar, P. D. (2023). Is industry 5.0 a human-centered approach? A systematic review. *Processes, 11*(1), 193.

Anthopoulos, L., Kazantzi, V. (2022). Urban energy efficiency assessment models from an AI and big data perspective: Tools for policy makers. *Sustainable Cities and Society, 76*, 103492.

Aysan, A. F., Bergigui, F., Disli, M. (2021). Using blockchain-enabled solutions as SDG accelerators in the international development space. *Sustainability, 13*(7), 4025.

Blanco-Novoa, Ó., Fernández-Caramés, T. M., Fraga-Lamas, P., Castedo, L. (2017). An electricity price-aware open-source smart socket for the internet of energy. *Sensors, 17*(3), 643.]

Brown, M. T. (2018). AI for earth: Innovating to mitigate human impact on the environment. *Trends in Ecology and Evolution, 33*(10), 669–672.

Cearley, D. W. (2020). *Top 10 Strategic Technology Trends for 2020*. Gartner: Stamford, CT, USA.

Chandan, A., John, M., Potdar, V. (2023). Achieving UN SDGs in food supply chain using blockchain technology. *Sustainability, 15*, 2109. https://doi.org/10.3390/su15032109

Chegini, H., Naha, R. K., Mahanti, A. (2021). Thulasiraman, P. process automation in an IoT–Fog–Cloud ecosystem: A survey and taxonomy. *IoT, 2*, 6.

Chen, L., Msigwa, G., Yang, M., Osman, A. I., Fawzy, S., Rooney, D. W., Yap, P. S. (2022). Strategies to achieve a carbon neutral society: A review. *Environmental Chemistry Letters, 20*(4), 2277–2310.

Dineva, K., Atanasova, T. (2021). Design of scalable IoT architecture based on AWS for smart livestock. *Animals, 11*, 2697.

El Jaouhari, A., Azari, M., Arif, J., El Farouk, I. I., Jawab, F., Moufad, I. (2022, May). IoT for the future of sustainable supply chain management in industry 4.0: A systematic literature review. In: *2022 14th International Colloquium of Logistics and Supply Chain Management (LOGISTIQUA)* (pp. 1–6). IEEE.

Elavarasan, R. M., Nadarajah, M., Pugazhendhi, R., Sinha, A., Gangatharan, S., Chiaramonti, D., Abou Houran, M. (2023). The untold subtlety of energy consumption and its influence on policy drive towards Sustainable Development Goal 7. *Applied Energy, 334*, 120698.

Fatima, Z., Tanveer, M. H., Waseemullah, Zardari, S., Naz, L. F., Khadim, H., Ahmed, N., Tahir, M. (2022). Production plant and warehouse automation with IoT and industry 5.0. *Applied Sciences, 12*(4), 2053.

Floridi, L., Cowls, J., Beltrametti, M., Chatila, R., Chazerand, P., Dignum, V., and Thiran, P. (2018). AI4People—An ethical framework for a good AI society: Opportunities, risks, principles, and recommendations. *Minds and Machines, 28*(4), 689–707.

Fraga-Lamas, P., Lopes, S. I., Fernández-Caramés, T. M. (2021). Green IoT and edge AI as key technological enablers for a sustainable digital transition towards a smart circular economy: An industry 5.0 use case. *Sensors, 21*(17), 5745.

Fregonese, G. (2023). Towards more sustainable Global Value Chains: The supporting role of the Block chain and Allied Technologies.

Furstenau, L. B., Rodrigues, Y. P. R., Sott, M. K., Leivas, P., Dohan, M. S., López-Robles, J. R.,and Choo, K. K. R. (2023). Internet of things: Conceptual network structure, main challenges and future directions. *Digital Communications and Networks*, *9*(3), 677–687.

Ghobakhloo, M., Iranmanesh, M., Foroughi, B., Tirkolaee, E. B., Asadi, S., Amran, A. (2023). Industry 5.0 implications for inclusive sustainable manufacturing: An evidence-knowledge-based strategic roadmap. *Journal of Cleaner Production*, *417*, 138023.

Ghobakhloo, M., Iranmanesh, M., Mubarak, M. F., Mubarik, M., Rejeb, A., Nilashi, M. (2022). Identifying industry 5.0 contributions to sustainable development: A strategy roadmap for delivering sustainability values. *Sustainable Production and Consumption*, *33*, 716–737.

Gonzalo, A., Harreis, H., Altable, C. S., Villepelet, C. (2020). *Fashion's Digital Transformation: Now or Never*. Textile Future: Chicago, IL, USA.

Halkos, G., Gkampoura, E. C. (2021). Where do we stand on the 17 Sustainable Development Goals? An overview on progress. *Economic Analysis and Policy*, *70*, 94–122.

Heeks, R., Ospina, A. (2016). ICT for sustainable development: Defining a global research agenda. *Information Technologies and International Development*, *12*(2), 3–18.

Howell, A., Saber, T., Bendechache, M. (2023). Measuring node decentralization in blockchain peer to peer networks. *Block chain: Research and Applications*, *4*(1), 100109.

Ilmola-Sheppard, L., Hukkinen, J. (1997). Technology and sustainable development: Myth or reality? *Technology Analysis and Strategic Management*, *9*(3), 297–309.

Ivanov, D. (2023). The industry 5.0 framework: Viability-based integration of the resilience, sustainability, and human-centricity perspectives. *International Journal of Production Research*, *61*(5), 1683–1695.

Javaid, M., Haleem, A., Singh, R. P., Khan, S., Suman, R. (2021). Blockchain technology applications for industry 4.0: A literature-based review. *Blockchain: Research and Applications*, 2(4), 100027.

Jiewu, L., Xiaofeng, Z., Zhiqiang, H., Kailin, X., Qiang, L., Xin, C., Du. (2023). Manu chain II: Block chained smart contract system as the digital twin of decentralized autonomous manufacturing toward resilience in industry 5.0. *IEEE Transactions on Systems, Man, and Cybernetics: Systems*, *53*(8), 4715–4728. https://doi.org/10.1109/TSMC.2023.3257172

Kasinathan, P., Pugazhendhi, R., Elavarasan, R. M., Ramachandaramurthy, V. K., Ramanathan, V., Subramanian, S., Kumar, S., Nandhagopal, K., Raghavan, R. R. V., Rangasamy, S., Alsharif, M. H. (2022). Realization of sustainable development goals with disruptive technologies by integrating industry 5.0, society 5.0, smart cities and villages. *Sustainability*, *14*(22), 15258.

Liu, L., Mishra, A. R. (2022). Enabling technologies challenges of green Internet of Things (IoT) towards sustainable development in the era of Industry 4.0. *Technological and Economic Development of Economy*, *30*(2), 1–32.

Masood, I., Daud, A., Wang, Y., Banjar, A., Alharbey, R. (2024). A blockchain-based system for patient data privacy and security. *Multimedia Tools and Applications*, *83*, 60443–60467.

Matsunaga, F., Zytkowski, V., Valle, P., Deschamps, F. (2022). Optimization of energy efficiency in smart manufacturing through the application of cyber–physical systems and industry 4.0 technologies. *Journal of Energy Resources Technology*, *144*(10), 102104.

Mbunge, E., Muchemwa, B., Jiyane, S., Batani, J. (2021). Sensors and healthcare 5.0: Transformative shift in virtual care through emerging digital health technologies. *Global Health Journal*, 5, 169–177.

Merrell, I. (2022). Block chain for decentralized rural development and governance. *Block chain: Research and Applications*, *3*(3), 100086.

Miglani, A., Kumar, N. (2021). Blockchain management and machine learning adaptation for IoT environment in 5G and beyond networks: A systematic review. *Computer Communications*, *178*, 37–63.

Miranda, J., Mäkitalo, N., Garcia-Alonso, J., Berrocal, J., Mikkonen, T., Canal, C., Murillo, J. M. (2015). From the Internet of Things to the internet of people. *IEEE Internet Computing*, *19*(2), 40–47.

Miraz, M. H., Ali, M., Excell, P. S., Picking, R. (2018). Internet of nano-things, things and everything: Future growth trends. *Future Internet*, *10*, 68.

Mourtzis, D., Angelopoulos, J., Panopoulos, N. (2022). A literature review of the challenges and opportunities of the transition from industry 4.0 to society 5.0. *Energies*, *15*(17), 6276.

Nahavandi, S. (2019). Industry 5.0—A human-centric solution. *Sustainability*, *11*, 4371.

Nambiar, V. K., Sridharan, R. (2016). Sustainability, innovation, and technology transfer. In: *Handbook of Research on Sustainable Development and Economics* (pp. 47–67). IGI Global.

Nastic, S., Truong, H. L., Dustdar, S. (2015). SDG-Pro: A programming framework for software-defined IoT cloud gateways. *Journal of Internet Services and Applications*, *6*, 1–17.

Omolara, A. E., Alabdulatif, A., Abiodun, O. I., Alawida, M., Alabdulatif, A., Alshoura, W. H., Arshad, H. (2022). The internet of things security: A survey encompassing unexplored areas and new in-sights. *Computer Security*, *112*, 102494.

Oscar, L. S., Saiz, M. D., Santos, Zarzuelo, J. M. (2023). Decentralized industry 5.0: Blockchain's contribution to industrial sustainability, resilience, and human-centricity. *Dyna*, *6*, 8 https://doi.org/10.6036/10804

Papadis, N., Tassiulas, L. (2020). Blockchain-based payment channel networks: Challenges and recent advances. *IEEE Access*, *8*, 227596–227609.

Parmentola, A., Petrillo, A., Tutore, I., De Felice, F. (2022). Is blockchain able to enhance environmental sustainability? A systematic review and research agenda from the perspective of Sustainable Development Goals (SDGs). *Business Strategy and the Environment*, *31*(1), 194–217.

Patil, A., Shardeo, V., Dwivedi, A., Moktadir, M. A., Bag, S. (2023). Examining the interactions among smart supply chains and carbon reduction strategies: To attain carbon neutrality. *Business Strategy and the Environment*, *33*(2), 1227–1246.

Pérez-Expósito, J. P., Fernández-Caramés, T. M., Fraga-Lamas, P., Castedo, L. (2017). VineSens: An eco-smart decision-support viticulture system. Sensors, *17*(3), 465.

Qin, M., Su, C. W., Lobonţ, O. R., Umar, M. (2023). Blockchain: A carbon-neutral facilitator or an environmental destroyer. *International Review of Economics and Finance*, *86*, 604–615.

Rada, M. (2018). *Industry 5.0-from Virtual to Physical*. LinkedIn. 7 March 2018. Available online: https://www.linkedin.com/pulse/industry-50-from-virtual-physical-michael-rada [Accessed 3 February 2022].

Rashed, A. H., Shah, A. (2021). The role of private sector in the implementation of sustainable development goals. *Environment, Development and Sustainability, 23,* 2931–2948.

Rosca, M. I., Nicolae, C., Sanda, E. and Madan, A., 2021. Internet of Things(IoT) and sustainability. In: Pamfilie, R., Dinu, V., Tăchiciu, L., Pleşea, D., Vasiliu, C., (Eds.), In:*7th BASIQ International Conference on New Trends in Sustainable Business and Consumption* (pp. 346–352). Foggia, Italy, 3-5 June 2021.Bucharest: ASE https://doi.org/10.24818/BASIQ/2021/07/044

Sachs, I. (1999). Sustainable development: Concepts and rankings. In: *Elements for a Theory of Sustainable Development* (pp. 1–35). Springer.

Sachs, J. D. (2001). The role of technology in sustainable development. In: *Globalization and the Developing World: An Integrated Approach* (pp. 227–252).

Saikia, B. (2023). Industry 5.0–Its role toward human society: Obstacles, opportunities, and providing human-centered solutions. In: *Fostering Sustainable Businesses in Emerging Economies: The Impact of Technology* (pp. 109–126). Emerald Publishing Limited.

Saqib, N. A., AL-Talla, S. T. (2023). Scaling up security and efficiency in financial transactions and blockchain systems. *Journal of Sensor and Actuator Networks, 12*(2), 31.

Schwab, K. (2017). *The Fourth Industrial Revolution.* World Economic Forum.

Shari, N. F. M., Malip, A. (2023). Block chain-based decentralized data dissemination scheme in smart transportation. *Journal of Systems Architecture, 134,* 102800.

Singh, D. (2023). Internet of Things. In: *Factories of the Future: Technological Advancements in the Manufacturing Industry* (pp. 195–227).

Stephens, M., Vashishtha, H., Wagner, D. N. (2023). *AI Enabled Business: A Smart Decision Kit.* IAP.

Suárez-Albela, M., Fraga-Lamas, P., Fernández-Caramés, T. M., Dapena, A., González-López, M. (2016). Home automation system based on intelligent transducer enablers. *Sensors, 16*(10), 1595.

Sujith, A., Sajja, G. S., Mahalakshmi, V., Nuhmani, S., Prasanalakshmi, B. (2021). Systematic review of smart health monitoring using deep learning and artificial intelligence. *Neurosciences Informatics, 2,* 100028.

Tallat, R., Hawbani, A., Wang, X., Al-Dubai, A., Zhao, L., Liu, Z., Min, G., Zomaya, A. Y., Alsamhi, S. H. (2023). Navigating industry 5.0: A survey of key enabling technologies, trends, challenges, and opportunities. *IEEE Communications Surveys and Tutorials,* 26(2), 1080–1126.

Tiwari, S., Bahuguna, P. C., Walker, J. (2022). Industry 5.0: A macro perspective approach. In: *Handbook of Research on Innovative Management Using AI in Industry 5.0* (pp. 59–73). IGI Global.

Trivedi, S. (2023). Block chain framework for insurance industry. *International Journal of Innovation and Technology Management, 20*(6), 2350034.

Tyagi, A. K., Dananjayan, S., Agarwal, D., Thariq Ahmed, H. F. (2023). Blockchain—Internet of Things applications: Opportunities and challenges for industry 4.0 and society 5.0. *Sensors, 23*(2), 947.

UN Global Pulse. (2019). AI for the SDGs: UN global pulse report on AI for sustainable development goals.

Verma, A., Bhattacharya, P., Madhani, N., Trivedi, C., Bhushan, B., Tanwar, S., Sharma, G., Bokoro, P., Sharma, R. (2022). Blockchain for industry 5.0: Vision, opportunities, key enablers, and future directions. *IEEE Access*, *10*, 69160–69199.

Vithanage, N. N. N., Thanthrige, S. S. H., Kapuge, M. C. K. P., Malwenna, T. H., Liyanapathirana, C., Wijekoon, J. L. (2021) . A secure corroboration protocol for Internet of Things (IoT) devices using MQTT version 5 and LDAP. In: *Proceedings of the 2021 International Conference on Information Networking (ICOIN)* (pp. 837–841), Jeju Island, Korea, 13–16 January 2021.

Wang, S., Wan, J., Li, D., Zhang, C. (2016). Implementing smart factory of industry 4.0: An outlook. *International Journal of Distributed Sensor Networks*, *12*(1), 3159805.

Wenhua, Z., Qamar, F., Abdali, T. A. N., Hassan, R., Jafri, S. T. A., Nguyen, Q. N. (2023). Block chain technology: Security issues, healthcare applications, challenges and future trends. *Electronics*, *12*(3), 546.

World Economic Forum, Internet of Things Guidelines for Sustainability, Future of Digital Economy and Society System Initiative. 2018. Available online: http://www3.weforum.org/docs/IoTGuidelinesforSustainability.pdf

Yang, S., Yang, D., Shi, W., Deng, C., Chen, C., Feng, S. (2023). Global evaluation of carbon neutrality and peak carbon dioxide emissions: Current challenges and future outlook. *Environmental Science and Pollution Research*, *30*(34), 81725–81744.

Zheng, K., Zheng, L. J., Gauthier, J., Zhou, L., Xu, Y., Behl, A., Zhang, J. Z. (2022). Block chain technology for enterprise credit information sharing in supply chain finance. *Journal of Innovation and Knowledge*, *7*(4), 100256.

CHAPTER 3

Unveiling the Ecological Effects of Digital Entrepreneurship: A Route to Sustainable Commercial Activities

SAYONEE DAS

Uttaranchal University, Dehradun, Uttarakhand, India

ABSTRACT

This chapter talks about how new technology and taking care of the environment are connected. It looks closely at how starting businesses online affects the environment. Many companies now operate online, using computers and the Internet. But have you ever thought about how this affects the environment? The chapter looks at how starting a business online can either help or harm the planet. It says that while technology has many benefits, it can also cause problems for the environment. For example, running servers and computers use a lot of energy. This energy mostly comes from burning fossil fuels, which release harmful gases into the air. Also, making all the gadgets we use, such as computers and phones, creates a lot of pollution. This pollution harms the environment. But the chapter also says that technology can help the environment too. The chapter says that businesses should start using eco-friendly strategies. This could mean using renewable energy to power their servers or using recycled materials in their products. It could also mean finding ways to reduce waste or pollution. Overall, the chapter urges businesses to think about the planet when using technology.

Industry 5.0: Innovation and Adoption for Sustainability.
Tilottama Singh, Pradeep Suri, and Zulkarin Jahangir (Eds.)

DOI: 10.1201/9781779641878-3

3.1 INTRODUCTION

Digital entrepreneurship is more than just a way to make money online. It is a big force that changes not only our economy but also our environment. When businesses use digital tools such as computers and the Internet, they can do things faster and better. But all this technology also has a downside: it can harm the environment.

Using digital technology means using a lot of energy. Think about all the servers and computers needed to run websites and online services (Adams, 2021). These machines need power to work, and most of that power comes from burning fossil fuels such as coal and oil. Burning fossil fuels releases harmful gases into the air, which can lead to climate change and other environmental problems.

But it is not just about the energy used to run digital technology. Making all those gadgets such as computers and phones also creates a lot of pollution. Factories that make these gadgets produce waste and emissions that harm the environment.

So, while digital entrepreneurship helps businesses grow and innovate, it also creates environmental challenges that we cannot ignore. That is why we need to understand how digital entrepreneurship affects the environment.

To deal with these challenges, businesses need to change how they operate. They need to think about the environment when they use digital technology. This means finding ways to reduce energy use and pollution. It also means using materials that are less harmful to the planet.

Businesses can do this by adopting sustainable practices. This postulates the use of alternate forms of energy such as solar power or wind power to serve their servers. It instructs on using second-hand materials on your product and deciding on how to minimize garbage.

But it is not just about what businesses do. It is also about how they think. Companies need to see environmental issues as important parts of their business strategies. Instead of seeing them as obstacles, they should see them as opportunities for long-term success.

By taking care of the environment, businesses can make sure that digital entrepreneurship is sustainable in the long run. This means finding ways to balance economic growth with environmental responsibility. It is not an easy task, but it is necessary if we want to create a better future for our planet.

In conclusion, digital entrepreneurship has a big impact on both our economy and environment. While it brings many benefits, it also creates challenges that we need to address. By adopting sustainable practices and changing how we think about the environment, businesses can ensure that digital entrepreneurship is not only profitable but also good for the planet.

3.2 LITERATURE REVIEW

The examination of the literature goes in-depth on the current research that addresses the effects of digital entrepreneurship on the ecosystem. It explores several questions, including the amount of energy consumed, the volume of waste produced, and the level of carbon emitted. This chapter serves as a firm basis for comprehending the effects of digital operations on the environment by synthesizing important findings picked from various studies. When we talk about digital entrepreneurship, we are talking about businesses that operate online. These businesses use computers, the Internet, and other digital tools to sell products or services. But all this technology has a cost to the environment.

For example, running servers and computers use a lot of energy. This energy mostly comes from burning fossil fuels, which release harmful gases into the air (Evans and Patel, 2019). Making all the gadgets we use also creates a lot of pollution. This pollution harms the environment.

When grouping various investigations, this review aids us in realizing its entirety. It considers how digital activities alter the ecology of energy use, generation of wastes, and elimination of carbon. This chapter provides a succinct version of these facts, thereby creating a perception about environment-friendly online business. One important aspect of this review is its focus on specific environmental factors such as energy, waste, and carbon footprints. By highlighting these factors, it helps us to see the complex relationships between digital business and the environment. This allows us to understand both the challenges and opportunities that come with digital entrepreneurship.

This review also acts as a guide for academics, researchers, and industry professionals interested in understanding the environmental effects of digital entrepreneurship (Williams and Martinez, 2017). By summarizing key findings and laying the groundwork for further study, it promotes ongoing discussion and exploration of this important topic.

Now, let us look at the methodology used in this research. To understand the environmental impact of digital entrepreneurship, the researchers collected data from digital firms and studied real-life cases. This approach gives us a detailed look at how digital business practices affect the environment.

Next, the chapter delves into specific measures and methods used to evaluate the environmental impact of digital entrepreneurship. It looks at things such as energy consumption, waste generation, and carbon emissions. By examining these footprints, we can better understand the environmental implications of digital business.

Finally, the chapter explores sustainable business practices in digital entrepreneurship. As the need for sustainability grows, digital entrepreneurs are faced with choices. This section examines strategies such as using energy-efficient technologies and reducing waste to minimize the environmental footprint of digital businesses.

In conclusion, this literature review provides a comprehensive look at how digital entrepreneurship affects the environment. By summarizing key findings, exploring methodologies, and discussing sustainable practices, it offers valuable insights for anyone interested in understanding the environmental impact of digital business.

3.3 CASE STUDIES

To understand how sustainable business practices can be applied in the digital realm, let us explore some real-life case studies.

3.3.1 GOOGLE'S RENEWABLE ENERGY COMMITMENT

If wind and solar energy projects are invested in, carbon footprint reduction follows as well as renewable energy infrastructure growth, thereby reducing theirs by also speaking for the former, one need not look further than tech labyrinth called Google. Find out how a giant with a conscience Google practices sustainability through this brief description. Three years ago, Google confirmed that they managed to buy as much green power as they needed globally making this corporation 100% self-sufficient regarding electricity consumption, which is an instance reflecting how much cleaner planet Earth can get if only more such measures would take place. Though this initiative has been applauded universally, there is no doubt that such commendations are rightly placed because there would be no digital services without it which depend heavily on these warehouses of data or some other type of online platform where all your information gets stored digitally under such tags as "private" or "public."

3.3.2 SALESFORCE'S GREEN CLOUD INITIATIVE

Salesforce, a cloud-based software company, launched its Green Cloud Initiative to reduce the environmental impact of its operations. One notable aspect of this initiative is their focus on energy-efficient data centers. Salesforce paired up with Equinix through which they seek to create a green data center adopting energy sources that are renewable and with systems of cooling that use less power. Salesforce optimizes the way its center relies on power hence less carbon emission, making it possible for digital businesses to take on sustainability practices within their establishments.

3.3.3 ETSY'S CARBON NEUTRAL SHIPPING

Etsy, an online marketplace for handmade and vintage goods, implemented a carbon offset program to make its shipping operations carbon neutral. Through partnerships with carbon offset projects, Etsy calculates the carbon emissions generated by shipping its products and invests in projects that reduce or capture an equivalent amount of carbon elsewhere. This initiative allows Etsy to mitigate the environmental impact of shipping while providing customers with eco-friendly shipping options. It serves as an example of how digital businesses can take responsibility for their carbon footprint throughout their supply chain.

3.3.4 PATAGONIA'S WORN WEAR PROGRAM

Patagonia is an outdoor clothing store. They have a program called Worn Wear that promotes sustainability and reduces textile wastage. In the Worn

Wear program, they want their clients to fix, reuse, or recycle their clothes. This can be done through repair services or trade-ins where clothes can be deposited for new ones. Patagonia even has a website where you can resell your old Patagonia items instead of buying new ones as well as prolong their duration in use. The promotion of circular economies is, therefore, in line with this as it helps reduce costs.

These case studies illustrate how digital businesses can integrate sustainable practices into their operations and supply chains. By prioritizing renewable energy, optimizing data center efficiency, offsetting carbon emissions, and promoting circular economy principles, these companies set examples for others to follow in building a more sustainable future.

3.4 CHALLENGES AND OPPORTUNITIES

Implementing sustainable practices in digital entrepreneurship is not always easy. Let us take a look at some of the challenges digital entrepreneurs face, along with examples of areas where innovation can make a big difference.

3.4.1 ENERGY CONSUMPTION

Digital entrepreneurs face a significant challenge in the form of high-power consumption by data centers and other digital facilities (Yang and Chen, 2019). They are required to consume substantial amounts of energy to operate servers and manage web applications. As an example, well-known streaming services such as Netflix or YouTube utilize enormous amounts of power for video storage and delivery throughout the globe. On the other hand, there exist possibilities for improvement through efficient electronic mechanisms and the use of sustainable power options. Some examples include Facebook and Apple investing in solar and wind power to drive their data centers even as they also make more efficient servers and cooling units to cut down on carbon emissions.

3.4.2 E-WASTE

Among other obstacles encountered in this context, is the matter of disposal of electronic waste, that is e-waste. The e-waste is a result of the discard of digital gadgets such as tablets, smartphones, and laptops. Once discarded, they find their way into landfills, a practice that exposes the environment to hazardous chemicals. Old mobile devices such as smartphones fall under this category as millions of them are discarded every year, thus contributing to the increasing problem of e-waste. Nevertheless, there is still potential for new ways of recycling and renovating these gadgets. Businesses such as Gazelle and Decluttr are good at purchasing and fixing used smartphones, so that they can be reused by someone else. In addition, they also destroy certain computer parts, which would lead to the continued presence of e-waste material in the Earth.

3.4.3 SUPPLY CHAIN TRANSPARENCY

Digital entrepreneurs often rely on complex supply chains to source materials and manufacture products. Ensuring transparency and ethical sourcing throughout these supply chains can be challenging. For example, it is difficult for companies to trace the origin of minerals used in electronic devices, such as cobalt and lithium, which are often mined under dangerous or exploitative conditions. There are chances to innovate in blockchain technology that would offer clear and unchangeable records about past transactions that took place in any supply chain. Take an example, some firms such as IBM or Walmart have utilized blockchain to follow the source of foodstuffs guaranteeing that they are safe for consumption and that the process used to obtain them is ethical. Blockchain can also help in following minerals back to their origins where they were extracted for use in making electronic gadgets, hence encouraging responsible mining.

Digital entrepreneurs can start the way to sustainability by recognizing these challenges and pointing out where there is room for novelty (Zhang and Wang, 2021). They can reduce environmental effects and help to make a more sustainable future by reducing their environmental footprints through using green technologies when it comes to investing in energy-efficient technologies, recycling e-waste, and ensuring to remain open in supply activities.

3.5 TECHNOLOGIES THAT ENABLE SUSTAINABLE PRACTICES

In a world of fast-changing technology today, new digital opportunities are available to entrepreneurs who wish to go green. Let us take a look at how technologies that are still emerging and which include environmentally friendly development methods through which they can be utilized by digital entrepreneurs such as artificial intelligence (AI) and blockchain among others can bring about sustainability in the field.

3.5.1 ARTIFICIAL INTELLIGENCE

AI possesses the capability of revolutionizing multiple aspects of digital enterprise such as resource management, energy efficiency, and waste minimization. For example, AI-powered algorithms can optimize energy consumption in data centers by dynamically adjusting server workloads based on demand. This helps reduce energy waste and lowers operational costs for businesses. Additionally, predicting demand, optimizing transportation routes, and identifying opportunities for waste reduction are some of how AI can enhance supply chain management. For example, AI algorithms are used by Walmart to analyze sales data and optimize inventory levels to minimize excess stock levels and reduce wastage. Also, automation driven by AI enhances business processes, thus eliminating the need for manual labor as well as workflows based on papers, hence resulting in increased efficiency. Consequently, this not only increases efficiency but also minimizes waste.

3.5.2 BLOCKCHAIN TECHNOLOGY

Blockchain technology is a suitable solution for enhancing supply chain transparency and sustainability since it operates as a decentralized and transparent tool to track and verify transactions. By putting down every transaction in a permanent ledger, the network provides an opportunity to trace everyone to the source. Take the fashion industry as an example, VeChain and Provenance are some of the companies that apply blockchain for tracing materials' sources and to guarantee that they are sourcing ethically. Blockchain gives buyers evidence-based data on how goods are made making it easier for them to select goods that meet their needs, which also promotes sustainability in businesses. Moreover, blockchain can facilitate the creation of carbon offset markets, where companies can buy and sell carbon credits to offset their carbon emissions. This incentivizes businesses to reduce their environmental footprint and invest in renewable energy projects.

Digital entrepreneurs can enable sustainability through such emerging technologies as AI and blockchain in business ventures they engage in.

These technologies provide novel responses to all the environmental challenges one faces be it in energy use optimization or transparent supply chain enhancement and carbon emission reduction. For a future that can be sustained, it is necessary to adopt green practices that are supported by emerging technologies as digital business startups grow.

3.6 POLICY IMPLICATIONS

When it comes to digital entrepreneurship and environmental concerns, the existing regulatory framework plays a crucial role in shaping business practices. Let us explore the current regulatory landscape and potential recommendations for enhancing environmental sustainability.

3.6.1 *EXISTING REGULATORY FRAMEWORK*

The regulatory framework governing digital entrepreneurship varies across different countries and regions. While some jurisdictions have specific laws addressing environmental issues, others may have limited or fragmented regulations in this area. For instance, the Sweeping Waste Equipment [Waste Electrical and Electronic Equipment (WEEE)] Law of the European Association exercises such as gathering, retrieving, and right dumping of electronic waste. Again, the Restriction of Hazardous Substances (RoHS) Directive stops specific pernicious substances employed in electronic equipment and comes up with strategies to address the e-waste. In contrast, countries such as the United States may have less comprehensive regulations governing electronic waste management, with individual states implementing their recycling programs and initiatives.

3.6.2 *EFFICACY OF CURRENT LAWS*

Despite the existence of some environmental regulations, the efficacy of current laws in addressing environmental concerns in digital entrepreneurship may be limited. Enforcement mechanisms, compliance requirements, and monitoring systems may vary, leading to inconsistencies and loopholes in implementation. For instance, while the WEEE Directive aims to reduce electronic waste, compliance rates among digital businesses may vary, and enforcement actions against noncompliant companies may be insufficient. Additionally, the rapid pace of technological innovation may outpace regulatory updates, making it challenging for laws to keep up with emerging environmental risks associated with digital entrepreneurship.

3.6.3 RECOMMENDATIONS FOR REGULATORY ENHANCEMENTS

There may be regulatory enhancements in digital entrepreneurship to promote ecocompliant businesses. This is achieved through enforcing compliance with more severe punitive measures or increased surveillance mechanisms, thereby discouraging air pollution by ensuring that the involved companies release minimalistic levels of greenhouse gases into the atmosphere.

The process of promoting Extended Producer Responsibility (EPR) involves holding digital enterprises accountable for the complete life cycle of what they make, ranging from production to disposal, which could pressure them into designing environmentally friendly goods and supporting the establishment of recycling facilities as well as waste disposal methods.

- Incentivizing ecolabeling and certification: Implementation of ecolabeling schemes and certification programs can assist consumers learn which goods and services maintain sustainability environmentally. Such incentives can be used to encourage digital companies to adopt green practices to distinguish themselves in the market.
- Stimulating collaboration and sharing of knowledge: Stimulating collaboration is important in digital entrepreneurship and this can be achieved when governments and civil societies work together and with the same token business because it enhances the flow of best practices based on technological developments and findings from research projects about sustainability in the environment within the entrepreneurial process.

Take for instance, the European Union may want to think about extending present directives, so that they incorporate digital products and services, while on the other hand, the United States may decide to look into an Act of Congress, which would make nationwide all electronic waste management practices.

In conclusion, there is a need to improve the regulative framework concerning digital entrepreneurship to attend to ecological problems sustainably, so that a supportive platform for digital age business operations that are environment-friendly can be created; regulators are supposed to enhance compliance strategies, lobby for product life extension via company buy-back schemes, and adopt ecolabeling while at the same time promoting cooperation.

3.7 FUTURE OUTLOOK

Let us break down and analyze what the future holds for digital entrepreneurship regarding environmental sustainability.

3.7.1 EVOLVING CUSTOMER EXPECTATIONS

More people become aware of issues about the environment, more customers worry about how products and services affect it. They are now preferring organizations that promote conservation and eco-friendly activities. To illustrate, some individuals would rather purchase goods from firms that recycle their stuff, lessen their carbon prints, or endorse green power (Jasen et al., 50).

3.7.2 EMERGING TECHNOLOGIES

Advancements in technology, such as AI, blockchain, and the Internet of Things (IoT), offer new opportunities for digital entrepreneurship to enhance environmental sustainability. For instance, AI can optimize energy usage and resource allocation, while blockchain can improve supply chain transparency and traceability. IoT devices can monitor environmental parameters and enable more efficient resource management.

3.7.3 POTENTIAL SHIFTS IN THE REGULATORY LANDSCAPE

Governments and regulatory bodies may introduce new laws and regulations to address environmental concerns in digital entrepreneurship. These regulations could include stricter emission standards, waste management requirements, or incentives for sustainable practices. For example, governments may implement carbon pricing mechanisms or offer tax breaks for companies that invest in renewable energy.

3.7.4 ALIGNMENT OF STRATEGIES WITH SUSTAINABLE PRACTICES

To succeed in the years that are to come, digital companies should make sure that they follow sustainable practices in their strategies, which means that they would have to design products with consideration given to the environment, manufacture them with a green supply chain, and market them using environmentally conscious marketing strategies. Touching every area of their functional frameworks, including everything from manufacturing lines, and assembling points down to transportation methods, these ensure that all materials are produced in an eco-friendly manner when considering how technological developments are important (Clark). For instance, this may entail creating green products, adopting energy-saving devices, or using recycle process that reduces garbage while at the same time maximizing the usage of resources available on Earth. In predicting these upcoming changes and making the first move to sustainability of the environment, these digital entrepreneurs can get a head start over others who have not started yet. They will find a new group of environmentally correct buyers and also take part in making our world less endangering. A case in point is the positioning of Tesla, which has established itself as a pacesetter regarding green transportation through the production of electrically powered cars besides plunging money into alternative power sources. In the same way, Patagonia and other organizations have developed a helpful presence among consumers by taking the lead in incorporating ecological accountability and moral corporate behaviors into their operations.

3.8 CONCLUSION

In conclusion, this research by Sayonee Das sheds light on the crucial relationship between digital entrepreneurship and environmental sustainability. By examining how technology impacts our planet, the chapter emphasizes the importance of adopting eco-friendly strategies in business operations.

Throughout the chapter, we have seen how digital entrepreneurship can both help and harm the environment. While digital tools offer efficiency and innovation, they also contribute to energy consumption, waste generation, and carbon emissions. However, there are opportunities for businesses to mitigate these negative impacts by embracing sustainable practices.

From renewable energy commitments to carbon offset programs, the examples provided showcase how digital businesses can lead the way in environmental stewardship. Prioritizing sustainability in their operations can help companies minimize their environmental impact, attract eco-friendly customers, and improve their chances of long-term prosperity. Looking ahead, the future of digital entrepreneurship hinges on addressing environmental concerns. As customer expectations evolve and emerging technologies offer new possibilities, businesses must align their strategies with sustainable practices. Digital entrepreneurs can smooth out a way to a more sustainable commercial future by adopting eco-friendly innovations and calling for regulatory advancements. More importantly, this study brings light to the significance of a more sustainable economy through green growth enabled by the cooperation of digital entrepreneurs, policymakers, and consumers. We, humans, must unite forces to design a future in which digital innovation and environmental responsibility are closely interconnected to develop a world that we will enjoy living in for future generations.

KEYWORDS

- **digital entrepreneurship**
- **carbon footprint**
- **waste footprint**
- **energy footprint**
- **sustainability**
- **environmental impact**
- **technological enablers**
- **sustainable business practices**
- **policy implications**

REFERENCES

Adams, S. (2021). Digital entrepreneurship and sustainability: A review of the literature. *Journal of Sustainable Business*, *35*(2), 87–104.

Evans, R., Patel, N. (2019). The carbon footprint of digital entrepreneurship: A comparative analysis. *Environmental Science & Technology*, *37*(3), 210–225.

Williams, K., Martinez, E. (2017). Greening digital entrepreneurship: Strategies for sustainability. *Journal of Sustainable Business Practices*, *25*(4), 301–315.

Yang, J., Chen, L. (2019). Environmental implications of digital entrepreneurship: A meta-analysis. *Journal of Environmental Management*, *42*(1), 55–68.

Zhang, Q., Wang, S. (2021). Sustainable digital entrepreneurship: Opportunities and challenges. *Journal of Business Ethics*, *19*(2), 123–137.

CHAPTER 4

Sustainable Stays: Assessing Traveler's Satisfaction in a Post-COVID-19 Era

MEGHA AGGARWAL,[1*] MANISH BADONI,[2] and BABITA RAWAT[1]

[1]*Uttaranchal Institute of Management, Uttaranchal University, Dehradun, India*

[2]*Uttaranchal School of Hotel & Hospitality Management, Uttaranchal University, Dehradun, India*

*Corresponding author

ABSTRACT

By analyzing the steps hotels have taken to ensure that guests are safe, the current study aims to analyze consumer satisfaction and their intent to return to the hotel establishment that conclusively provides a sustainable solution to the hospitality industry. The study was conducted in Mussoorie, often referred to as "The Queen of Hills," which is the most popular tourist destination in India, to determine if visitors were still seeking sufficient safety measures to mitigate the risk of the pandemic. The study identifies the factors with the help of principal component analysis and factor analysis. Further, the study uses the multiple regression model to develop a relationship between the factors, followed by structural equation modeling to analyze the results. According to the study's findings, customers are most satisfied with hotels that emphasize cleanliness and hygiene, and they expressed intentions to return during their future trips. The study also revealed that tourists are keen to visit domestic destinations, provided that both the public sector and private property owners adhere to the necessary safety and sanitation standards. Customers who have been at the same

Industry 5.0: Innovation and Adoption for Sustainability.
Tilottama Singh, Pradeep Suri, and Zulkarin Jahangir (Eds.)

DOI: 10.1201/9781779641878-4

hotel before do not increase their degree of satisfaction and have apparently noticed the COVID-19 risk. The customers are reluctant to revisit the hotel because of least security measures, and this attitude of tourists shows a path block in the growth of the hospitality industry as a whole.

4.1 INTRODUCTION

COVID-19, during the underlying episode in Wuhan, the infection and sickness were ordinarily alluded to as "coronavirus" or "Wuhan coronavirus" and were highly contagious in nature.[1] COVID-19 pandemic, being an unprecedented situation, disrupted every aspect of life throughout the globe. All the organizations, workplaces, and transport were shut down to curb the menace of the coronavirus. A total closure implies a setup to all business' economies and the life of common men. The hospitality industry and the travel industry, being client-driven and manpower driven, suffered the most (Alam and Parveen, 2021). Other ventures such as FMCG, banking, training, or IT were given little unwinding, yet the hospitality or the travel industry, as most times connected with recreation and leisure, could not get any unwinding. Like other countries, the Indian Government, to safeguard the existence of residents of India, declared a total lockdown on March 21, 2020, bringing everything to a halt except emergency services.

India is one of the favorites among domestic and foreign travelers due to its rich cultural, historical, and religious sights with a unique ideology of Atithi Devo Bhav[2] (guests are to be considered as "Bhagwan"). The Indian travel industry creates job opportunities for lots of job seekers. The travel industry is assessed to have straightforwardly contributed 6.8% to the gross domestic product in 2019 and 8% to the work of the country in 2020 (IBEF Report).[3] If the indirect contribution of the tourism industry will be included, the percentage share would go up to 5.2% and 15.3%, respectively. Nonetheless, the pandemic has not just undermined the development possibilities of the area yet, in addition, caused a financial haul, which the area is requiring some investment to recuperate (Alam and Parveen, 2021).

---

[1]https://www.who.int/health-topics/coronavirus#tab=tab_1

[2]https://en.wikipedia.org/wiki/Atithi_Devo_Bhava

[3]https://www.ibef.org/industry/tourism-hospitality-india

4.2 LITERATURE REVIEW

Studies in this short pandemic period have analyzed various issues in the travel and tourism industry. It has been identified the COVID-19 responses by the government and industry in South Africa. Study reveals that due to rapid changes in policy and regulations by the government in tourism sector, it creates differences between stakeholders (Rogerson and Rogerson, 2020). A study has identified two types of limitation factors for tourists: (1) internal and (2) external. External factors include safety and security, lack of information, distance, and transportation (Suhud, 2022). It is also identified the risk perceptions of travelers by taking South Africa and Zimbabwe as competing destinations, where tourists consider South Africa as a safe destination compared to Zimbabwe, though risk perceptions have a negative impact on the brand image of the specific destination post-COVID-19 (Mandina and Preez, 2022). Similar factors have been identified in previous research (Kozak et al., 2007; Page et al., 2012; Sharifpour et al., 2014; Sönmez et al., 1999; Williams and Baláž, 2015) that risk perceptions of travelers are the biggest reason for the reduction in tourist numbers or arrivals if there is some pandemic or virus attack. The author studied the impact of COVID-19 on MICE tourism in Rwanda and found that MICE tourism is negatively affected due to COVID-19 not only in Rwanda but around the globe (Rwigema, 2020). The visit for MICE tourism has been replaced by virtual meetings and events, which is unfavorable for the tourism industry to survive. Tourist also cancels their plans to any destination if there is some threat to their health, as in the cases of pandemics (Nicoll, 2006) and concern for their safety have been identified as a major factor while traveling (Crotts, 2003).

Due to low COVID-19 cases, various economies have started to entertain tourists in a very restrictive and planned manner. A study by Sucheran (2022) identified the economic impact of COVID-19 on South African guesthouses, noting a decline in revenue generation within this specific sector of tourism in South Africa. Guesthouses have started identifying the strategies for the future to address and resolve the issues related to the COVID-19 pandemic. Being the first country to spread the virus, China has also become the first to entertain tourists with strategies such as slow and smart tourism with reconsideration of service designs (Wen et al., 2021). Tourists are becoming confident while traveling their own country with shorter distances, the places that are not too far from their home-based stations (Enger et al., 2020). Similar results have been found that tourists are willing to travel to destinations within the nation for the speedy recovery of their tourism sector post-COVID-19 and also to become a support by creating positive publicity (Bama and Nyikana, 2021). Guided tour packages or tours with big groups are being avoided by tourists, and they are preferring groups with a minimum of 10 members with ease of travel (Enger et al., 2020). According to a survey (Eric Jones, 2023), 87% travelers wish to travel their own country, and 56% travelers wish to go on holiday with their own vehicles. Whereas 65% and 55% of travelers were willing to travel to beaches, restaurants, and bars, respectively, if social distancing norms were fulfilled.

So, after the second and third waves of COVID-19,[4] people have started to get back to their normal life, but this turning point has reached an advanced stage of "new normal." In India, destinations are also trying to transform according to the "new normal" phenomenon for better and safer customer experiences. Tourists have started moving out and exploring (Dileep and Nair, 2021), and a term called revenge travel (Agrawal et al., 2023) is getting very common among people in India to take shorter breaks or for workcations rather than vacations. In India, destinations such as Shimla, Mussoorie, Nainital, Dalhousie, Auli, Dhanaulti, and Dharamshala are attracting tourists from Delhi, Uttar Pradesh, Haryana, Punjab, Rajasthan, and Bihar due to their proximity and carry weightage among other nearby destinations. This study specifically focuses on the destination, Mussoorie, to identify the satisfaction of the travelers with their hotel stay if COVID-19 guidelines are being properly followed. Hotels with high cleanliness and hygiene would result in high customer satisfaction and increase the likelihood of guests intending to return for their next stay.

Due to the decline of COVID-19 cases after the third wave of coronavirus, the Government of India and the Uttarakhand Government relieved[5] the COVID-19 guidelines partially, which attract tourists from nearby places to visit the capital of Uttarakhand, Dehradun, and Mussoorie. During the opening ceremony of Food Festival 2021, organized at Mussoorie, Mr Dilip Jawalkar[6] stated that *due to Xmas and New Year, there might be a huge footfall of tourists in Mussoorie, and tourists across the world can taste the local cuisines of Uttarakhand (Pahadi Pakwaan). Tourism*

---

[4]https://www.mha.gov.in/notifications/circulars-covid-19

[5]994/USDMAV792(2020) TC-2

[6]https://uttarakhandtourism.gov.in/sites/default/files/document/type/press_0.pdf

Department has made special arrangements for the same. The number of tourists traveling to Mussoorie was 12.29 lakhs in the year 2021. After the COVID-19 relaxation, the destination experienced more than 11 lakh tourists[7] from February 2022 to April 2022. Destinations in Uttarakhand, such as Mussoorie, are dependent on the tourism sector and face huge losses during COVID-19, have tried to lift their economy by getting the maximum number of tourist arrivals.

Mussoorie, a hill station in Uttarakhand, known as the "*Queen of the Hills*," and is considered one of the top hill stations of India, is famous among travelers as a weekend gateway due to less travel distance from the National Capital Region, that is, Delhi. Mussoorie is considered as a perfect summer destination with scenic views and a home of personalities such as Mr Ruskin Bond.[8] The destination has various places of attraction such as religious places, Gun Hill, Company Bagh, Kempty Fall, Landour, and Mall Road for shopping and food. With a distance of around 35 km from the capital city of Uttarakhand, Dehradun, Mussoorie offers a variety of activities, including trekking, ziplining, angling, and winter sports. Around the destination, travelers also seek to visit places in nearby cities such as Dehradun and Dhanaulti.

4.3 RESEARCH METHODOLOGY

4.3.1 RESEARCH GAP AND PROBLEM STATEMENT

Numerous studies have been done in the past focusing on COVID-19 and its various impacts on different industries. There are also researches based on identifying the post-COVID-19 impact on customer satisfaction and their willingness to adopt the change. A similar study has been done by Bama and Nyikana (2021) with respect to South African domestic destinations. The authors have used a mixed methodology of quantitative and qualitative analysis by taking all stakeholders as respondents for the study. Other various authors (Kaushal and Srivastava, 2021; Patil et al., 2020; Patel et al., 2020; Vasant Jagdale et al., 2021; Wangmo and Dubey, 2022) have studied the post-COVID-19 effect on Indian tourism and their learnings and implementations. Though few studies (Borges et al., 2022;

[7]https://www.india.com/news/india/dehradun-mussoorie-uttarakhand-unlock-latest-update-record-over-11-lakh-tourists-in-last-three-month-after-govt-relaxes-covid-curbs-hotel-cinema-halls-malls-markets-5342896/

[8]https://uttarakhandtourism.gov.in/index.php/destination/mussoorie

Matiza and Slabbert, 2022; Rather and Najar, 2022; Yagmur and Aksu, 2021) have identified the role of new habit, perceived risk while traveling and staying at hotels but still in the context of country India, the researcher is unable to find any study based on post-COVID-19 customer satisfaction in hotel stay specifically in India and in the region of Uttarakhand. In this study, factors related to COVID-19 have been identified through previous studies, that is, fulfillment of COVID-19 guidelines, tactics adopted by hotels to attract travelers, and travelers' previous experience of stay have been taken to analyze customer satisfaction and their intention to return to the same hotel on their next travel.

4.3.2 RESEARCH VARIABLES AND MODEL

The current study talks about the satisfaction of customer post-COVID-19 with respect to their stay at hotels. Thus, the satisfaction of the customers or travelers has been kept as a dependent variable, which will generate a feeling of intent to revisit the destination, whereas other variables such as hotels are fulfilling guidelines on cleanliness and hygiene-related issues; hotels are using various publicity tactics to attract customers during post-COVID-19, and customers have some previous experience of staying in a specific hotel while selecting have been kept as independent variables for the current study. On the basis of the above variables, following model has been designed to analyze and verify, as shown in Figure 4.1.

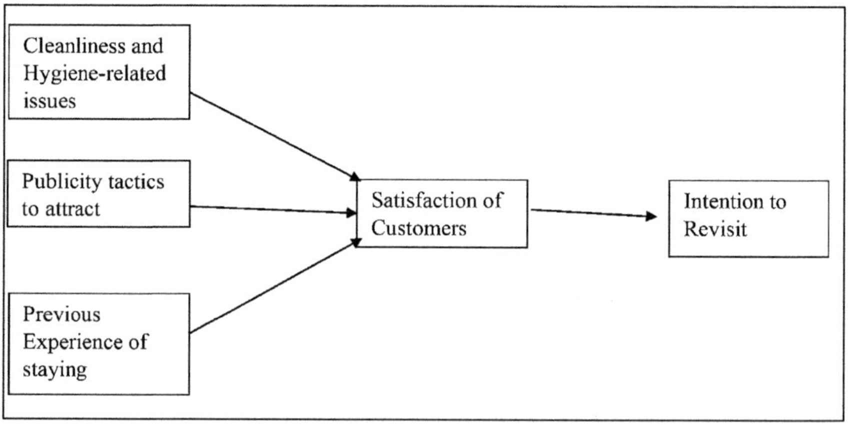

FIGURE 4.1 Research model.

Source: Authors' compilation.

4.3.3 HYPOTHESIS DEVELOPMENT

The current study discusses whether customer gets satisfaction during their stay in hotels with respect to the COVID-19 protocol or not. The purpose of this study is to examine the variations in customer satisfaction based on whether hotels are adhering to COVID-19 protocols. The following three null hypotheses have been designed for the same:

1. H1: There is no significant difference in customer satisfaction if hotels fulfill the COVID-19 protocol.
2. H2: There is no significant difference in customer satisfaction if hotels publicize their COVID-19-related norms.
3. H3: There is no significant difference in customer satisfaction in hotels that customers have already experienced in terms of stay.

4.3.4 RESEARCH INSTRUMENT

The study uses a self-administered structured questionnaire divided into two sections. The first section is related to the demographics of the respondents such as age, gender, education, and the information related to their number of travels, especially in the post-COVID-19 period, that is, from July 2022 to June 2023. The purpose of this question is to know the number of travel respondents did in post-COVID-19 duration because after the second wave, Indian tourists became reluctant to stay at home and they started traveling to nearby destinations that could be reached using their own vehicles. The majority of the respondents shared that avoiding public transport was a big help to explore nearby places and cover the maximum destinations of the specific place.

The second segment of the questionnaire relates to the information about variables, including norms of COVID-19 followed by hotels, self-identified or guided by governing bodies, discounts offered by hotels due to a smaller number of tourists, and the satisfaction of the respondent post-COVID-19 by the services related to COVID-19. A Likert scale ranging from 1 (strongly disagree) to 5 (strongly agree) has been used for data collection with 25 items. After pilot testing with 20 respondents, a few questions, which found ambiguous or unclear or repetitive in nature, have been reduced to 13, comprising all related variables.

4.3.5 SAMPLING AND DATA COLLECTION

The current study specifically collected data through one city of Uttarakhand, Mussoorie, and convenience and simple random sampling have been used to collect data. Travelers have been approached to fill out the questionnaire in hard form, but many of the travelers denied it due to time or inconvenience. Due to the paucity of time and nonresponse of the travelers, it was difficult to fill out the questionnaire, and after approaching around 400 travelers, only 220 filled out the questionnaire. The data was collected from July 2023 to August 2023.

4.4 DATA ANALYSIS AND RESULTS

SPSS 25 and AMOS 25 were used for analysis purposes. The responses received in an Excel sheet, which has been converted to an SPSS file after cleaning the data in terms of missing information, and coding and labeling of variables have been done accordingly. As stated earlier, out of 222, only 157 responses were found suitable and included in the analysis.

The descriptive statistics of the data show that most of the tourist belongs between the ages of 25–35 (57%), and 35–45 (34%) years, and the rest belong to the age group of 45 years and above. The gender distribution shows that the majority of the respondents were females (55%). Over 70% of respondents had a postgraduate degree, and 30% respondents had an undergraduate degree. Out of 157 respondents, 70% respondents were from the nearby states of Mussoorie, including Uttarakhand, Uttar Pradesh, Delhi, Haryana, Punjab, and Rajasthan. This high number of respondents from nearby states indicates that tourists prefer nearby places rather than any other place far from their residing city.

4.4.1 RELIABILITY ANALYSIS

The stability, consistency, and dependability of a set of measurements or a questionnaire can be analyzed with the help of reliability analysis. It is used to determine the reliability or consistency of a measurement instrument or a set of data. Cronbach's alpha is the most frequently utilized measure of internal consistency, particularly for scales and questionnaires. It evaluates the effectiveness of a set of items in gauging a singular concept or attribute.

TABLE 4.1 Reliability Analysis

Reliability Statistics		
Cronbach's Alpha	**Cronbach's Alpha-Based on Standardized Items**	**Number of Items**
0.913	0.924	13

Source: Authors' compilation.

Table 4.1 shows the value of Cronbach's alpha; for the total scale, it was 0.913, which is above the acceptable threshold (0.70) as recommended by Hair et al. (2006). The value of Cronbach's alpha after deleting some of the items from the questionnaire will remain greater than 0.900 after which is also acceptable for the questionnaire used in this analysis.

4.4.2 *DIMENSION REDUCTION OR FACTOR ANALYSIS*

Principal component analysis (PCA) with varimax rotation was used for the identification of the factors with their loadings, and a few items were not found suitable for the construction of factors and, hence, removed from the final analysis. It was found that the Kaiser–Meyer–Olkin (KMO) value was 0.898, and a significant Bartlett's test of sphericity strongly confirmed the use of factor analysis to extract the factors that are assumed to affect the satisfaction of travelers post-COVID-19.

There was a total of 13 items in the questionnaire for the identification of the factors, and any value of a factor loading with or above 0.5 has been taken for further analysis for which 12 items were found suitable for the analysis (Q6 has been removed due to values less than 0.5).

By using PCA, four factors were extracted and explained about 68% of the variance. Any factors that have eigenvalues greater than one were taken, and the rest were removed in the final analysis. With the help of PCA, the following four factors have been extracted that influence the satisfaction of travelers post-COVID-19, as shown in Table 4.2.

4.4.3 *MULTIPLE REGRESSION ANALYSIS: THE DEVELOPMENT*

After factor reduction, multiple regression analysis was used for the current study to identify the relationship between the dependent variable and the independent variables. The developed equation for the multiple regression model is given below:

$$\text{SATFAC} = \alpha + \beta_1\text{CLHY} + \beta_2\text{PUBCOV} + \beta_3\text{PRVEXP} + \varepsilon$$

TABLE 4.2 Results of Rotated Component Matrix

Item	Factor 1	Factor 2	Factor 3	Factor 4
Cleanliness and hygiene-related issues (CLHY)				
The hotels you stayed have adequate equipment such as sanitizers and temperature testing machine to follow hygiene	0.788			
The physical facilities such as rooms, food, dining hall, and others were visually appealing in terms of cleanliness and neatness	0.716			
Employees were well dressed and appeared to be neat and clean	0.750			
The hotels you stayed have adequate facilities for social distancing	0.806			
Previous experience of customers while selecting the hotel (PRVEXP)				
You have chosen this hotel due to your familiar terms with the manager or staff		0.660		
You have chosen this hotel due to your previous stay experience		0.650		
Publicity tactics by hotels post-COVID-19 (PUBCOV)				
You have chosen this hotel due to special discounts due to COVID-19			0.565	
You have chosen this hotel because of the publicity of health and safety measures by the hotel, especially after COVID-19			0.776	
You have chosen this hotel because it follows the corona-related health guidelines issued by the government			0.797	
Satisfaction of customers post-COVID-19 (SATFAC)				
You believe that you receive the worth of your money after choosing this hotel in post-COVID-19 period				0.804
Overall, you are satisfied with all facilities provided by the hotel				0.835
You can revisit the hotel again on your next trip				0.808

Source: Authors' compilation.

Where three different factors, cleanliness and hygiene-related issues (CLHY), publicity tactics by hotels post-COVID-19 (PUBCOV), and previous experience of customers while selecting the hotel (PRVEXP) have been identified as independent variables, IV 1, IV 2, and IV 3, respectively. The dependent variable (DV) in this study is satisfaction of customers post-COVID-19 (SATFAC). α is denoted as the constant term, and β is the coefficient of variables where ε has been taken as the residual.

Before attempting multiple regression analysis, assumptions related to the analysis have been checked, and it was found that all assumptions of multiple regression analysis have been met except one, residual values are normally distributed. However, as only extreme deviations from normality are likely to have a significant impact on findings, the results can be said to be valid in this case because the distance of the value from the line is not too far.

4.4.4 MULTIPLE REGRESSION MODEL: THE ANALYSIS

The model that has been designed for the analysis of post-COVID-19 satisfaction by travelers by using hotel, the variables such as cleanliness and hygiene-related issues, post-COVID-19 publicity tactics, and previous experience of the customers are affecting the satisfaction of the travelers, which can decide their intention to revisit the place. Tables 4.3and 4.4 show the values of each variable by which the equation has been developed.

TABLE 4.3 Regression Summary

R	*R* Square	*F*	Sig.	Durbin–Watson
0.803	0.644	92.364	0	1.969

TABLE 4.4 Regression Beta Coefficient

Regression Coefficients	*B*	Sig.	Tolerance	VIF
Constant	2.22	0.001		
Clean and hygiene	0.538	0	0.579	1.72
Publicity	0.112	0.05	0.453	2.2
Previous travel experience	0.031	0.594	0.657	1.52

Source: Authors' compilation.

The value of *R* square is 0.644, which indicates that approximately 64% of the variation in the dependent variable is explained by the independent variables. Value of *F*-test with significant *p*-value indicates that the model is significant for predicting the dependent variable based on identified independent variables. Thus, the multiple regression model with all four indicators/variables produced $R^2 = 64\%$, $F\ (4, 157) = 92.36$, $p \leq 0.05$. It can be seen from Table 4.4 that IV 1 and IV 2 contribute much to the model with significant *p*-values, but IV 3, the previous experience of travelers, does not contribute any significant impact post-COVID-19 satisfaction ($B = 0.031, p = 0.594$). So, the other two factors, (1) cleanliness and hygiene (IV 1) and (2) publicity tactics (IV 2), are significantly contributing to the model with significant *p*-values. Thus, null hypotheses, H_1 and H_2, are being rejected in this analysis, but H_3 cannot be rejected due to an insignificant *p*-value. Further, the multiple regression model can be summarized below:

$$\text{SATFAC} = \alpha + \beta_1\text{CLHY} + \beta_2\text{PUBCOV} + \beta_3\text{PRVEXP} + \varepsilon$$

$$\text{SATFAC} = 2.22 + 0.538\text{CLHY} + 0.112\text{PUBCOV} + 0.031\text{PRVEXP}$$

All the variables have been used for the model, though the previous experience indicator (IV 3) did not contribute much to the model.

4.4.5 STRUCTURAL EQUATION MODELING MODEL RESULTS

To analyze the structural relationship between variables, structural equation modeling (SEM) has been performed with the help of AMOS 25, and model fit identification has been done. Figure 4.2 and Table 4.5 show the results and analysis.

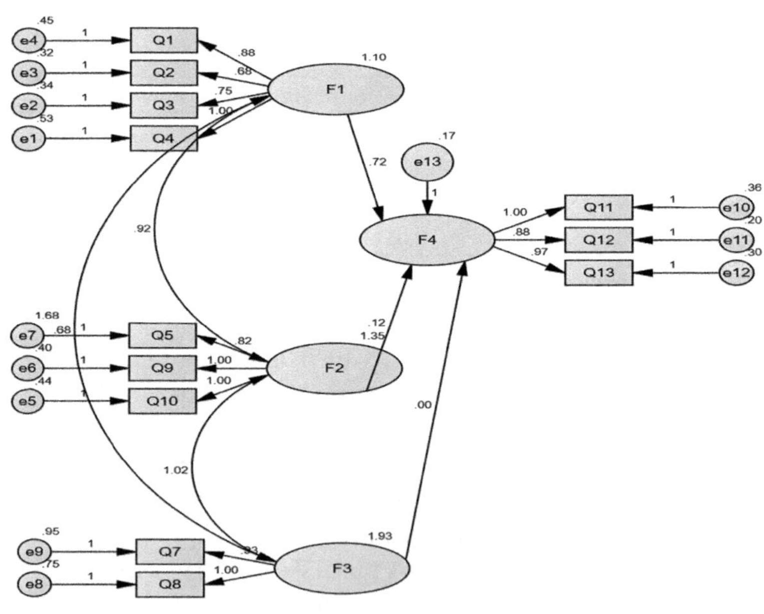

FIGURE 4.2 AMOS results.

4.4.6 MODEL FIT ANALYSIS: RESULTS AND DISCUSSION

The quality of fit of the model is adequate, with CMIN = 117.04, DF = 48, with $p = 0.000$. Table 4.5 shows the results of SEM.

TABLE 4.5 SEM Results

CMIN/DF	RMR	GFI	CFI	RMSEA
2.438	0.106	0.894	0.945	0.096

Source: Authors' compilation.

Table 4.5 shows CMIN/DF (2.438) for the default model is less than 3, which indicates an acceptable fit. RMR stands for root mean square residual, and smaller the RMR value, better the fit. In this case, it is 0.10, but here the value of interest is GFI, where 1 represents the perfect fit. In this model, the value of GFI is 0.89, closer to 0.9, indicating a reasonable fit of the model.

Table 4.5 also shows the value of interest is CFI, which is close to 0.95; thus, again the model can be said to be a reasonable fit. The further value of the interest is RMSEA, where values higher than 0.1 are considered poor. The values between 0.08 and 0.1 are considered borderline and a reasonable fit. In this model, the value of RMSEA is 0.96, defining the model as a reasonable fit.

At the end, it can be said that the model is a reasonable fit with the analyzed factors, and customer satisfaction can be achieved with proper hygiene and by fulfilling the COVID-19 protocols. The planned and rigorous marketing and publicity strategies to communicate these hygiene policies to customers also affect the satisfaction of the travelers, whereas any previous experience of the customer with similar hotels has less effect on customer satisfaction. The higher satisfaction of travelers will lead to creating their intention to revisit a similar hotel on their next travel to a similar destination.

4.5 IMPLICATIONS TO THE INDUSTRY

Coronavirus has impacted life in every space and also put a pause on daily activities. Governments of every country also restrained people from traveling to any social places such as restaurants, cinema halls, or events (Yadav and Mohsin Qureshi, 2021). Schools and colleges were also being shut down due to the virus, and maximum operations were done with the help of information technology tools. The tourism industry has also faced a declining trend in terms of footfalls of travelers, revenue generation, employment, etc. But just after the relaxation of local and national governments, people have started to move out of their houses for some fun or recreational activities to get rid of the dullness of the pandemic. To gain customers' maximum satisfaction and as the industry adaptation, the safety protocols and procedures should be better aligned with travelers' expectations and their concerns post-COVID-19. This study is suitable for enhancing the traveler's confidence in choosing accommodation options by providing insights into the effectiveness of safety measures implemented by hotels. Hotels with superior safety measures could gain a competitive edge in the market and showcase their commitment to safety and customer satisfaction through the marketing and rebranding of their name.

This research is also helpful for policymakers and industry stakeholders to develop a standard safety practice to be maintained in the post-COVID-19 era. The hospitality and tourism industry could revive again by instilling confidence in travelers to resume travel and stay in hotels. The public health authorities also try to understand the factors of traveler's satisfaction with safety measures and should develop and spread awareness regarding safer travel practices. This study also contributes to the broader body of knowledge in hospitality and tourism management, specifically in the situations of postpandemic recovery and adaptation. This study further proposes the adoption of robust safety measures as a strategic approach to safeguard the long-term viability of the hotel industry, minimizing the potential for future outbreaks.

4.6 CONCLUSION

The current study discusses the satisfaction of travelers in their hotel stay with a specific destination of Uttarakhand, Mussoorie, which in turn can define their intention to revisit the hotel. The study found that hotels that are fulfilling the COVID-19 protocols completely and maintaining their cleanliness and hygiene are able to satisfy their customers with proper marketing strategies and by using modern technologies to ensure safety from COVID-19. Similar findings have been identified in a study (Woyo and Ukpabi, 2022) that the transition and efficiency of tourism has been increased due to smart tourism, and destinations are providing safe places to the travelers in South Africa after COVID-19.

On the other hand, any previous experience with the hotel did not affect their satisfaction level during the post-COVID-19 duration. Although the period of COVID-19 is in a declining phase, however, this pandemic has given a new area to the tourism industry, which includes proper sanitization, cleanliness, and hygiene-related issues to be taken care of well by hotels even with fewer occupancies. This trend of domestic tourism or using private vehicles might not have a long journey in this industry, but travelers are seeking places to visit and identifying the possible options with less risk, which is helpful for policymakers to redesign the image of the destination by including such factors (Mkwizu, 2021). Thus, having a less risky place of stay with proper communication channels by hotels is increasing satisfaction among travelers.

The study also has some limitations, which can become the base of further research. The study includes only one destination, Mussoorie, for the reason that the destination is in close proximity to places such as Delhi, Punjab, Haryana, and Uttar Pradesh; thus, the maximum number of travelers belong to these regions only. The current study includes the factors related to COVID-19 protocols with a specific focus on staying at hotels. Travelers from other regions of the country might demand other factors such as law and order safety mechanisms, learning a new culture, safety while dining, and destination safety. The current study only has taken the travelers viewpoint with regard to the COVID-19 protocol and its fulfillment. Other studies can also include other stakeholders of the tourism industry to identify their opinion about COVID-19 measures and the speedy recovery of the industry. Future research can be done by including more destinations to identify customer satisfaction and their intention to revisit as a comparable study with varied destinations. Studies could also compare satisfaction with two or more destinations with similar factors or by including more factors. Finally, more studies can be done to identify the post-COVID-19 scenario in countries such as India, a major tourist destination for both domestic and foreign travelers.

KEYWORDS

- **tourism**
- **Mussoorie**
- **COVID-19**
- **customer satisfaction**
- **safety measures**
- **hotel accommodation**
- **post-COVID-19 risk**
- **travel satisfaction**

REFERENCES

Agrawal, C., Duggal, T., Gupta, P. (2023). *Revenge Travel* (pp. 72–91). https://doi.org/10.4018/978-1-6684-8726-6.ch005

Alam, M., Parveen, R. (2021). Covid-19 and tourism. *International Journal of Advanced Research*, *9*(4), 788–804. https://doi.org/10.21474/IJAR01/12766

Bama, H. K. N., Nyikana, S. (2021). The effects of COVID-19 on future domestic travel intentions in South Africa: A stakeholder perspective. *African Journal of Hospitality, Tourism and Leisure*, *10*(1), 179–193. https://doi.org/10.46222/ajhtl.19770720-94

Borges, A. P., De Almeida, A. L., Vieira, E., Beluccio, M. (2022). Travelling and discovering new destinations after the Covid-19's lockdown: The role of routines and habits. *Enlightening Tourism. A Pathmaking Journal*, *12*(1), 70–93. https://doi.org/10.33776/et.v12i1.5345

Crotts, J. C. (2003). Theoretical perspectives on tourist criminal victimisation: [Reprint of original article published in v.7, no.1, 1996: 2-9.]. *Journal of Tourism Studies*, *14*(1), 92–98.

Dileep, M. R., Nair, B. B. (2021). COVID-19 and the future of tourism: Back to normal or reformation? *International Journal of Hospitality & Tourism Systems*, *14*(special issue), 10–14.

Enger, W., Saxon, S., Suo, P., Yu, J. (2020). *The-way-back-What-the-world-can-learn-from-Chinas-travel-restart-after-COVID-19-vF2*.

Eric Jones. (2023, April 14). *Survey: Traveling and Entertainment in 2020 Amid COVID-19 Part III—Events & Vacation Locations*. The Vacationer.

Hair, J. F., Black, W. C., Babin, B. J., Anderson, R. E., Tatham, R. L. (2006). *Multivariate Data Analysis*, 6th edn.

Kaushal, V., Srivastava, S. (2021). Hospitality and tourism industry amid COVID-19 pandemic: Perspectives on challenges and learnings from India. *International Journal of Hospitality Management*, *92*, 102707. https://doi.org/10.1016/j.ijhm.2020.102707

Kozak, M., Crotts, J. C., Law, R. (2007). The impact of the perception of risk on international travellers. *International Journal of Tourism Research, 9*(4), 233–242. https://doi.org/10.1002/jtr.607

Mandina, S. P., Preez, E. A. Du. (2022). Travellers' risk perceptions and intentions to visit African destinations amidst COVID-19: The case of brands South Africa and Zimbabwe. *African Journal of Hospitality, Tourism and Leisure, 11*(3), 975–995. https://doi.org/10.46222/ajhtl.19770720.270

Matiza, T., Slabbert, E. (2022). The destination media profile and tourist travel intentions: The mediating effect of covid-19 induced perceived risk. *Advances in Hospitality and Tourism Research (AHTR), 10*(4), 580–604. https://doi.org/10.30519/ahtr.943356

Mkwizu, K. (2021). Re-defining domestic tourism in the new normal: A literature review. In: *Proceedings of the 3rd Africa-Asia Dialogue Network (AADN) International Conference on Advances in Business Management and Electronic Commerce Research* (pp. 38–40). https://doi.org/10.1145/3503491.3503496

Nicoll, A. (2006). Personal (non-pharmaceutical) protective measures for reducing transmission of influenza—ECDC interim recommendations. *Weekly Releases (1997–2007), 11*(41). https://doi.org/10.2807/esw.11.41.03061-en

Page, S., Song, H., Wu, D. C. (2012). Assessing the impacts of the global economic crisis and swine flu on inbound tourism demand in the United Kingdom. *Journal of Travel Research, 51*(2), 142–153. https://doi.org/10.1177/0047287511400754

Patil, A., Naik, S., Patil, D. Y. (2020). *A Study on Impact of Lockdown on Tourism and Post Lockdown Expectations of Tourists*. http://publishingindia.com/atithya/

Patel, P. K., Sharma, J., Kharoliwal, S., Khemariya, P. (2020). The effects of Nobel corona virus (Covid-19) in the tourism industry in India. *International Journal of Engineering Research And, V9*(05). https://doi.org/10.17577/IJERTV9IS050618

Rather, A. H., Najar, A. H. (2022). An empirical assessment of tourist's risk perception and its impact on destination loyalty: A study of Kashmir Valley. *Enlightening Tourism. A Pathmaking Journal, 12*(2), 627–660. https://doi.org/10.33776/et.v12i2.6957

Rogerson, C. M., Rogerson, J. M. (2020). Covid-19 tourism impacts in South Africa: government and industry responses. *GeoJournal of Tourism and Geosites, 31*(3), 1083–1091. https://doi.org/10.30892/gtg.31321-544

Rwigema, P.-C. (2020). Impact of covid-19 pandemic to Meetings, Incentives, Conferences and Exhibitions (MICE) tourism in Rwanda. *Strategic Journal of Business & Change Management, 7*(3). https://doi.org/10.61426/sjbcm.v7i3.1679

Sharifpour, M., Walters, G., Ritchie, B. W., Winter, C. (2014). Investigating the role of prior knowledge in tourist decision making. *Journal of Travel Research, 53*(3), 307–322. https://doi.org/10.1177/0047287513500390

Sönmez, S. F., Apostolopoulos, Y., Tarlow, P. (1999). Tourism in crisis: Managing the effects of terrorism. *Journal of Travel Research, 38*(1), 13–18. https://doi.org/10.1177/004728759903800104

Sucheran, R. (2022). The COVID-19 pandemic and guesthouses in South Africa: Economic impacts and recovery measures. *Development Southern Africa, 39*(1), 35–50. https://doi.org/10.1080/0376835X.2021.2003758

Suhud, U. (2022). Edith Cowan University. In: *The Grants Register 2023* (pp. 432–433). Palgrave Macmillan: UK. https://doi.org/10.1057/978-1-349-96053-8_427

Vasant Jagdale, D., Ganatra, H., Chowdhury, F., Rogers, J., Samuel, S., Tomar, A., Dutta Roy, A., Latollari, E., Hussain, A., Bagch, S., Panchani, M., Singh, S., Manimekalai, K., Iswarya, B. (2021). Article impacts of covid-19 pandemic on tourism industry of India academic Editors: G M Omar. *Asian Microeconomic Review*, *1*(1), 39–45. https://doi.org/10.5281/zenodo.5573943

Wangmo, A. P., Dubey, A. (2022). An assessment of COVID-19 and revival strategies for tourism: A study of Sikkim. *Indian Journal of Applied Hospitality & Tourism Research*, *14*, 33–44.

Wen, J., Kozak, M., Yang, S., Liu, F. (2021). COVID-19: Potential effects on Chinese citizens' lifestyle and travel. *Tourism Review*, *76*(1), 74–87. https://doi.org/10.1108/TR-03-2020-0110

Williams, A. M., Baláž, V. (2015). Tourism risk and uncertainty. *Journal of Travel Research*, *54*(3), 271–287. https://doi.org/10.1177/0047287514523334

Woyo, E., Ukpabi, D. (2022). Building resilient smart cities for sustainable Urban tourism in Africa post-COVID-19 pandemic. In: *Information and Communication Technologies in Tourism 2022* (pp. 256–267). Springer International Publishing. https://doi.org/10.1007/978-3-030-94751-4_23

Yadav, S. K. S., Mohsin Qureshi, M. (2021). Impacts of covid-19 on Indian travel & tourism industry. *International Journal of Trade and Commerce-IIARTC*, *9*(2), 310–318. https://doi.org/10.46333/ijtc/9/2/5

Yagmur, Y., Aksu, A. (2021). Model to determine the perceived risks of tourists staying at hotels where the halal tourism concept is present: Antalya region example. *Enlightening Tourism. A Pathmaking Journal*, *11*(2), 428. https://doi.org/10.33776/et.v11i2.5165

CHAPTER 5

Sustainable Innovation Framework: Lean Six Sigma Approach Toward Industry 5.0

SHRISH SINGH[1*] and MAMTA CHAWLA[2]

[1]*School of Leadership and Management (SLM), Manav Rachna International Institute of Research and Studies, Faridabad, India*

[2]*Amity Global Business School, Amity University, Noida, Uttar Pradesh, India*

*Corresponding author

ABSTRACT

The development of value with a human-centered design is one of Industry 5.0's goals. Lean Six Sigma (LSS) has been effectively employed for optimizing manufacturing activities in the quality field. LSS can perfectly combine more human processes with an improved outcome-oriented environment in a field that values human contribution. The major strength of lean philosophy prioritizes sustainable development while simultaneously focusing on a people-oriented approach, taking care of both the consumers and employees. Moreover, both Industry 5.0 and the well-appreciated techniques of LSS place a strong emphasis on innovation as a means of reducing variations in the process yet achieving iterative improvements in the processes over time. Thus, to achieve process excellence, this chapter proposes a sustainable innovation framework using inductive and integrative methods, which advocates adoption of LSS.

Industry 5.0: Innovation and Adoption for Sustainability.
Tilottama Singh, Pradeep Suri, and Zulkarin Jahangir (Eds.)

DOI: 10.1201/9781779641878-5

5.1 INTRODUCTION

Over the past 10 years, new terminologies in operations management like Smart Factory, Cyber-Physical Production System (CPPS), and Industry 4.0 have emerged because of the continued growth of the concept and potential for integrating information and communication technology (ICT) and the recently introduced technology termed as industrial internet of things (IIoT) into manufacturing processes and systems. An accepted definition is as follows for a rapid separation between the myriad of terminology: The IIoT refers to the technology foundation that uses ICT (Boyes et al., 2018). A long-term approach of complete integration of value-driving strategies known as CPPS, resulting in the Smart Factory, having a production system that uses an IIoT technology base (Lee et al., 2017). Industry 4.0 explains the financial effects of the previously mentioned methods. Since there is no official consensus governing this set of definitions, the terms are occasionally used interchangeably.

In 2011, the phrase "Industry 4.0" was first used in Germany. It serves as shorthand for the fourth industrial revolution, which is brought about by the integration of ICT into manufacturing processes (Veile et al., 2020). First industrial revolution based on the mechanization working on steam-powered machinery, second industrial revolution resulting in mass production using electrical production lines, and automation as third industrial revolution which was based on IT and microcontrollers together provided bases toward Industry 4.0. The foundation of Industry 4.0 as an improved platform for the industry can be explained as the Smart Factory concept, wherein the manufacturing systems are horizontally embedded in the distributed value-adding networks, which further can be managed in real time, as well as vertically connected to business processes within organizations (Pascual et al., 2019). Smart Factories make it possible to meet specific consumer needs and enable the profitable production of even one-off items. Key components of the Smart Factory include (Mabkhot et al., 2018; Osterrieder et al., 2020):

1. **Product and Service Providers:** Product and service manufacturing integrated into a single product or service provider,
2. **Smart Personalized Products:** With the help of CPPS, and
3. **High Level of Collaboration:** Collaborative manufacturing and other value-adding business processes through production networks.

Another crucial element of Industry 4.0 along with Smart factory concept is CPPS, which has the following features (Goetz et al., 2017; Rojas and Rauch, 2019):

1. **Smart Products:** Smart products are those that are tailored specifically to the demands of the client, are instantly identifiable, as well as located, with a previous record of preference and usage, present, and alternate paths to the customer.
2. **Single-Item Production:** Single-item manufacturing should be planned and controlled by CPPS to be cost-effective as well as serial mass production.
3. **Production with Flexible Plans:** As CPPS is required to manage the manufacture of single-item items, it should handle production without rigid plans.
4. **Energy Efficiency:** CPPS serves as the foundation for decisions that lead to more environmentally friendly manufacturing and decision-making processes.
5. **Cyber Security:** Since the CPPS's most vulnerable point is cloud computing, it needs to be protected from hacker attacks.
6. **Production Networks:** All processes that add value together with product development and manufacturing are supported by mutually networked CPPSs.

The COVID-19 pandemic drove the disruption in the value chains across the globe resulting in a reconsideration of the existing global supply-chain paradigms (Bacchetta et al., 2021). It is even though the Industry 4.0, a new industrial platform, had a huge impact and attracted the attention of the academic community and became one of the most popular subjects in recent years. The robustness of the production system, with people as its most important but also most susceptible component, became one of the new paradigms' key focal points. An entirely new paradigm has emerged from an Industry 4.0 workshop held by the European Commission in 2020, highlighting a step toward a new era of Industry 5.0 (Zizic et al., 2022; Mourtzis et al., 2022), which is the integration of a sustainable, human-centric, and resilient industry.

Progress toward Industry 4.0 has always been viewed as an evolution rather than a revolution. Hence, the progress toward Industry 5.0 was anticipated to occur shortly, but it was preponed due to the COVID-19 pandemic. So, in the face of global challenges such as climate change, pandemics, and economic uncertainties, the concept of resilience has gained significant importance across industries. Industry 5.0, often considered as the next phase of industrial evolution, emphasizes the harmonious collaboration between humans and machines (Lasi et al., 2014). Unlike its predecessors, focusing solely on automation and efficiency (Industry 4.0), Industry 5.0 places human values, creativity, and adaptability at the forefront. It acknowledges the irreplaceable role of human intuition, empathy, and problem-solving abilities in complex decision-making processes. The fundamental pillars of Industry 5.0 are depicted in Figure 5.1.

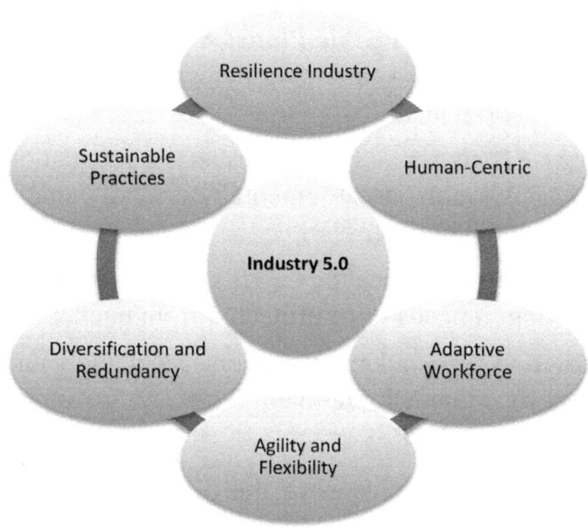

FIGURE 5.1 Pillars of Industry 5.0.

Resilience in Industry: Resilience here refers to the capacity of organizations to adapt, recuperate, and thrive in the face of disruptions (Ahmadi et al., 2020). These disruptions may arise from various sources such as natural disasters, supply chain interruptions, or unexpected market shifts. Traditional approaches to resilience often involve redundancy, risk mitigation strategies, and contingency planning (Sikula et al., 2015). However, Industry 5.0 introduces a new dimension to resilience by emphasizing human–machine symbiosis.

Human Centricity: Beyond providing goods and services for profit, these fundamental components further expand the purposefulness of industry, incorporating subordering and employing human-centric application of technology, not the other way around. The approach demands to place human demands and interests at the focal point of the manufacturing systems and processes. At the core of Industry 5.0 lies the recognition of human potential and the need to harness it alongside technological advancements (Ibarra-Esquer and López-de-Ipiña, 2020). By integrating human intelligence, creativity, and emotional capabilities with automation and data-driven processes, Industry 5.0 fosters a more resilient operational framework (Ivanov, 2023). For instance, cobots which are collaborative robots have been designed as coworkers for humans, augmenting productivity while ensuring protection and adaptability in dynamic environments.

Agility and Flexibility: Resilient industries require agility and flexibility to respond effectively to changing circumstances. Industry 5.0 facilitates this agility by empowering workers with the tools and autonomy to make real-time decisions (Schumacher et al., 2016). Fast insights are made possible by progressive analytics and AI-steered systems, which enable businesses to quickly adapt production procedures, allocate resources optimally, and satisfy changing demands. Furthermore, decentralized decision-making strengthens resilience against system-wide failures by reducing reliance on centralized control (Adana et al., 2024).

Adaptive Workforce: A resilient industry requires an adaptive and skilled workforce capable of responding to evolving challenges. Industry 5.0 emphasizes continuous learning and upskilling through human–machine collaboration (Ibarra-Esquer and López-de-Ipiña, 2020). By integrating augmented reality, virtual reality, and through dynamic and captivating training simulations, employees learn new skills and information. Furthermore, collaborative platforms help employees share information and solve problems together, which promotes an innovative and resilient culture (Carmeli et al., 2021).

Divergence and Redundancy: Redundancy and divergence are encouraged by Industry 5.0 as essential components of resilience. Businesses use modular production systems and distributed networks in place of centralized manufacturing facilities and single-source suppliers (Prajapati et al., 2021). This decentralization mitigates the risk of disruptions by spreading resources across multiple locations and suppliers. Furthermore, additive manufacturing technologies enable on-demand production, reducing lead times and enhancing supply chain resilience (Belhadi et al., 2022).

Sustainable Practices: Sustainable means to satisfy various interests and demands of the working generation without endangering the interest of future generations. In other words, approaches toward sustainable development of circular processes through inevitable 3Rs—reuse, repurpose, and recycle natural resources. Resilience and sustainability are inherently interconnected concepts (Pisano, 2012). Industry 5.0 promotes sustainable practices by optimizing resource utilization, minimizing waste, and reducing environmental impact (Lasi et al., 2014). Smart manufacturing processes enable real-time monitoring of energy consumption, emissions, and other sustainability metrics, allowing companies to recognize spaces and verticals for enhancement and execute corrective measures promptly. Furthermore, the integration of renewable energy sources and circular economy principles enhances the long-term resilience of industrial operations (Pan and Hashemizadeh, 2023). Higher levels of robustness in industrial production are being developed to provide resilience against interruptions in global value and supply chains.

Industry 5.0 represents a paradigm shift toward a more resilient and sustainable industrial ecosystem. By embracing human centricity to collaborative and technological innovation, Industry 5.0 harnesses the full potential of both human intelligence and technological advancements (Adel, 2022; Mourtzis et al., 2022). This holistic approach not only enhances operational efficiency and productivity but also strengthens the capacity of industries to adapt and thrive in the face of uncertainty and disruptions. As we continue to navigate through an increasingly complex and interconnected world, Industry 5.0 offers a roadmap toward a sustainable future.

5.2 PRINCIPLES OF LSS

Organizations are constantly striving to improve their operational efficiency, reduce costs, and enhance customer satisfaction. Lean Six

Sigma (LSS) has emerged as a powerful methodology that combines the principles of Lean manufacturing and Six Sigma quality management to achieve these objectives. As Industry 5.0 emphasizes on value creation and human-centered design, which can be achieved through LSS. LSS has been effectively used to improve manufacturing activities in the quality field. LSS can perfectly combine more human processes with a better result environment in a field that values people. The lean philosophy prioritizes sustainable development, employs a people-oriented strategy, and gives attention to both customers and staff. The Six Sigma methodology and the evolved industrial wave both place a strong emphasis on innovation as a means of reducing process variation and achieving iterative production process improvements for meeting sustainable development objectives.

LSS integrates the Lean philosophy of waste reduction and efficiency improvement, employing statistical exactitude and problem-solving techniques. While Lean focuses on eradicating activities which are not adding any value and thus streamlining different processes, the focus of Six Sigma is to minimize variability, imperfections, and defectiveness to achieve consistent quality outcomes (George, 2002). By combining these methodologies, LSS stipulates a well-structured approach to identify and eliminate process inefficiencies, thereby enhancing overall performance and customer satisfaction. According to Dahlgaard-Park et al. (2013), more studies are being done on tools and methods used in the quality management framework, such as lean, the Toyota production system, and Six Sigma. The waste in the organization is successfully eliminated using various lean tools.

LSS techniques are applied across various industries and functional areas to drive continuous improvement and achieve operational excellence. These techniques enable organizations to:

- Identify and eliminate process inefficiencies, defects, and waste.
- Improve product and service quality, consistency, and reliability.
- Enhance customer satisfaction and loyalty by delivering products matching or exceeding customer expectations.
- Optimize resource utilization, reduce costs, and increase profitability.
- Foster a culture of constant amendment, improvements and productivity enhancement, innovation, and employee empowerment and engagement.

5.2.1 KEY PRINCIPLES OF LSS

1. Customer Focus: LSS emphasizes understanding and meeting customer requirements by delivering products and services that provide value and meet quality standards (George, 2002).
2. Continuous Improvement: The methodology promotes a culture of continuous improvement, encouraging organizations to seek incremental enhancements in processes, products, and services.
3. Data-Driven Decision-Making: Here, LSS supports statistical analysis of data to ascertain sources and causes of issues and problems, measure process performance, and make data-based calculated-wise decisions.
4. Employee Involvement: Employees' participation and engagement in the process improvements, throughout the organizational hierarchy, fosters ownership, commitment, and innovation, driving sustainable improvement efforts (George, 2002).
5. Systematic Problem Solving: Well-structured problem-solving methodologies of LSS-DMAIC (define-measure-analyze-improve-control) and define, measure, analyze, design, and verify (DMADV) (discussed in the next section) are capable of systematically addressing process issues and achieving desired outcomes.

5.2.2 METHODOLOGIES AND TOOLS: LSS

1. Value stream mapping (VSM) can be created using lean concepts and methodologies. To find waste, inefficiencies, and non-value-added procedures in their product development processes, manufacturing businesses have adopted VSM approaches. As a visual instrument, VSM can analyze and optimize the movement of information and materials within a value stream or a process. Thus, supporting to identify any bottlenecks, possible waste of resources and efforts, and scope and opportunities for improvement, enabling organizations to streamline operations and reduce lead times (Rother and Shook, 2003).

2. Root Cause Analysis is a technique used to identify the underlying causes of problems or defects within a process. By systematically investigating contributing factors, organizations can implement effective corrective and preventive actions to address root causes and prevent recurrence (Pyzdek and Keller, 2014).
3. Statistical Process Control comprises monitoring and controlling the possible process variation utilizing statistical tools and methods to ensure consistent quality outcomes. By establishing control charts and process capability indices, organizations can detect deviations from target values and take corrective action to maintain process stability and performance (Montgomery, 2012).
4. Just-in-time as an LSS tool describes the successful planning and execution of events to provide a finished product in lean manufacturing, inventory level to dispatching. Assembling lines encounter issues because manufacturing lines frequently employ a push mechanism. By creating a Kanban system, the push system was changed to a pull system.
5. DMADV is a methodology used to design new processes or products. It consists of five phases: Define: Set project objectives and goals as per customer needs and business focus and prerequisites. Measure: Identify and measure critical-to-quality characteristics to ensure alignment with customer expectations. Analyze: Analyze process options and design alternatives to develop the optimal solution. Design: Develop and implement the new process or product design based on the analysis and evaluation conducted. Verify: Validate the effectiveness of the new design through pilot testing and implementation (Pyzdek and Keller, 2014).
6. DMAIC is another structured LSS approach of problem-solving employed to improve the quality of prevailing processes and enhance production processes. It consists of five phases: DMAIC Define: Clearly define project goals, objectives, and customer requirements. Measure: Quantify process performance and collect relevant data to identify areas for improvement. Analyze: Analyze data to identify root causes of defects, inefficiencies, or variation. Improve: Implement solutions to address identified issues and optimize process performance. Control: Establish controls to

sustain improvements and monitor process performance over time (Pyzdek and Keller, 2014). For implementation, the improvement team members must be trained and skilled because it primarily depends on statistical tools and methodologies. According to Sutphin et al. (2015), the DMAIC may take longer than three months, depending on the size and duration of the project. The complexity of the problem and the technique that needs to be modified will determine this. It also requires a time and resource investment.

7. Quick changeover/Single-Minute Exchange of Die system has the core premise to transfer as many changeover tasks as feasible to "external" processes, conducted while the equipment (internal) remains operational, while also simplifying and optimizing the remaining steps. Quick Changeover is the time required to change or replace a piece of machinery between producing two subsequent production lots. Through its effective implementation, machines can be reconfigured expeditiously, thus enhancing line flexibility and productivity.
8. The Plan-Do-Check-Act advocates that the iterative method should be repeatedly applied, particularly at the outset of process improvement endeavors, especially for the activities involving repetitive processes. Plan: Identify a problem or opportunity and devise a plan for continuous improvement. Formulate a hypothesis regarding potential issues. Do: Implement the plan through a small-scale test, enabling the measurement of results and a deeper understanding of the hypothesis. Check: Evaluate whether the problem has been effectively addressed. Act: If the initial test proves successful, scale up the solution for broader implementation.
9. Takt time represents the pace required to produce a product to satisfy customer demand. In manufacturing, takt serves as a critical gauge of production efficiency relative to customer requirements. Takt time is determined by dividing the total production time by the customer demand. The available production time refers to the duration required to complete the product from inception to completion, excluding intervals such as worker breaks, scheduled maintenance, and shift changeovers.

$$Takt\ Time = \frac{Available\ time\ working\ per\ shift}{Rate\ of\ customer\ demand}$$

Implementing takt time into production operations allows for: Quick identification of bottlenecks: By easily pinpointing stations slowing down the production line. Efficient localization of under-performing stations: Enabling prompt intervention and optimization measures.

10. Poka-Yoke also known as Error-Proofing is a methodology designed to prevent product defects by anticipating, rectifying, and signaling human errors as they occur. Derived from Japanese terminology meaning "error" and "machine operator," poka-yoke encompasses any mechanism within a process aimed at minimizing the occurrence of mistakes, thereby empowering individuals and processes to achieve accuracy from the outset.
11. Similarly, there is a concept Jidoka which interprets that in cases of technical issues, a machine or system automatically deactivates itself (Powell, 2024). Monitoring systems, such as those employing sensors or detectors, are employed for this task. When production or machine faults occur, the machine is capable of self-detection. Consequently, it initiates its own shutdown, hence why Jidoka is referred to as "autonomation," combining "autonomy" and "automation."
12. Kanban concept, originating from Japan, refers to a signal, often in the form of a card or board, utilized within a process. Integral to the JIT methodology, Kanbans—whether physical or electronic—indicate the need to reorder inventory, progress a unit, or proceed to the subsequent stage in a process.
13. Kaizen technique translates to "improve for the better," is a Japanese term encapsulating the concepts of "change" and "goodness." A Kaizen Event, also termed a Rapid Improvement Event, typically lasts from 1 to 5 days and involves key process participants focusing on resolving a narrowly scoped process improvement opportunity. Noteworthy distinctions between Kaizen and standard workshops include extensive event planning, explicit leadership approval for change, and implementation of agreed-upon improvements before the event's conclusion.

5.3 SUSTAINABLE INNOVATION FRAMEWORK (SIF)

Innovation which is futuristic and contributes toward social well-being termed as sustainable innovations results in new global markets, including smart specialization of specific regions and supporting governments comfortable with the long-term perspective for policy action (Boons et al., 2013). SIF can be defined as "a framework that places innovation management by the industry for a sustainable environment" (Varadarajan, 2017). The idea of blending innovation with sustainability can be viewed as an illustrious way to approach the paradigm envisioned in Industry 5.0. The SIF shows how compactly human intellect and robots can work together. Employees can develop their skills and abilities, along with using their intelligence and talents, to focus more on value-adding tasks and activities in the manufacturing process. Hence, providing high services to clients and contributing toward process efficiencies, by involving individuals in production with IT-enabled cobots.

Figure 5.2 illustrates the internal framework of SIF and the integrate LSS with Industry 5.0. Incepting from the creative generating stage, it is transformed into improvement through a series of linked activities known as innovation management. Creative management is required at every stage of the innovation process, which entails building networks rather than lines. To reach a higher level of creativity, this method incorporates a variety of functional domains and disciplines. Throughout the process, they are combined to govern three perspectives: individual, organizational, and technical resources. They can implement game-changing innovations contributed through involving LSS, which dramatically boosts business performance and productivity in line with the relevant requirements of the customer and society at large while taking care of the ecosystem simultaneously.

According to Antony et al. (2016), LSS can be used for a variety of innovations, including incremental, process, and capability innovations. Customers can anticipate more individualized goods and services in Industry 5.0, which is only possible with greater consumer involvement in the process of designing goods and services. Mass production of mass customization is possible with the inclusion of ICT in production and delivery. The ideal conclusion is to build the design for Six Sigma role structure (DFSS). The focus of DFSS is on creative endeavors (Hwang et al., 2018). To identify things that please customers, find innovations, and link them, products must be brought to market. It is not always easy, and innovation needs to be integrated into every phase of the process. In addition to technology, innovation also relates to how things are organized, how people interact with one another, and how new tools and tactics are used. To improve the present product development process, we try to incorporate innovation. Customer empowerment is wherein customers can modify certain product characteristics through mass customization in Industry 5.0 at the same time maintaining costs at par or close to mass manufacturing levels. Thus, the improved manufacturing processes enable businesses manufacturers to develop interchangeable components that can be assembled in a variety of ways to create products that are both affordable and tailored to the demands of individual customers. On top of it, improvements are eco-friendly and progressive toward reducing wastage of time, efforts, and resources, and carbon emissions.

To explore possibilities and applications of new technology, Industry 5.0 reinstates the human-centric approach. As a result, the LSS collaborative management idea can be seen as an Industry 5.0 enabler. First, a lean mind-set prioritizes people, both customers and staff. Lean methodologies foster an egalitarian culture where each employee has a voice and their ideas matter, which is consistent with Industry 5.0's human-centered mission to increase the sustainability of employee technology systems. Second, Six Sigma strives to increase output quality by locating and eliminating flaws and faults as well as lowering the variability of both business and manufacturing procedures (Antony et al., 2012). Finally, from the standpoint of LSS, the emphasis is on sustainability along with innovation throughout various business processes, comprising of eradicating variability, providing better-value, lowering intricacy, and enhancing precision with efficiency to support human prosperity.

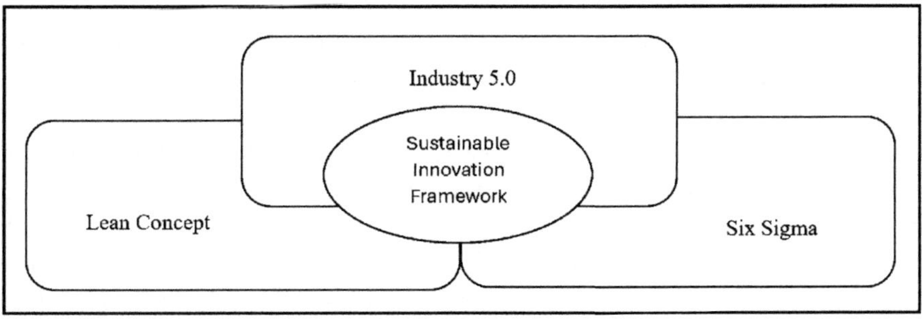

FIGURE 5.2 Sustainable innovation framework.

5.4 GUIDELINE FOR APPLYING LSS 5.0 TOOLS

SIF supports the inclusion of LSS as supporting Industry 5.0, which is a sequential path. To reach the target, Figure 5.3 represents a path toward application of Lean Six Sigma to achieve Industry 5.0.

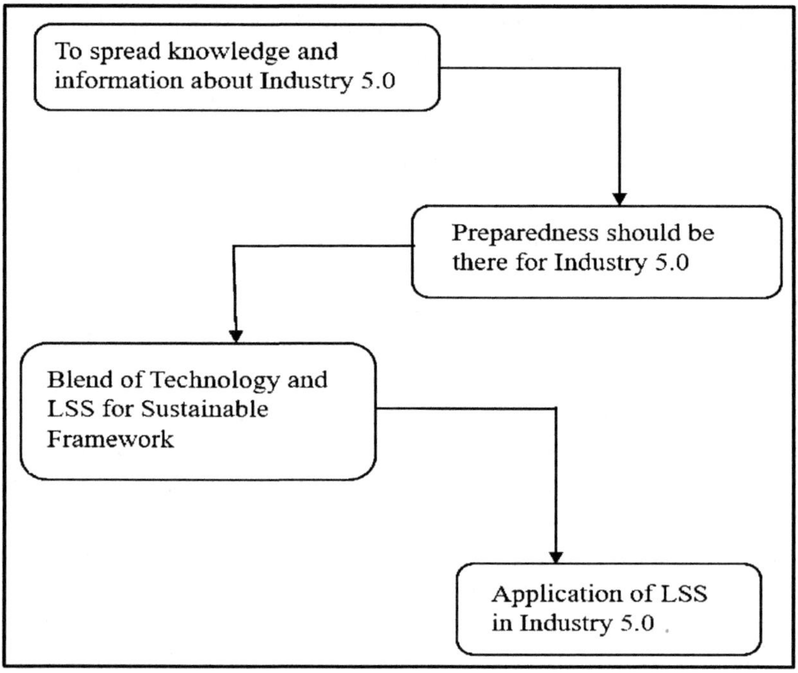

FIGURE 5.3 LSS application.

5.5 AWARENESS OF INDUSTRY 5.0

Innovation is inevitable in every sphere of the business and thus industry and manufacturing. Even though Industry 5.0 is still a young concept, some early academic works outline its main features suggestive to the industry. It is theorized that Industry 5.0 would lead toward a rediscovered and extended industry purpose, beyond producing things and services for profit, after examining literature and considering the future. As already discussed, the inclusion of different elements of Industry 5.0 leads toward a robust future. Instead of examining the potential for developing technology to improve efficiency, emphasis on fundamental human requirements is at the focal point, resulting in adopting 3Rs to reduce waste and address its effects on the environment (Memon, 2010). To reemphasize on the focus in the present context, resilience refers to building a strong industrial production system that can support all essential infrastructures in an emergency and better resist disturbances.

5.6 READINESS FOR INDUSTRY 5.0

Human existence cannot survive without ongoing progress. So, there are revolutionary changes taking place in both the planet and the many industrial sectors. Industry 5.0's change is unavoidable. It is no longer possible to turn back the technological revolution. Everything in the modern world is computerized. Instead of concentrating on robotics, and to progress toward Industry 5.0, through integration and alignment of machines and humans, adjustments are implemented and workplace improvements are introduced. Individuals currently can communicate with robots through chat, information requests from virtual assistants, and other means. Industry 5.0 is the future for many businesses at a time of creation of new jobs; its changed skill requirements, requirement of personalized communication, products and services, and automated equipment, resulting in delighted customers, supporting the ecosystem through resource conservation and sustainable development.

5.7 PROVISION OF REQUIRED TECHNOLOGIES FOR INDUSTRY 5.0

By fusing enabling improved latest information and communication technologies like Internet of Everything, big data analytics, cobots, materials, and smart supplies which have embedded sensors and trackers, combining with human cognitive abilities of creativity and innovation; business and industries can simply improve output and offer products, services, and solutions demanded by customers. Industry 5.0, which utilizes these latest technologies, new methods, and technique, develops a cutting-edge production model based on the interaction of robots and humans. It enhanced productivity and performance in a world where intelligent robots collaborate with people, and smaller enterprises and individuals are better able to automate jobs than ever before.

5.8 AWARENESS OF LSS

Senior corporate executives are advocating and implementing LSS to enhance their operations in greater numbers. Before they may start to experience the benefits and advantages of continuous process improvement, these businesses must comprehend the core components of DMAIC. The company will be able to turn into an LSS company, where Six Sigma tools can be applied at every level, by using DMAIC methods and other suitable tools and techniques. Senior management must put in a lot of work in the early phases of this process. In order to guarantee that the LSS is done as intended and that the assigned actions are finished on time, it is essential for businesses to create a high level of discipline. Projects will progress, with the needed level of support, as individuals find they can fulfil their duties with better efficacy, allowing them to develop personally and ensure their organizations remain competitive.

5.9 APPLICATION OF LSS 5.0 TOOLS

In Industry 5.0, the integration of human capital with cutting-edge technology is the key aspect of transforming the operations of any firm. Industry 5.0 is not only a revolutionary vision but also a logical extension of Lean and Six Sigma's attempts to reduce waste and unpredictability. When LSS 5.0 tools are employed, the user will be made aware of more waste and faults right away. Manufacturers should assist their LSS teams, e.g., Kaizen teams in their attempts to develop real-world uses for LSS 5.0 tools as part of the transition process, in addition to deploying Industry 5.0. The implication of each of the tools in terms of appropriateness needs to be deliberated and the right combination of technology and human, with output and performance monitoring only can pave the way for an improved and sustainable resilient industry.

5.10 CONCLUSION

Industry 5.0 emphasizes on the consistent humanization of the latest technologies. ICT-driven solutions like usage of big data, bots, and smart sensors and the proliferation of related technologies are examples of recently developed solutions that can ease the workers and employees, and comforting task and situations while making a secure setting for them to do their work. A well-liked LSS technique included in Industry 5.0 is proposed to be successful in lowering waste and process variability. Advanced analytics can boost LSS project performance due to faster data collection and analysis, which can further be utilized for improving decision-making and problem solving. A sustainable environment will foster incremental innovation, which can result through integration of LSS into Industry 5.0. Lean speeds up Six Sigma to fix issues and enhance procedures more quickly and effectively. Thus, the SIF will be more useful for this application in a human-centric, resilient, and sustainable environment, as this chapter underlines. The infusion of technology that has altered how businesses approach mass personalization without any additional cost is one of the major advances. The success of SIF depends upon critical factors like industry readiness, availability of infrastructure and technologies, along with the top management's consistent support and vision. Application of SIF in the industry needs to be further investigated and can be compared with the results of other similar proposed frameworks/ approaches focused on Industry 5.0.

KEYWORDS

- **Industry 5.0**
- **Lean Six Sigma**
- **sustainable innovation framework**

REFERENCES

Adana, S., Manuj, I., Herburger, M., Cevikparmak, S., Celik, H., Uvet, H. (2024). Linking decentralization in decision-making to resilience outcomes: A supply chain orientation perspective. *The International Journal of Logistics Management*, *35*(1), 256–280.

Adel, A. (2022). Future of industry 5.0 in society: Human-centric solutions, challenges and prospective research areas. *Journal of Cloud Computing*, *11*(1), 40.

Ahmadi, H., Seuring, S., Mäntymäki, M. (2020). Understanding resilience in industry: A literature review and research agenda. *International Journal of Production Economics*, *222*, 107520.

Antony, J., Gijo, E. V., Childe, S. J. (2012). Case study in Six Sigma methodology: Manufacturing quality improvement and guidance for managers. *Production Planning & Control*, *23*(8), 624–640.

Antony, J., Setijono, D., Dahlgaard, J. J. (2016). Lean Six Sigma and Innovation—An exploratory study among UK organisations. *Total Quality Management & Business Excellence*, *27*(1–2), 124–140.

Bacchetta, M., Bekkers, E., Piermartini, R., Rubinova, S., Stolzenburg, V., Xu, A. (2021). *COVID-19 and Global Value Chains*. World Trade Organisation.

Belhadi, A., Kamble, S. S., Venkatesh, M., Jabbour, C. J. C., Benkhati, I. (2022). Building supply chain resilience and efficiency through additive manufacturing: An ambidextrous perspective on the dynamic capability view. *International Journal of Production Economics*, *249*, 108516.

Boons, F., Montalvo, C., Quist, J., Wagner, M. (2013). Sustainable innovation, business models and economic performance: An overview. *Journal of Cleaner Production*, *45*, 1–8.

Boyes, H., Hallaq, B., Cunningham, J., Watson, T. (2018). The industrial internet of things (IIoT): An analysis framework. *Computers in Industry*, *101*, 1–12.

Carmeli, A., Levi, A., Peccei, R. (2021). Resilience and creative problem-solving capacities in project teams: A relational view. *International Journal of Project Management*, *39*(5), 546–556.

Dahlgaard-Park, S. M., Chen, C.-K., Jang, J.-Y., Dahlgaard, J. J. (2013). Diagnosing and prognosticating the quality movement—A review on the 25 years quality literature (1987–2011). *Total Quality Management & Business Excellence*, *24*(1–2), 1–18.

George, M. L. (2002). *Lean Six Sigma: Combining Six Sigma with Lean Speed*. McGraw-Hill: New York.

Goetz, S., Keitzel, G., Klocke, F. (2017). Going smart—CPPS for digital production. In: *Industrial Internet of Things: Cybermanufacturing Systems* (pp. 401–422).

Hwang, G. H., Lee, K. C., Seo, Y. W. (2018). How does Six Sigma influence creativity and corporate performance through exploration and exploitation. *Total Quality Management & Business Excellence*, *29*(13–14), 1668–1684.

Ibarra-Esquer, J. E., López-de-Ipiña, D. (2020). Human-centric approaches in industry 4.0: A review. *Journal of Ambient Intelligence and Humanized Computing*, *11*(2), 605–622.

Ivanov, D. (2023). The industry 5.0 framework: Viability-based integration of the resilience, sustainability, and human-centricity perspectives. *International Journal of Production Research*, *61*(5), 1683–1695.

Lasi, H., Fettke, P., Kemper, H.-G., Feld, T., Hoffmann, M. (2014). Industry 4.0. *Business & Information Systems Engineering*, *6*(4), 239–242.

Lee, C. K. M., Zhang, S. Z., Ng, K. K. H. (2017). Development of an industrial Internet of things suite for smart factory towards re-industrialization. *Advances in Manufacturing*, *5*, 335–343.

Mabkhot, M. M., Al-Ahmari, A. M., Salah, B., Alkhalefah, H. (2018). Requirements of the smart factory system: A survey and perspective. *Machines*, *6*(2), 23.

Memon, M. A. (2010). Integrated solid waste management based on the 3R approach. *Journal of Material Cycles and Waste Management*, *12*, 30–40.

Montgomery, D. C. (2012). *Introduction to Statistical Quality Control*. John Wiley & Sons: Hoboken, NJ, USA.

Mourtzis, D., Angelopoulos, J., Panopoulos, N. (2022). A literature review of the challenges and opportunities of the transition from industry 4.0 to society 5.0. *Energies*, *15*(17), 6276.

Osterrieder, P., Budde, L., Friedli, T. (2020). The smart factory as a key construct of industry 4.0: A systematic literature review. *International Journal of Production Economics*, *221*, 107476.

Pan, Y., Hashemizadeh, A. (2023). Circular economy-based assessment framework for enhancing sustainability in renewable energy development with life cycle considerations. *Environmental Impact Assessment Review*, *103*, 107289.

Pascual, D. G., Daponte, P., Kumar, U. (2019). *Handbook of Industry 4.0 and Smart Systems*. CRC Press.

Pisano, U. (2012). Resilience and sustainable development: Theory of resilience, systems thinking. *European Sustainable Development Network (ESDN)*, *26*, 50.

Powell, D. J. (2024). Artificial intelligence in lean manufacturing: Digitalization with a human touch? *International Journal of Lean Six Sigma*, *15*(3), 719–729.

Prajapati, P., Sankar, S., Tiwari, M. K. (2021). Industry 5.0—A comprehensive review. *Journal of Manufacturing Systems*, *61*, 760–781.

Pyzdek, T., Keller, P. A. (2014). *The Six Sigma Handbook: A Complete Guide for Green Belts, Black Belts, and Managers at All Levels*. McGraw-Hill Education: New York.

Rojas, R. A., Rauch, E. (2019). From a literature review to a conceptual framework of enablers for smart manufacturing control. *The International Journal of Advanced Manufacturing Technology*, *104*, 517–533.

Rother, M., Shook, J. (2003). *Learning to See: Value Stream Mapping to Add Value and Eliminate Muda*. Lean Enterprise Institute: Cambridge, MA, USA.

Schumacher, A., Erol, S., Sihn, W. (2016). A maturity model for assessing industry 4.0 readiness and maturity of manufacturing enterprises. *Procedia CIRP*, *52*, 161–166.

Sikula, N. R., Mancillas, J. W., Linkov, I., McDonagh, J. A. (2015). Risk management is not enough: A conceptual model for resilience and adaptation-based vulnerability assessments. *Environment Systems and Decisions*, *35*, 219–228.

Sutphin, P. D., Reis, S. P., McKune, A., Ravanzo, M., Kalva, S. P., Pillai, A. K. (2015). Improving inferior vena cava filter retrieval rates with the define, measure, analyze, improve, control methodology. *Journal of Vascular and Interventional Radiology*, *26*(4), 491–498.

Varadarajan, R. (2017). Innovating for sustainability: A framework for sustainable innovations and a model of sustainable innovations orientation. *Journal of the Academy of Marketing Science*, *45*, 14–36.

Veile, J. W., Kiel, D., Müller, J. M., Voigt, K. I. (2020). Lessons learned from Industry 4.0 implementation in the German manufacturing industry. *Journal of Manufacturing Technology Management*, *31*(5), 977–997.

Zizic, M. C., Mladineo, M., Gjeldum, N., Celent, L. (2022). From industry 4.0 towards industry 5.0: A review and analysis of paradigm shift for the people, organization and technology. *Energies*, *15*(14), 5221.

CHAPTER 6

Exploring the Industry 5.0 Toward a Greener Future: The Role of Sustainable Practices and Technological Innovations

VANSHIKA KAKKAR[1,*] and BABITA RAWAT[2]

[1]*Operational Risk & Analytics Analyst, BNY, India*

[2]*Uttaranchal Institute of Management, Uttaranchal University, Dehradun, Uttarakhand, India*

*Corresponding author

ABSTRACT

Industry 5.0 methodically suggests viewing industrial workers at the center of industrial production to create a relationship where both industry and society benefit. Industrial 5.0 encompasses technical advancements, sustainability, human-centered design, and sustainable development goal resilience. The next big wave is Industry 5.0, and industry insiders are looking forward to it. The urgency of finding quick fixes for a sustainable and affluent future is increased by the changes in the global environment. Additionally, in our methodical investigation of Industry 5.0 toward a sustainable future, we address the significance of technology advancements and sustainable practices. The idea of Industry 5.0 has been put out in an effort to reinstate human importance in the appropriate use of technology. This study proposed to analyze a systematic literature review based on what is currently known about this subject. For 14 years (2009–2023),

Industry 5.0: Innovation and Adoption for Sustainability.
Tilottama Singh, Pradeep Suri, and Zulkarin Jahangir (Eds.)

DOI: 10.1201/9781779641878-6

191 research publications were analyzed for this study. The results make it possible to describe innovation via Industry 5.0 by highlighting the necessity of fusing technology and sustainable practices. Originality and worth: The research achieved a variety of goals, including technologically driven innovation and sustainable innovation.

6.1 INTRODUCTION

The society we live in today's world is voluntary, uncertain, capricious, intricate, and ambiguous to an extraordinary and growing degree. Consequently, there are an increasing number of worldwide difficulties that society must confront across all domains and levels to maintain stability and prosperity, be it technical, economic, social, or environmental. Era 5.0, which centers innovation and technological change on the human being, can and must use technology to help improve human well-being, societal problem-solving, and quality of life. Industry 5.0 is the organizational capability across the whole chain that contributes distinct intellect to the organization's unique processes. According to the literature, Industry 5.0 may integrate sustainability into company's operations and enhance corporate sustainable development (Lasi et al., 2014). Since research on sustainable performance and Industry 5.0 is still in its infancy, experts and scholars are working to comprehend the findings while considering specific sustainable metrics in various settings and proportions, such as societal, ecological, and financial ones (Aslam et al., 2020; Xu et al., 2021). Consequently, the goal of this chapter is to facilitate the transition to sustainable growth in Era 5.0, drawing on a thorough analysis of the literature. (1) To look at the implications on the important Industry 5.0 technical advances is the primary objective of this study; (2) to evaluate the role that Industry 5.0 automation will play; and (3) to assist in identifying the opportunities, challenges, and any loopholes that will allow further investigation developments and play in facilitating sustainable development.

The environmentally friendly evolution goals of Industry 5.0, as outlined in the Europid authority 2022's Industry 5.0 mission, include the comprehensive advancement of the social, ecological, and political pillars of sustainability as well as the underlying micro-objectives of each. Two major disputes are linked to Industry 5.0, which serves as a fresh approach to achieving sustainable competitiveness. First, only few studies have been conducted on Industry 5.0's parameters, guiding concepts, and operational aspects. Second, the practical implications of Industry 5.0's potential to advance sustainability remain obscure. The current investigation aims to fill in these knowledge voids. Different explanations on why Industry 5.0 is so common have been put forth by researchers. Özdemir and Hekim (2018) presented Industry 5.0 as a progressive phase and a gradual upgrade to Industry 4.0 which the Industry 5.0 may provide. To deal with the inadequacies of the Industry 4.0 revolution network, balanced innovation is needed. This study suggested that, in addressing the difficulties associated with using the phrase "Industry 5.0," it is better to think of Industry 5.0 as a logical development from Industry 4.0 rather than as a replacement or substitute (Müller, 2020). Industry 5.0 is not the same as Industry 4.0 in the following ways:

- Sustainable development is valued by Industry 5.0.
- Industry 5.0 strengthens human-centered approaches to technology advancement.
- Industry 5.0 drives technical innovation (e.g., intelligent green power systems).
- Industry 5.0 prioritizes stakeholders in the governance of technology, the expansion of innovation, and sustainability performance management.

Green marketing is a critical tool for increasing a company's competitiveness because it views environmental considerations as primary strategic goals and drivers of the company's competitive values (Chung, 2020). Goals for sustainability and Industry 5.0 resilience aim to promote innovation and the general prosperity of the community (Enang et al., 2023). Organizations may achieve sustainable development and meet their ecological and economic objectives with the aid of eco-friendly supply chain oversight strategies (Das, 2018). Businesses to achieve their objectives, companies with green supply chains are constantly concentrating on persuading their suppliers to adopt eco-friendly practices and be mindful of the environment linked to performance (Heras-Saizarbitoria et al., 2011). Industry 5.0 outlines how to properly leverage technological progress by using clever technology to predict the optimal ideal configuration of a savvy culture in the future. This chapter's argument is organized as follows: A review of the literature on Industry 5.0, technical advancements, and long-term growth in terms of sustainability is given in the next section. The discussions or conclusions for conducting the review paper are expounded in Section 6.3. Section 6.44 presents the limitations and future study. Considering this, a framework for innovation management is put up and examined. Future research directions and a conclusion are provided. A structured literature review is performed to:

- Determine the fundamental areas to conceptualize the key elements of the business, use Business 5.0.
- Provided insight into the main forces behind the creation of Innovation 5.0.
- Identify ramifications and recommendations for more research.

6.2 LITERATURE REVIEW

The Industry 5.0 phenomenon is explained in this particular section that aligns with long-term effectiveness, particularly in terms of sustainable practices, environmental factors, and technological innovations.

6.2.1 INDUSTRY 5.0 AND SUSTAINABLE PERFORMANCE

An organization's ability to execute sustainably is essential to achieving a better, more dynamic, and long-lasting future (Weber et al., 2014; Emas, 2015; Wichaisri and Sopadang, 2018). Academics assert that Industry 5.0 has made an equitable contribution to several facets of sustainable performance (Grabowska et al., 2022). A fundamental competency of financial institutions is sustainable performance, notwithstanding the obstacles it encounters, including poverty, inequality, environmental uncertainty, variations in the atmosphere, and moral dilemmas. The sustainable development of the financial perspectives-related operations is linked to enduring growth. By delivering justice and managing stakeholders properly, businesses aim to build long-lasting and lost-lasting connections with their many stakeholders, including suppliers, purchasers, and society. A number of professionals and academic specialists have studied sustainable performance in many fields to highlight its significance on a global scale. For example, Webber's study assessed the impact of sustainable performance on the finance sector (Weber et al., 2014). The impact of energy and the environment on long-term sustainability was identified by Rosen and Dincer (2001). Businesses should concentrate on Industry 5.0

technologies, which have a favorable correlation with sustainable development, in order to attain sustainable performance (Ghobakhloo et al., 2022; Waheed et al., 2022). To confirm the significance of Industry 5.0 in relation to total business performance, another study was carried out (Ghobakhloo et al., 2022). According to a survey, Industry 5.0 is important for companies' business strategies (Alexa et al., 2022). Few research has shown how Industry 5.0 affects the long-term success in finance-related industries, which is the primary objective of this study. Even though sustainable performance was the subject of separate discussions, experts affirmed its significance globally. However, there has not been much research done in the past to look at how sustainable performance and practices interact. Furthermore, a variety of environmentally and ecologically friendly methods have been connected to sustainable performance in the literature (Reyes-Rodríguez et al., 2016; Phan et al., 2020).

6.2.2 MEDIATING ROLE OF SUSTAINABLE PRACTICES

The organizational aspects of technology advancements and the implementation of environmentally friendly practices are intricate procedures that need advanced training and the development of resources utilized by humans (Del Brío and Junquera, 2003). Implementing eco-friendly marketing strategies can improve sustainability. However, organizational characteristics that are supportive of green marketing strategies are necessary for their adoption (Chung, 2020). It is simpler for organizations to embrace green marketing strategies when appropriate and supporting organizational variables are present, as per the framework of technology, organization, and environment (Kumar, 2015). Green technologies require sustainable behaviors (Tsai and Liao, 2017). Adoption of environmentally friendly, sustainable methods is positively impacted by the caliber of human resources (Singh et al., 2017). Adoption of eco-friendly methods is also influenced by the support of the organization. Staff members who are driven have a higher propensity to embrace green behaviors and practices. Thus, it is simple to implement green practices when there is organizational support. The presence of pertinent elements helps to make the implementation of sustainable green action. A key aspect in adopting green strategies is organizational characteristics (Hwang et al., 2016). Finance is a crucial component of the organization that makes green innovation easier to embrace (Aboelmaged and Hashem, 2019). Green goods and process are the transformation that provides organizations with a competitive edge competing in the external environment. Still, a lot of it depends on internal circumstances. The effectiveness of organizational elements determines the acceptance and use of eco-friendly innovations (Chong and Olesen, 2017; Zhang et al., 2020).

6.2.3 MEDIATING ROLE OF ENVIRONMENTAL FACTORS

As stated by the Environmental Protection Agency (EPA, 2020), it is the process of preserving the environment by considering a number of variables, such as greenhouse gas emissions, water quality, plant and animal life, air quality, and hazardous waste. A little amount of research on Electronic National Vendor Declaration (ENVD) from the standpoint of Industry 5.0 has validated the importance of ENVD globally (Waheed et al., 2022). The industry model's good reaction to environmental sustainability was concluded by another expert. Numerous investigations were made that have examined sustainability's impact from various angles worldwide (Bai et al., 2020; Ukko et al., 2022; Yadav et al., 2020). Nonetheless, research has shown that Industry 5.0 has affected the ecological durability, which is a sign of sustainable performance. Technological innovation will increase an organization's capacity to implement sustainable green practices in an unpredictable climate (Rothenberg and Zyglidopoulos, 2007).

The adoption of environmentally friendly sustainable techniques is significantly influenced by environmental conditions. Sustainable green techniques are increasingly being used mostly due to environmental laws (Aboelmaged, 2018). When a business decides to implement green supply chain methods, certain environmental conditions must be met (Hwang et al., 2016). Adopting green innovations is really facilitated by pertinent environmental conditions, and these strategies can only be effective in an externally encouraging atmosphere. The environmental elements must provide support for the use of green marketing methods (Chung, 2020; Kumar, 2015). Green innovation adoption is pushed by environmental variables within organizations. Organizations are being pushed to switch from old times to green technologies in order to survive due to the emergence of environmental challenges (Khan et al., 2018; Mohd Saudi et al., 2019). Sustainability is the result of green advertising campaigns (Papadas et al., 2017). By implementing these procedures, environmental problems are addressed.

These procedures deal with environmental issues and have a connection to sustainable performance. Performance and sustainability have been taken into consideration while developing and testing the idea of green marketing in the previous years (Hasan and Ali, 2015; Lam and Li, 2019).

6.2.4 MEDIATING ROLE OF TECHNOLOGICAL INNOVATIONS

Industry 5.0 focuses on the importance of the necessity of "machines," or the pinnacle of technology, to foster innovation and competition as well as to consider the effects of technology on society and the environment. As long as technical innovation is used to achieve its ideals, Industry 5.0 will remain a technological phenomenon (Müller, 2020). Organizations often embrace technology that shows higher economic gains and better performance, and the perceived value of the innovation is a relative advantage (Rogers, 2010; Tornatzky and Klein, 1982). Adoption of innovations is influenced by their relative advantages, which also have a good correlation with environmentally friendly, sustainable development. To achieve sustainability goals, using green marketing techniques is made simple by the existence of technical aspects (Karjaluoto and Vaccaro, 2009). Green innovation adoption is facilitated by suitable and pertinent technology aspects (Chong and Olesen, 2017; Zhang et al., 2020). Technology plays a major part in developing green innovation methods. The technical aspects' worth is contingent upon their interoperability with other technologies required for environmentally conscious advances (Zhang et al., 2020). Industry 5.0 depends on automation that anticipates virtual and machine–human interaction. It could be challenging to meet Industry 5.0's claimed sustainability goals (Xu et al., 2021). To achieve its primary goals of humans, resilient, and sustainable practices, Industry 5.0 additionally is dependent on the most innovative technology developments indicated under the reference model's emerging technologies (Müller, 2020).

6.3 CONCLUSION

Data from 2009 to 2023 were retrieved from Scopus for the present investigation. To examine the role that green finance and financial technology play in achieving the goals of sustainable economic growth, a conceptual framework was established. The following will be the sequence in which the study findings and discussion are presented: Prior to presenting findings of Industry 5.0, basic information about how many reports are there in the field will be provided, with a focus on human-centric, sustainable, and resilient publications. To attain sustainable performance, we examined Industry 5.0's function as well as the mediating roles of environmental, technical, and sustainable practices. While many publications found in the "Industry 5.0" domain are focused on technology, the keywords for those articles fell into three groupings. The goal of the study is to identify themes pertaining to "environment," "humans," "sustainability," or "resilient." The terms "innovation," "technologies," and "process," which are associated with segments, are equally appropriate in the matter of the new idea. One of the important considerations influencing how businesses function in the market as is they pursue sustainability under Industry 5.0 as a process-based methodology.

Industry 5.0 is created with the assistance of Industry 5.0 technology. If the greatest method to promote smart technologies in enterprises is through sociotechnical circumstances, businesses are concentrating on process innovation and optimization. In today's competitive world, Industry 5.0 offers unique capabilities that help businesses become quicker, smarter, and more competitive. To guarantee that the programmed manufacturing revolution under Industry 5.0 is in line with long-term goals, a fundamentally diverse set of enabling measures is needed to meet the specific and hopeful aims. To achieve equilibrium between economic expansion along with the advancement of social and environmental progress, Industry and stakeholders must work together interactively to achieve Industry 5.0. Integrating stakeholders also enable social contributors to assume a front-runner position in implementing the many functional economic pillars, like cooperative consumerism, cooperative economy, and eco-friendly developments, which holds significant significance and insights for accomplishing Industry 5.0's goals to maintain ecological and social equity. The following terms are coupled with the adjective "sustainable": economy, development, product, process, manufacturing, and Industry 5.0. Resilience is synonymous with adaptability. The term "resilient" defines numerous elements of businesses, and most crucially, it connects with technology.

Defining the characteristics of the new Industrial 5.0 paradigm for economies, society, and industries is seen to be important. Through responsible business, sustainability, and corporate social responsibility, technology may help organizations enhance their corporate image in addition to helping to meet the Global Goals which are created by the UN. The consideration of potential downsides and unforeseen effects is crucial for achieving environmental sustainability. To provide a comprehensive approach to environmental sustainability, factors like the energy and resource needs for proper use of technology should also be considered. The Internet of Things presents a significant opportunity to improve environmental sustainability through smart resource management and device and system connections. Several scientific studies highlight the necessity of incorporating sustainability and the vital significance that people play in the creation of future manufacturing. Social and environmental demands form the cornerstone of recently developed concept creation, while economic sustainability forms the cornerstone of evolution. The three development pillars which are focusing on humans, sustainability, and resilient development are fundamental to Industry 5.0.

Advanced developments might lead toward eco-friendly operations in organizations. The kind of innovation that impacts specific resources and company operations is one of the most important factor (Sampson, 2007). This will result in sustained performance and a greater competitive edge (Chege and Wang, 2020a). All organizations are expected to invest in technology advancements to implement environmentally friendly practices, including green supply chains, green finance, green marketing, and green innovation, to achieve sustainability. Governments should develop practical plans for fostering green partnerships, giving priority to sustainability projects, and sharing sustainability expertise to strengthen a focus on sustainability that involves all stakeholders and accelerate the Industry 5.0 transition. As an alternative, governments are urged to actively support long-term business leadership, which can entail empowering and assisting businesses in their pursuit of long-term profits and the transition to sustainable business models. Lastly, and perhaps most crucially, the roadmap suggests that supporting eco-innovation is part of the crucial role governments play in enabling Industry 5.0. Governments can accomplish this specific aim in a variety of ways, such as by encouraging funding for environmentally friendly research.

6.4 LIMITATIONS AND FUTURE STUDY

First, the authors of the chapter suggest more investigation into social aspects, governance, and the creation of the Industry 5.0 idea in the framework of ecological balance. The present investigation has several practical limitations that might be considered in future investigations. Second, the search range of publications is restricted because this study only looks at one database, Scopus. Further search databases, such as "Web of Science," "Google Scholar," and "IEEE Xplore," may be considered throughout the document analysis process. Third, we consider articles that have only been published in English, leaving out anything that may be examined later. Fifth, to increase the efficacy of green financing, policymakers must plan an ecological revelation backdrop and enhance the guidance given to state governments by the Indian government.

KEYWORDS

- **Industry 5.0**
- **green finance**
- **sustainability**
- **technological innovations**
- **sustainable future**

REFERENCES

Aboelmaged, M. (2018). The drivers of sustainable manufacturing practices in Egyptian SMEs and their impact on competitive capabilities: A PLS-SEM model. *Journal of Cleaner Production, 175*, 207–221. https://doi.org/10.1016/J.JCLEPRO.2017.12.053

Aboelmaged, M., Hashem, G. (2019). Absorptive capacity and green innovation adoption in SMEs: The mediating effects of sustainable organisational capabilities. *Journal of Cleaner Production, 220*, 853–863. https://doi.org/10.1016/j.jclepro.2019.02.150

Alexa, L., Pîslaru, M., Avasilcăi, S. (2022). From industry 4.0 to industry 5.0—An overview of European Union enterprises. In: *Sustainability and Innovation in Manufacturing Enterprises* (pp. 221–231).

Aslam, F., Aimin, W., Li, M., Ur Rehman, K. (2020). Innovation in the era of IoT and industry 5.0: Absolute innovation management (AIM) framework. *Information, 11*(2), 124.

Bai, C., Dallasega, P., Orzes, G., Sarkis, J. (2020). Industry 4.0 technologies assessment: A sustainability perspective. *International Journal of Production Economics, 229*, 107776.

Chege, S. M., Wang, D. (2020a). The influence of technology innovation on SME performance through environmental sustainability practices in Kenya. *Technology in Society, 60*, 101210. https://doi.org/10.1016/j.techsoc.2019.101210

Chong, J. L. L., Olesen, K. (2017). A technology-organization-environment perspective on eco-effectiveness: A meta-analysis. *Australasian Journal of Information Systems, 21*. https://doi.org/10.3127/ajis.v21i0.1441.

Chung, K. C. (2020). Green marketing orientation: Achieving sustainable development in green hotel management. *Journal of Hospitality Marketing and Management, 29*(6), 722–738. https://doi.org/10.1080/19368623.2020.1693471

Das, D. (2018). The impact of sustainable supply chain management practices on firm performance: Lessons from Indian organizations. *Journal of Cleaner Production, 203*, 179–196. https://doi.org/10.1016/J.JCLEPRO.2018.08.250

Del Brío, J. Á., Junquera, B. (2003). A review of the literature on environmental innovation management in SMEs: Implications for public policies. *Technovation, 23*(12), 939–948. https://doi.org/10.1016/S0166-4972(02)00036-6

Emas, R. (2015). The concept of sustainable development: Definition and defining principles. *Brief for GSDR, 2015.*

Enang, E., Bashiri, M. and Jarvis, D. (2023). Exploring the transition from techno centric industry 4.0 towards value centric industry 5.0: A systematic literature review. *International Journal of Production Research, 61*(22), 7866–7902, https://doi.org/10.1080/00207543.2023.2221344.

EPA. (2020). Retrived from "Our Mission and What We Do." [Accessed 27 June 2020]

Ghobakhloo, M., Iranmanesh, M., Mubarak, M. F., Mubarik, M., Rejeb, A., Nilashi, M. (2022). Identifying industry 5.0 contributions to sustainable development: A strategy roadmap for delivering sustainability values. *Sustainable Production and Consumption, 33*, 716–737.

Grabowska, S., Saniuk, S., Gajdzik, B. (2022). Industry 5.0: Improving humanization and sustainability of Industry 4.0. *Scientometrics, 127*(6), 3117–3144.

Hasan, Z., Ali, N. A. (2015). The impact of green marketing strategy on the firm's performance in Malaysia. *Procedia—Social and Behavioral Sciences, 172*, 463–470. https://doi.org/10.1016/J.SBSPRO.2015.01.382

Heras-Saizarbitoria, I., Landín, G. A., Molina-Azorín, J. F. (2011). Do drivers matter for the benefits of ISO 14001? *International Journal of Operations and Production Management, 31*(2), 192–216. https://doi.org/10.1108/01443571111104764

Hwang, B.-N., Huang, C.-Y., Wu, C.-H. (2016). A TOE approach to establish a green supply chain adoption decision model in the semiconductor industry. *Sustainability, 8*(2), 168. https://doi.org/10.3390/su8020168

Karjaluoto, H., Vaccaro, V. L. (2009). B2B green marketing and innovation theory for competitive advantage. *Journal of Systems and Information Technology, 11*(4), 315–330. https://doi.org/10.1108/13287260911002477

Khan, S. N., Busari, A. H., Abdullah, S. M., Mughal, Y. H. (2018). Followership moderation between the relationship of transactional leadership style and employees' reactions towards organizational change. *Polish Journal of Management Studies, 17*(1), 131–143. https://doi.org/10.17512/pjms.2018.17.1.11

Kumar, P. (2015). Green marketing innovations in small Indian firms. *World Journal of Entrepreneurship, Management and Sustainable Development, 11*(3), 176–190. https://doi.org/10.1108/WJEMSD-01-2015-0003

Lam, J. S. L., Li, K. X. (2019). Green port marketing for sustainable growth and development. *Transport Policy, 84*, 73–81. https://doi.org/10.1016/J.TRANPOL.2019.04.011

Lasi, H., Fettke, P., Kemper, H. G., Feld, T. and Hoffmann, M. (2014). Industry 4.0. *Business and Information Systems Engineering, 6*, 239–242.

Mohd Saudi, M. H., Sinaga, O., Gusni, G., Zainudin, Z. (2019). The effect of green innovation in influencing sustainable performance: Moderating role of managerial environmental concern. *International Journal of Supply Chain Management, 8*(1), 303–310. http://www.excelingtech.co.uk/. [Accessed 24 April 2021]

Müller, J. (2020). *Enabling Technologies for Industry 5.0: Results of a Workshop with Europe's Technology Leaders*. European Commision. https://doi.org/10.2777/082634

Özdemir, V., Hekim, N. (2018). Birth of industry 5.0: Making sense of big data with artificial intelligence, "the internet of things" and next-generation technology policy. *Omics: A Journal of Integrative Biology, 22*(1), 65–76.

Papadas, K. K., Avlonitis, G. J., Carrigan, M. (2017). Green marketing orientation: Conceptualization, scale development and validation. *Journal of Business Research, 80*, 236–246. https://doi.org/10.1016/j.jbusres.2017.05.024

Phan, T. T. H., Tran, H. X., Le, T. T., Nguyen, N., Pervan, S., Tran, M. D. (2020). The relationship between sustainable development practices and financial performance: A case study of textile firms in Vietnam. *Sustainability, 12*(15), 5930. https://doi.org/10.3390/su12155930

Reyes-Rodríguez, J. F., Ulhøi, J. P., Madsen, H. (2016). Corporate environmental sustainability in Danish SMEs: A longitudinal study of motivators, initiatives, and strategic effects. *Corporate Social Responsibility and Environmental Management, 23*(4), 193–212. https://doi.org/10.1002/csr.1359

Rogers, E. M. (2010). *Diffusion of Innovations*, 4th edn., Simon and Schuster: New York. https://www.simonandschuster.com/books/Diffusion-of-Innovations-4th-Edition/Everett-M-Rogers/9781451602470.

Rosen, M. A., Dincer, I. (2001). Exergy as the confluence of energy, environment and sustainable development. *Exergy, An International Journal, 1*(1), 3–13.

Rothenberg, S., Zyglidopoulos, S. C. (2007). Determinants of environmental innovation adoption in the printing industry: The importance of task environment. *Business Strategy and the Environment, 16*(1), 39–49. https://doi.org/10.1002/bse.441

Sampson, R. C. (2007). R&D Alliances and firm performance: The impact of technological diversity and alliance organization on innovation. *The Academy of Management Journal, 50*(2), 364–386. https://doi.org/10.2307/20159859

Singh, S., Tan, B. C., Khan, N. (2017). Driving green practice adoption in restaurant sector: Restaurant operator's perspectives—MMU institutional repository. In: *Proceedings of the 5th International Conference on Innovation and Entrepreneurship* (ICIE 2017) (pp. 219–226). Academic Conferences and Publishing International Limited. http://shdl.mmu.edu.my/7108/.

Tornatzky, L. G., Klein, K. J. (1982). Innovation characteristics and innovation adoption-implementation: A meta-analysis of findings. *IEEE Transactions on Engineering Management, EM-29*(1), 28–45. https://doi.org/10.1109/TEM.1982.6447463

Tsai, K.-H., Liao, Y.-C. (2017). Innovation capacity and the implementation of eco-innovation: Toward a contingency perspective. *Business Strategy and the Environment, 26*(7), 1000–1013. https://doi.org/10.1002/BSE.1963

Ukko, J., Saunila, M., Nasiri, M., Rantala, T. (2022). The importance of sustainability engagement in small businesses supplier collaboration. *Sustainable Development, 30*(1), 1–9.

Waheed, A., Alharthi, M., Khan, S. Z., Usman, M. (2022). Role of industry 5.0 in leveraging the business performance: Investigating impact of shared-economy on firms' performance with intervening role of i5.0 technologies. *SAGE Open, 12*(2), 21582440221094608.

Weber, O., Diaz, M., Schwegler, R. (2014). Corporate social responsibility of the financial sector—Strengths, weaknesses and the impact on sustainable development. *Sustainable Development, 22*(5), 321–335.

Wichaisri, S., Sopadang, A. (2018). Trends and future directions in sustainable development. *Sustainable Development, 26*(1), 1–17.

Xu, X., Wang, C., Zhou, P. (2021). GVRP considered oil-gas recovery in refined oil distribution: From an environmental perspective. *International Journal of Production Economics, 235*, 108078.

Yadav, G., Kumar, A., Luthra, S., Garza-Reyes, J. A., Kumar, V., Batista, L. (2020). A framework to achieve sustainability in manufacturing organisations of developing economies using industry 4.0 technologies' enablers. *Computers in Industry, 122*, 103280.

Zhang, Y., Sun, J., Yang, Z., Wang, Y. (2020). Critical success factors of green innovation: Technology, organization and environment readiness. *Journal of Cleaner Production, 264*, 121701. https://doi.org/10.1016/j.jclepro.2020.121701

CHAPTER 7

Consonance of Finance: Nurturing an Inclusive, Resilient, and Sustainable Ecosystem

RITUPARNA CHOUDHURY* and SUMEET GUPTA

School of Business, University of Petroleum and Energy Studies (UPES), Dehradun, Uttarakhand, India

*Corresponding author

ABSTRACT

The indeterminate scenario of financial ecosystem is shifting toward the determinate landscape of holistic finance with networking across the globe. Being in a world of multipolar and robust nature, we are confronted with many unforeseen challenges such as evolving economic challenges, social inequalities, and environmental threats, which demands attention. This chapter offers an exhaustive analysis of the complex elements vital for creating an inclusive, resilient, and sustainable financial ecosystem. Furthermore, it will establish a regulatory framework to elucidate the interplay between inclusivity, resilience, and sustainability within the ecosystem of finance. Realizing these principles in actual policy and practice necessitates worldwide cooperation to fulfill the objectives set forth by the United Nations Sustainable Development Goals—envisaging a more affluent, just, and environmentally sustainable future. Such endeavors will pave the way for enhanced financial well-being.

7.1 INTRODUCTION

The world is recognized for its workforce diversity, which must be inclusive in nature, not ostracism. It has been already proven that a diverse

Industry 5.0: Innovation and Adoption for Sustainability.
Tilottama Singh, Pradeep Suri, and Zulkarin Jahangir (Eds.)

DOI: 10.1201/9781779641878-7

workforce collaborates more efficiently, innovatively, and creatively, and are at a competitive advantage. Also, they are more equipped to solve complex problems. Furthermore, as workers increasingly want for inclusive and inviting workplaces, organizations that prioritize diversity and inclusion are more likely to draw in and keep top talent. The future of diversity is one in which we recognize and celebrate our differences while also finding common ground and shared values. Industry 5.0 is the next version of Industry 4.0 with the preferment of artificial intelligence (AI) and machine learning. With further use of virtual reality and augmented reality technologies, immersive experiences can be created that foster empathy and understanding of different behaviors and their perspectives. The trending robot, that is COBOT (cooperative robot and collaborative robots) is the autonomic robot which is used to conduct a more independent and comprehensive technique of manufacturing in locations where humans are unable to operate. There are devices to optimize productivity and human effectiveness in Industry 5.0 (Butt, 2020).

However, certain challenges limit the enhancement of Industry 5.0. The probable concern would be AI algorithms that can have biased decision-making skills if they are not designed and trained properly with diversity in their artificial mind. It is better to reinforce inclusivity and diversity rather than existing inequalities within the financial ecosystem. So, we need to have intersections to address the equitable, inclusive, resilient, and sustainable spaces that value the diverse experiences and perspectives of all individuals.

As per the Organisation for Economic Co-operation and Development's (OECD) Economic Outlook, Volume 2023, the real GDP growth rate will decelerate to 6.3% in FY 2023–2024 and 6.1% in the following year with a strong outcome revealed in FY 2022–2023. Further, the report says that the main hurdles to stumble down are eradication of poverty, mitigating the changes in the climate, and income generation which requires huge efforts and standards to accomplish an inclusive, resilient, and sustainable ecosystem of finance.

7.2 INDUSTRY 5.0

The digital industry emerged in the time preceding Industry 5.0. Technological developments like the Industrial Internet of Things and the fusion of AI and big data have created a new class of technology that can provide businesses with information based on data. This has consequently led to the development of procedures like business and operating intelligence that produce models that employ technology in an effort to make decisions that are more precise and less ambiguous. But the focus of Industry 4.0 has been on process automation with a minimum of human intervention. Humans have been displaced in various situations and placed in a situation where they must compete with technology. The term "Industry 5.0" refers to a revolution in which both humans and machines collaborated to increase manufacturing production methods and efficiency. Industry technologists, ideologies, and others argue that Industry 5.0 is forced to concentrate on the technology and human elements in the manufacturing systems. Industry 5.0 is thought to be the edge of the smart industry, interacting with both humans and robots. Social networks are used to facilitate communication between electrical components and people. Already, Industry 5.0 has brought about irrevocable changes. Through this approach, businesses can harness the power of ever-more-powerful machinery and highly skilled professionals to promote efficient, sustainable, and secure production. Industry 5.0 is a new paradigm for manufacturing that has implications for productivity, the economy, and business. It is not a passing trend. As a result, businesses who do not adapt their production to the factory 5.0 model will quickly become outdated and lose out on the advantages it offers in the marketplace (Adel, 2022).

7.2.1 BENEFITS OF INDUSTRY 5.0

7.2.1.1 COST OPTIMIZATION

The paradigm of Industry 5.0 builds upon prior advancements that have, since the inception of the First Industrial Revolution, facilitated the refinement of operational processes. The quest for business models that minimize resource consumption while maximizing profitability reaches its apex within the framework of Industry 5.0, wherein human and machine collaboration engenders optimal financial decision-making for enterprises.

7.2.1.2 GREENER SOLUTIONS

None of the aforementioned industrial evolutions have prioritized environmental protection as a primary objective. The advent of Industry 5.0 heralds the introduction of novel corporate technologies and heightened environmental awareness, thereby altering this trajectory. This evolution has fostered the development of sustainable policies whereby, for instance, the minimization of waste generation and its management become integral, and pervasive processes that simultaneously enhance organizational efficacy. This transformation aligns with the escalating demands of international organizations, governmental regulations, and consumer expectations.

7.2.1.3 PERSONALIZATION AND CREATIVITY

Technological advancements have historically failed to provide a level of personalization that adequately addresses consumer demands. The workforce associated with Industry 5.0 will harness technological capabilities while simultaneously creating opportunities to incorporate their unique insights, resulting in products that are expressly tailored to individual preferences.

Moreover, the automation achieved during Industry 4.0 emancipates employees from repetitive tasks, thereby allowing them to focus on devising more significant strategies or exercising their creative abilities.

Factory 5.0 has introduced a new role called Chief Robotics Officer. This person has expertise in the interaction between machines and operators, as well as knowledge in robotics and AI. The Chief Robotics Officer is responsible for making decisions related to these factors in the company. The training of employees will improve with the widespread adoption of virtual education. This approach reduces costs for companies and allows for safer training that prevents workers from being exposed to unnecessary risks during training. Interactive learning environments resulting from virtual education also boost communication and employee motivation. Industry 5.0 introduces collaborative robots, also known as "COBOTs." These robots are designed for simple and intuitive interaction with humans and consider processes like plant safety and goals. COBOTs act as apprentices, capable of observing a human's actions and replicating them to assist operators. Another necessary technology in Factory 5.0 is the expansion of digital twins. These are visual models of a product or process that can be better understood and tested. As processes become increasingly complex, suitable software is required to manage vast amounts of data and provide human operators with a space to interact with machines. Overall, the emergence of Industry 5.0 is expected to create numerous employment positions related to the interaction with robotic systems and AI, among other technologies.

7.3 PROBLEM STATEMENT

According to the European Union, Industry 5.0 complements Industry 4.0 by further embracing its inclusive, human-centric, resilient, and sustainable strategies. As per the OECD's Economic Outlook, Volume 2023, the real GDP growth rate will decelerate to 6.3% in FY 2023–2024 and 6.1% in the following year with a strong outcome revealed in FY 2022–2023. Further, the report says that the main hurdles to stumble down are eradication of poverty, mitigating the changes in the climate, and income generation, which requires huge efforts and standards to accomplish an inclusive, resilient and sustainable ecosystem of finance.

There is a requirement for having a financial ecosystem consisting of inclusivity, resilience, and sustainability leading toward financial well-being of the people.

7.4 FINANCIAL ECOSYSTEM

Unlike the scientific ecosystem, the financial ecosystem refers to the biotic and abiotic components that interact with each other and the surrounding environment. It is the chain of interactions and thus influences each other, thereby creating a dynamic and interrelated system. The biotic components include the participants, financial service providers, and the investors, while the abiotic components involve the financial institutions, financial markets, regulatory frameworks, technology, infrastructure, economic policies, and risk management and insurance. Even a slight difference in any part of the ecosystem will have a rippling effect throughout the system, influencing stability in financial condition. Other than the biotic and abiotic components, there are some extraneous variables that can have an impacting effect on the ecology of finance such as political events, technological innovations or cultural differences, and so on.

The financial ecosystem follows the laws of biology more than the laws of physics where the people act in their own self-interest, make mistakes, then rectify themselves. A well-functioning ecosystem is based on cloud computing platforms through which they can do the orchestration of the available products and services to provide security, integrity, and resilience within the system.

7.4.1 INCLUSIVITY IN FINANCIAL ECOSYSTEM

Inclusivity in the financial ecosystem means providing an equitable distribution of access to basic financial amenities and opportunities to every citizen regardless of their caste, creed, religion, background, economic status, or demographic attributes. To spread the essence of inclusivity, there must be diversity in the workforce who were once added to the vulnerable group and were not having the tools to involve themselves in the financial ecosystem. Inclusivity in the financial ecosystem is an economic imperative. It contributes to economic growth, reduces inequality, and fosters a more stable and resilient financial system. The policymakers, financial institutions, and other stakeholders play crucial roles in driving initiatives that enhance inclusivity in the financial ecosystem.

Inclusivity has three pillars that make it balanced in the financial ecosystem. They are access, usage, and quality. The access refers to the availability or the reach of the financial services, whether the services are obtainable in that vicinity or not. Usage means the utilization of funds or the financial services available in an efficient manner. It measures the frequency of using the available financial resources or services. Also, quality is a vital element of inclusivity which depends on the satisfactory position of the customers. Most of the inclusivity within the financial ecosystem hinges on the quality parameter of inclusion whether it is related to financial literacy and education or dispute resolution. According to Ehrbeck et al. (2012), the Consultative Group to Assist the Poor has explained how the government plays the role of rulemaking, increasing volumes, and promoting infrastructure for inclusivity in the financial ecosystem.

Importance of the facets of inclusivity within financial ecosystem is given below.

7.4.1.1 ACCESSIBILITY TO AMENITIES

Much of the work has been done on access part of financial inclusion by the government by taking many initiatives and thus implementing them. Starting from the Pradhan Mantri Jan Dhan Yojana, providing and ensuring access to basic banking facilities such as savings accounts and so on. Also, credit services like loans and microfinance that are easily available without much ado about the formal documents have enabled the business units as well as the individuals to finance, grow, and oversee the essential requirements. This led to an increase in the volume of accounts as well as account holders.

7.4.1.2 DIGITAL FINANCIAL INCLUSION

The ongoing evolution of online payment mode, the Unified Payment Interface, has emerged as a revolutionary action taken by the Government to promote inclusivity in a sustainable manner. Various application software have made the transactions process readily accessible. It has promoted in using mobile banking services and digital financial services to people living in remote or secluded regions in a way minimizing the geographical gap to financial inclusion. Also, offering creative and affordable substitutes for conventional banking, the Fintech technologies have embraced inclusivity and contributed to the economy.

7.4.1.3 FINANCIAL EDUCATION AND LITERACY

The government has taken many initiatives to educate people on financial concepts, budgeting, savings, and so on to foster financial behavior in them such that they can take crucial informed decisions. The vulnerable and underprivileged ones are offered training programs to embrace their financial literacy skills to become a part of this financial ecosystem.

7.4.1.4 MICROFINANCE AND SMALL BUSINESS SUPPORT

Programs and schemes like Emergency Credit Line Guarantee Scheme have delivered additional credit to more than 130,000 micro, small, and medium enterprises (MSMEs). Raising and Accelerating MSME performance program, MSME Self-Reliant India Fund, Udyam, e-Shram, National Career Service, and Aatamanirbhar Skilled Employee Employer Mapping portals are there to benefit the MSMEs. The launch of the Udyam Assist platform is done to bring the informal microenterprises under the formal ambit of MSMEs for availing the benefit of priority sector lending.

7.4.1.5 FINANCIAL INCLUSION POLICIES AND REGULATIONS

Enforcement of consumer protection laws instills the interests and rights of individuals having little knowledge of finance by dispute resolution which comes under the quality aspect of financial inclusion. The government needs to put more efforts on this section of inclusivity and make policy recommendations and regulations that are essential for maintaining the ecosystem of finance.

7.4.1.6 PARTNERSHIPS AND COLLABORATION

Synergy is maintained between the public and private sectors, including financial institutions, governments, and nonprofit organizations, to increase the outcomes of inclusivity. Community engagement can be done by understanding the community needs and providing them with tailored financial services, accordingly, fostering a sense of ownership and trust.

7.4.1.7 MEASURING IMPACT AND PROGRESS

The government should check the improvement or progress of their initiatives to address the gaps of inclusivity within the financial ecosystem. The ecosystem must have transparency and trust so that the individuals or business units continue to take financial services, thus balancing the financial ecosystem.

7.4.2 RESILIENCE IN FINANCIAL ECOSYSTEM

The ability of an individual, household, organization, or system to endure and then recover the financial losses or shocks, disruptions, or unexpected events is said to be resilience. Not only it involves the capacity to absorb and adapt to economic challenges but also to fight with uncertainties and cures itself from this disease without experiencing a significant decline in one's financial well-being. Withstanding the adverse financial phases of life and adapting to such challenges and then ultimately emerging as much stronger and equipped than before for future uncertainties is resilience. Resilience is crucial at various levels—individual, household, organizational, and systemic—because it enables entities to withstand, adapt to, and recover from challenges and disruptions.

Financial resilience has been studied extensively, with different definitions provided by various researchers. In an earlier study, Lusardi (2011) defined financial resilience as an individual's ability to access emergency funds from various sources when required. This definition is based on the life-cycle hypothesis model of consumption and savings, which posits that individuals save to ensure a smooth consumption pattern over their lifetime and to be prepared for unforeseen expenses or temporary income loss. However, other researchers like Salignac et al. (2019) have defined financial resilience as the ability of individuals to rely on their internal and external resources during adverse financial events. Internal resources refer to an individual's ability to manage their finances by saving and controlling expenses, while external resources refer to the availability of social support from family and friends.

According to Salignac et al. (2019), financial resilience cannot be determined solely by an individual's ability to manage financial shocks. Other factors such as context, structures, and support must also be taken into consideration. To understand financial resilience from the perspective of financial inclusion, they have developed a framework that encompasses four interconnected factors: economic resources, financial products and services, financial knowledge and behavior, and social capital. Similarly, Goyal et al. (2021) have studied financial resilience among Indians during the COVID-19 pandemic by incorporating individuals' reliance on both internal and external resources.

The concept of financial resilience, as defined by the OECD, is based on several critical factors, including effective control of one's finances, prudent expenditure, maintenance of a financial cushion, adept handling of financial shortfalls and stress, proactive financial planning, and an awareness of fraudulent activities. These factors primarily relate to an individual's ability to assess and use internal resources when confronted with financial challenges. This approach mostly corresponds to the economic resource factor analyzed by Salignac et al. (2019).

In recent years, there has been an increase in the number of studies focusing on financial literacy and its impact on financial resilience. Lyons and Kass-Hanna (2021) have conducted research to explore the role of financial literacy in promoting financial resilience. According to Lyons and Kass-Hanna (2021), higher levels of financial literacy promote improved saving habits, better risk management behaviors, and enhanced emergency preparedness, including the acquisition of life and health insurance. Other researchers found that higher financial literacy is associated with better emergency preparedness, reduced debt burdens, improved future planning, and increased saving and retirement planning.

The literature has extensively explored the role of financial inclusion in influencing financial resilience. Although the definition of financial inclusion varies among different studies, most of them define it in terms of comprehensive financial product holdings. Early studies focused on basic account access, while current studies assert that financial inclusion enables access to suitable and reasonable financial products and services. This, in turn, facilitates investment in various aspects, including business, education, and health. The unbanked population often engages in informal financing, which is more expensive compared with formal financing.

According to Lyons and Kass-Hanna (2021), having financial products related to savings, loans, and insurance is critical in making more strategic financial decisions that mitigate risk and prepare individuals to face financial shocks. Financially resilient individuals have lower financial vulnerability and are bank account holders. Based on the developmental economic perspective, financial inclusion is a catalyst that stimulates economic development. Countries with low financial inclusion often have higher poverty rates, greater income inequality, and lower economic growth. Lyons and Kass-Hanna (2021) affirmed that greater financial inclusion is linked to lower financial vulnerability. Therefore, enhancing financial inclusion can play a crucial role in promoting financial resilience among individuals and countries.

Salignac et al.'s (2021) paper presents a snapshot of financial resilience in Indonesia, offering the country's first comprehensive analysis of the prevalence and impact of financial exclusion in the region. The authors underscore the importance of financial resilience in contributing to economic development, highlighting its potential to inform and shape policy interventions aimed at promoting financial inclusion and reducing poverty in developing countries. The study's findings demonstrate that financial exclusion is a significant challenge in Indonesia, with many individuals lacking access to formal financial services. The authors argue that such exclusion can have far-reaching effects on economic growth and development, limiting opportunities for entrepreneurship, savings, and investment. The paper's emphasis on financial resilience offers a compelling framework for understanding the importance of access to financial services in developing countries. The authors conclude that promoting financial inclusion is essential for reducing poverty and achieving sustainable economic growth, and that policymakers must take active steps to address the challenges of financial exclusion in these regions. Overall, this research provides valuable insights into the importance of financial resilience and its potential role in shaping policy interventions aimed at promoting economic development in developing countries.

The importance of the facets of financial resilience is mentioned below.

7.4.2.1 EMERGENCY SAVINGS

We are aware of the COVID-19 pandemic situation where people were craving for cash-in-hand facilities. One must have emergency fund or contingency amount in case of uncertainties such as job loss or medical emergencies that may cover the living expenses for a specified period. This will help to lever the financial ecosystem and balance the circular economy of the country.

7.4.2.2 FINANCIAL LITERACY AND PLANNING

A good knowledge and understanding of financial concepts helps in making informed decisions about savings, investing, and managing debts. This will reduce financial losses to a certain extent. In the current world scenario, dependence on a single source of income should not be done. At least one must diversify their stream of income which will try to mitigate the instability in jobs or will prevent fluctuations in the economy. To be resilient one must be accustomed to identifying the financial risks. This includes having appropriate insurance coverage, understanding investment risks, and making informed financial decisions.

7.4.2.3 ADAPTABILITY AND FLEXIBILITY

Reevaluation of expenditure, new income opportunities, and taking vital decisions on long-term financial plans all depend on adaptability and flexibility in accepting the dynamic financial ecosystem.

7.4.2.4 GOVERNMENT AND INSTITUTIONAL SUPPORT

Benefits from supportive government policies, social safety nets, and institutional frameworks contribute to the overall economic stability and individual financial well-being.

7.4.3 SUSTAINABILITY IN FINANCIAL ECOSYSTEM

Encapsulation of sustainability surrounds the three subthemes of environmental, society, and governance (ESG). They encompass a wide range of matters such as carbon regulation, health and safety, financial inclusion, corruption, and bribery. To take responsibility for issues arising from these operations, sustainability and ESG must go hand in hand and thus require a company to take a more holistic approach. Many researchers have demonstrated a better economic outcome for business models that incorporate ESG. Sustainable development is described as "development that meets the needs of the present without compromising the ability of future generations to meet their own needs." In the financial ecosystem, sustainability may be defined as the integration of ESG into financial practices and decision-making. This may lead to long-term benefits toward financial well-being. Sustainability in the financial ecosystem is an evolving and dynamic concept that requires ongoing commitment, collaboration, and innovation from all stakeholders, including financial institutions, investors, regulators, and the broader community. Integrating sustainability principles into financial practices not only contributes to a more responsible financial system but also aligns with broader global efforts to address environmental and social challenges.

The importance of sustainability within the financial ecosystem is given below.

7.4.3.1 ENVIRONMENTAL CONSIDERATIONS

The more the promotion of green finance in environmentally sustainable projects, such as renewable energy, energy efficiency, and sustainable agriculture, the more will the environment be sustainable. Timely mitigation of risks involved with climate change including physical risks (e.g., extreme weather events) and transitional risks (e.g., regulatory changes impacting carbon-intensive industries) also leads to sustainability.

7.4.3.2 SOCIAL CONSIDERATIONS

Social considerations in integration to investment decisions to support companies and projects that prioritize social impact, ethical practices, and social justice help maintain balance not only in society but also in the financial ecosystem. Ensuring that financial services are accessible to all segments of the population, including marginalized and underserved communities, to promote inclusive economic growth will fulfill the quota of Sustainable Development Goals (SDG 10). Spreading financial education knowledge, training and development skills has made a step ahead toward sustainability.

7.4.3.3 GOVERNANCE CONSIDERATIONS

Emphasis is on transparent and accountable corporate governance practices to ensure that companies are managed in the best interests of stakeholders. Adoption of ethical behavior, fair labor practices, and smooth supply chain management will encourage ethical business practices which in a way will lead to accomplish SDG 8 and 16. Implementation of regulations mandate companies to disclose their ESG performance, promoting transparency and accountability. Also, recent development in economic conditions is done by encouraging financial institutions to integrate sustainability into their business models, especially banking institutions.

7.4.3.4 INNOVATION AND TECHNOLOGICAL CONSIDERATIONS

The revolutionary innovation of financial technology harnesses to develop innovative solutions that promote sustainability, including blockchain for transparent supply chains and digital tools for impact measurement.

7.4.4 RESEARCH GAP

There is a dearth in literature about the relation between financial ecosystem leading to financial well-being.

7.4.5 RESEARCH QUESTION

What is the impact of a financial ecosystem consisting of inclusivity, resilience, and sustainability on financial well-being?

7.4.6 RESEARCH OBJECTIVE

To establish a model to elucidate the interplay between inclusivity, resilience, and sustainability within the ecosystem of finance leading to financial well-being.

7.4.7 RESEARCH DESIGN

The research's design is descriptive. The relations that were fetched from the literature are as shown in Table 7.1.

TABLE 7.1 Relation of the Constructs Inclusivity, Resilience, Sustainability, and Financial Well-Being among Themselves from the Literature

Relation of Constructs	References
Inclusivity and financial well-being	Kanungo and Gupta (2021), Mukong and Amadhila (2021)
Resilience and financial well-being	Salignac et al. (2019), Hasler et al., (2023)
Sustainability and financial well-being	Munisamy et al. (2022), Shair et al. (2024)
Relation among Industry 5.0 components (human-centric, resilient, and sustainable)	

7.5 CONSONANCE OF INCLUSIVITY, RESILIENCE, AND SUSTAINABILITY WITHIN FINANCIAL ECOSYSTEM

The conceptual relationship between inclusivity, resilience, and sustainability within the financial ecosystem revolves around the idea of creating a well-balanced and robust system that promotes economic, social, and environmental well-being. How these concepts are interconnected is shown in Figure 7.1.

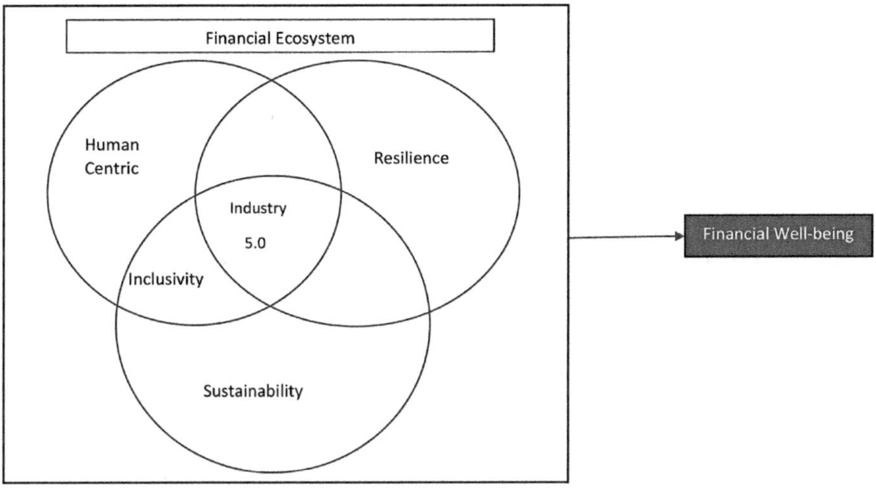

FIGURE 7.1 Relation between human-centric, resilience, inclusivity, and sustainability within financial ecosystem with financial well-being.

A sustainable financial system requires inclusivity, which means offering financial services to a wide range of people and companies. Inclusive financial systems help to mitigate risk and promote general stability. They help to diversify the financial ecosystem, increase its robustness, and promote economic growth. By encouraging investments in environmentally and socially conscious projects, a sustainable financial ecosystem considers environmental factors and promotes the planet's long-term health. Inclusive financial systems are the backbone of sustainability, resilience, and inclusivity. The interconnection between these concepts is vital for promoting positive feedback loops that support each other. Resilience is fostered by inclusivity, and it supports sustainability. Financial institutions that promote inclusive practices are more likely to adopt sustainable business models. Therefore, a well-crafted regulatory framework is crucial for promoting sustainability and inclusivity. Strong rules that prevent and mitigate systemic threats contribute to the long-term stability of inclusive and sustainable financial activities.

Financial well-being is a state of overall satisfaction, security, and comfort with one's financial situation. It encompasses various aspects of financial life such as income, employment stability, savings, debt management, budgeting, investments, insurance, financial literacy, retirement planning, social and emotional well-being, and access to resources and opportunities. It is essential for individuals to have a stable and sufficient income that covers basic living expenses, savings, an emergency fund, and appropriate insurance coverage to protect against unexpected events and mitigate financial risks. Creating a budget and making informed investment decisions are also crucial for long-term financial well-being. Furthermore, understanding basic financial concepts and having the ability to make informed financial decisions are important for an individual's long-term well-being. Planning for retirement ensures that individuals can maintain their desired lifestyle after they stop working. Finally, having access to education, job opportunities, and financial resources within the financial ecosystem contributes to overall financial well-being.

7.6 GLOBAL APPROACH TOWARD INDUSTRY 5.0

The global market for Industry 5.0 is brimming with key players such as ABB from Switzerland, KUKA from Germany, Universal Robots from Denmark, Nexus Integra from Spain, IBM, Microsoft, Rockwell Automation Inc., and Cisco Systems Inc. from the US, Siemens AG from Germany, and Schneider Electric from France. These players have been instrumental in driving digital transformation across various industries. In August 2021, UnifyTwin, Inc. (US) launched a new Industry 5.0 that promises to revolutionize the manufacturing sector. By integrating human and machine intelligence, UnifyTwin's Industry 5.0 eliminates digital blind spots and enables smart manufacturing operations. With UnifyTwin's solution, manufacturers can optimize their production processes, improve product quality, and enhance overall efficiency.

7.6.1 NEXUS INTEGRA

Nexus Integra is a platform that integrates industrial processes management and analysis, aiming to simplify the way these operations are carried out. As we move toward Industry 5.0, which merges sustainable production and personalization with advanced technologies like AI and robotics, Nexus Integra has become a crucial player in facilitating these changes. The reasons why Nexus Integra is implementing Industry 5.0 are shown in Table 7.2.

TABLE 7.2 The Working of Nexus Integra

Integration of Complex Systems	Data Management and Analytics
Nexus Integra is a comprehensive framework that facilitates the seamless integration of various industrial systems and data sources. It acts as the central nervous system of an enterprise's operations, in line with Industry 5.0's focus on interoperability between human workers and smart systems. For instance, consider a factory that produces custom bicycles, each of which is tailored to customer specifications such as color, size, and components, and must be produced sustainably. With Nexus Integra, all subsystems, from ordering to manufacturing, can communicate effectively, ensuring personalized output while optimizing the use of resources.	The platform is specifically designed to manage large amounts of data generated by connected devices (Internet of Things) in an industrial setting. It leverages big data analytics and AI algorithms to help businesses make informed decisions that align with the principles of Industry 5.0—combining efficiency with customization. For instance, a food processing plant can use Nexus Integra to continuously monitor the quality of ingredients across various batches and adjust recipes based on real-time customer feedback or supply availability without compromising safety standards.

7.6.2 ROCKWELL AUTOMATION

Rockwell Automation is the go-to solution for industrial automation. Their unique approach toward realizing the Industry 5.0 vision involves seamlessly blending advanced technologies with human touchpoints throughout the entire manufacturing process. Rockwell Automation's Industry 5.0 implementation processes are unparalleled, ensuring that businesses stay ahead of the curve (Table 7.3).

TABLE 7.3 The Working of Rockwell Automation

Collaborative Robotics (COBOTs)	Scalable Solutions
Rockwell Automation integrates COBOTs that work safely alongside humans, enhancing productivity and flexibility in tasks that require fine motor skills or human judgment. For instance, in automotive assembly lines, Rockwell's COBOTs help technicians install wiring harnesses with precision while identifying fitment issues that may arise.	Rockwell provides scalable automation solutions for different industries and segments, catering to the individualized paradigm of Industry 5.0. For example, craft breweries can use Rockwell's modular controls and equipment to increase production while preserving the manual oversight and craftsmanship behind their product.

7.7 IMPLEMENTATION OF INDUSTRY 5.0 IN INDIA

ABB, a Swiss-based company, is leading the way in sustainable and innovative technology solutions. In August 2022, the company launched a new smart factory in India that follows Industry 5.0 standards, resulting in a 40% increase in productivity due to human and COBOT collaboration. ABB's commitment to sustainability is evident in its acquisition of Sevensense, a Swiss startup known for its AI-enabled 3D vision navigation technology for autonomous mobile robots. With a sustainability agenda aligned with international standards and frameworks, ABB is enabling a low-carbon society, preserves resources, and promotes social progress for a net-zero future. ABB's culture of integrity and transparency extends throughout the value chain, ensuring that the company's actions are consistently aligned with its mission.

The Smart Power factory in Nelamangala, Bengaluru, is a prime example of ABB's commitment to innovation and sustainability. Equipped with one of the first Industry 5.0 production processes, the factory harnesses advanced collaborative robotics technology, AI, and advanced digitalization technologies. The factory produces ABB Smart Power's full range of low-voltage power equipment and energy management technologies to meet industry demand for increased reliability and energy savings. This includes commercial and residential buildings, infrastructure operators, utilities, and solar and wind power systems. The ABB Smart Power factory is over 8,400 m^2 (about the area of a Manhattan city block) and is equipped with "COBOTs," motors, and drives linked to the Internet of Things. The connected factory software highlights process optimization opportunities and schedules predictive maintenance to maximize productivity and efficiency. The result? An impressive 40% increase in productivity in the same space, with an energy efficiency of 15%, and a 50% improvement in the test automation ratio. ABB's commitment to sustainable and innovative technology solutions is evident in its actions. By investing in Industry 5.0 standards and acquiring Sevensense, ABB is leading the way in creating a sustainable, low-carbon society that promotes social progress for a net-zero future.

7.8 CONCLUSION

Industry 5.0 marks a significant shift from an emphasis on economic value to the prioritization of societal value and well-being. Within the financial ecosystem, fostering inclusivity, resilience, and sustainability is no longer solely a moral imperative but has become a strategic necessity for a thriving global economy. The interconnectivity of financial markets and the growing awareness of social and environmental challenges reinforce the significance of embracing these principles. The promotion of inclusivity is essential in ensuring that all individuals, regardless of their background or economic status, have access to financial services and opportunities. This not only reduces inequality but also enhances economic productivity by unlocking the full potential of diverse talents and perspectives. Building resilience within the financial ecosystem is critical to withstand various shocks and uncertainties. By implementing robust risk management practices, regulatory frameworks, and technological innovations, a more resilient financial system that can weather economic downturns and unexpected crises can be created. Sustainability is key for the long-term health of our planet and economies. Aligning financial activities with ESG criteria is vital to creating a more sustainable and responsible financial ecosystem. Investments in renewable energy, green technologies, and socially responsible enterprises can drive positive change and help address pressing global challenges. Thus, the pursuit of inclusivity, resilience, and sustainability within the financial ecosystem requires collaboration among various stakeholders, including governments, financial institutions, businesses, and individuals. Through prioritizing ethical practices, environmental stewardship, and social responsibility, a good financial ecosystem can be made that not only thrives economically but also contributes to the well-being of people and the planet. This holistic approach lays the foundation for a more equitable, adaptable, and sustainable future for generations to come.

KEYWORDS

- **financial ecosystem**
- **financial inclusion**
- **financial resilience**
- **sustainability**
- **financial well-being**

REFERENCES

Adel, A. (2022). Future of industry 5.0 in society: Human-centric solutions, challenges and prospective research areas. *Journal of Cloud Computing*, *11*(1), 40. https://doi.org/10.1186/s13677-022-00314-5

Butt, J. (2020). A strategic roadmap for the manufacturing industry to implement industry 4.0. *Designs*, *4*(2), 11. https://doi.org/10.3390/designs4020011

Ehrbeck, T., Pickens, M., Tarazi, M. (2012). *Financially Inclusive Ecosystems: The Roles of Government Today*. CGAP.

Goyal, K., Kumar, S., Rao, P., Colombage, S., Sharma, A. (2021). Financial distress and COVID-19: Evidence from working individuals in India. *Qualitative Research in Financial Markets*, *13*(4), 503–528.

Hasler, A., Lusardi, A., Yagnik, N., Yakoboski, P. (2023). Resilience and wellbeing in the midst of the COVID-19 pandemic: The role of financial literacy. *Journal of Accounting and Public Policy*, *42*(2), 107079. https://doi.org/10.1016/j.jaccpubpol.2023.107079

Industry 5.0: Towards more sustainable, resilient and human-centric industry. (2021, January 7). Research and Innovation. https://research-and-innovation.ec.europa.eu/news/all-research-and-innovation-news/industry-50-towards-more-sustainable-resilient-and-human-centric-industry-2021-01-07_en

Kanungo, R. P., Gupta, S. (2021). Financial inclusion through digitalisation of services for well-being. *Technological Forecasting and Social Change*, *167*, 120721. https://doi.org/10.1016/j.techfore.2021.120721

Lusardi, M. M. (2011). Older adults and their families. *Geriatric Physical Therapy-eBook*, *207*.

Lyons, A. C., Kass-Hanna, J. (2021). A multidimensional approach to defining and measuring financial literacy in the digital age. In: *The Routledge Handbook of Financial Literacy* (pp. 61–76).

Mukong, A. K., Amadhila, E. (2021). Financial inclusion and household wellbeing in namibia. *Southern African Business Review*, *25*. https://doi.org/10.25159/1998-8125/7866

Munisamy, A., Sahid, S., Hussin, M. (2022). Socioeconomic sustainability for low-income households: The mediating role of financial well-being. *Sustainability*, *14*(15), 9752. https://doi.org/10.3390/su14159752

Salignac, F., Hanoteau, J., Ramia, I. (2021). Financial resilience: A way forward towards economic development in developing countries. *Social Indicators Research*, *160*, 1–33. https://doi.org/10.1007/s11205-021-02793-6

Salignac, F., Marjolin, A., Reeve, R., Muir, K. (2019). Conceptualizing and measuring financial resilience: A multidimensional framework. *Social Indicators Research*, *145*(1), 17–38. https://doi.org/10.1007/s11205-019-02100-4

Shair, W., Hussain, S., Asif, M., Niamat, A. (2024). Financial inclusion, sustainability, and financial wellbeing in OIC and non-OIC countries. In: *Financial Inclusion, Sustainability, and the Influence of Religion and Technology* (pp. 164–185). https://doi.org/10.4018/979-8-3693-1475-3.ch008

CHAPTER 8

Empowering Communities: An Investigation into the Corporate Social Responsibility Efforts of Indian Public Sector Banks

HIMANI GREWAL

Department of Management, Moradabad Institute of Technology, Moradabad, Uttar Pradesh, India

ABSTRACT

The Indian banking sector has undergone a significant transformation, evolving from its colonial roots through various reforms, including nationalization and privatization, to reach its current state amid challenging business environments. The sector has shifted from traditional banking methods to modern practices. Indian public sector banks prioritize corporate social responsibility (CSR) programs that address issues such as healthcare, poverty reduction, self-employment training, financial literacy, infrastructure development, education, and environmental protection. Enhancing financial inclusion and providing services to underserved communities are the goals of these initiatives. This study aims to examine the diverse CSR initiatives undertaken by Indian public sector banks to contribute positively to society. It includes data from the annual reports of 18 public sector banks over a 9-year period, spanning from the financial year 2013–2014 to 2020–2021. The findings will enrich the existing body of knowledge on CSR practices in the Indian banking industry and provide

Industry 5.0: Innovation and Adoption for Sustainability.
Tilottama Singh, Pradeep Suri, and Zulkarin Jahangir (Eds.)

DOI: 10.1201/9781779641878-8

valuable insights for banks to formulate more strategic CSR policies, taking into account the activities of their competitors.

8.1 INTRODUCTION

Financial institutions are the economic agents committed to simultaneously purchasing and reselling financial instruments and contracts. Banks are a particular category of financial institutions that deal in financial resilience. They purchase the securities produced by borrowers and sell them to lenders. Banks have a major role in the growth of an economy (Roy, 2015). A business's financial existence is greatly influenced by banks. The bank gathers money from the populace and lends those savings to those in need, acting as a lubricant to keep the wheels of the economy turning. A thriving banking system mobilizes societal savings and uses them for investments and other beneficial purposes (Limbore and Mane, 2014). Due to their pivotal role in financing global economic and developmental activities, financial institutions, especially banks, have a major responsibility to support sustainable development. It is difficult to overstate the importance of banks acting as good corporate citizens in this setting, particularly in a developing nation like India (Chaudhury et al., 2012).

A strong banking system effectively channels individuals' savings into investments and other productive endeavors. The Indian banking industry has a long and rich history that includes changes from nationalization to privatization as well as customary banking techniques from the colonial era through the reform era (Sawant, 2020). The Indian banking sector has undergone a significant transformation and has now established a new position amid challenging business conditions. This journey reflects the breadth of its evolution, from the British era to the reform period, and illustrates the transition from nationalization to privatization. The Indian banking sector has evolved significantly and now occupies a new position amid challenging business conditions (Gauba, 2012).

Globally, corporate social responsibility (CSR) has grown tremendously (Galan, 2006), and banking is not an exception to that misguided trend (Bihari and Pradhan, 2011; Singh and Verma, 2017). The rise in popularity is attributed to the current era of globalization, which urges organizations of all sizes, whether large or small, to incorporate CSR into their strategies for community development and environmental sustainability. The banking sector, having faced the repercussions of severe international financial crises (Bihari and Pradhan, 2011; Rani and Khan, 2014) and amid intense competition, continues to play a crucial role in economic growth and is guiding the country in implementing CSR practices for the benefit of stakeholders (Fatma et al., 2020; Tran, 2014). In the 21st century, there has been a concerted effort to integrate economic, social, legal, and ethical principles to foster sustainable development. CSR is the commitment made by a business to work with its workers, their families, and the community in order to promote sustainable economic development. The stability of financial systems depends on the banking sector. Because it encourages the flow of money and ensures that resources are used wisely to support economic growth and development, it is essential to the state of our country's economy. CSR initiatives are aggressively pursued by Indian banks. The CSR practices framework, however, varies from bank to bank. And for the identification of this variation we get motivated to conduct this thematic analysis. As per the literature, a clear picture of Indian banks contribution toward CSR is missing, so keeping this objective into mind we have selected this methodology to provide a detailed analysis of the Indian banking industry's CSR practices. As per the content analysis and methodology adopted in the study, it has been observed that the Indian banking industry has identified its thematic areas where the banks could start star initiative programs. The chosen banks' CSR initiatives cover a broad range of endeavors such as rural development, community welfare, sustainability, skill development, sports promotion, healthcare, women's empowerment, and protection of the environment. The bank uses a sizeable portion of its annual profits to support CSR activities in a variety of ways. Participating in socially responsible activities gives banks more chances to stand out from rivals and raises the public perception of the caliber of their services. The Star Initiatives programs are helping society in multiple ways and also strengthening the credibility of the banks.

As per the literature, the studies based on CSR in the banking industry are mainly focused on the relationship between CSR and financial benefits (Sharma and Mani, 2013; Mansoor, 2016), CSR investments and economic growth (Sharma, 2011), CSR investments and brand image (Yuan et al., 2011), CSR investments and bank performance (Chen et al., 2018), and CSR investments and economic growth (Skare and Golja, 2014). However, a detailed analysis and discussion have not been yet identified in the literature. So to fulfill this gap, this study has been proposed. This study is focused toward identifying the various CSR programs running by Indian public sector banks (PSBs) in order to give back to the society. The study will provide a detailed analysis of the annual reports of the Indian banks. The study will add to the existing literature in order to analyze the CSR approach of the Indian banking industry and will also help the banking institutions to design their policies in a more strategic manner while analyzing the approaches of their competitors at the same time.

To the best of the author's knowledge, this is the first study to offer a thorough examination of the CSR initiatives carried out by Indian PSBs. The structure of the banking sector and the literature analysis are covered in the paper that follows, which focusses on CSR in the Indian banking sector. The study's goals and research methodology are covered in the next section. The study's conclusions are listed in the final part, which is then followed by some closing thoughts (Figure 8.1).

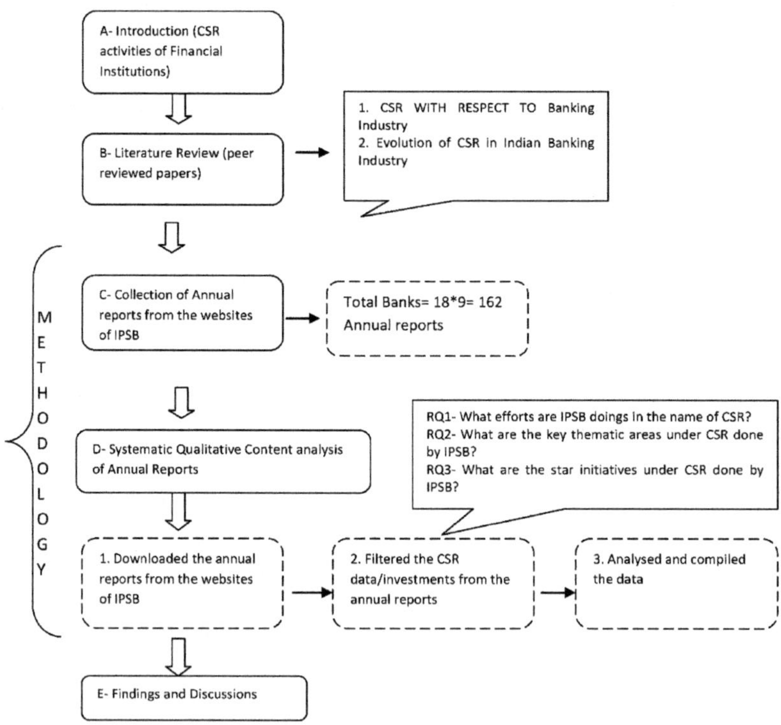

FIGURE 8.1 Methodology frame.

Source: Framed by the author.

IPSB—Indian public sector banks; CSR—corporate social responsibility.

8.2 LITERATURE REVIEW

With the rise of numerous multinational firms in the early 1970s, the phrase "corporate social responsibility" gained popularity. CSR is described as the deliberate taking into account of the public interest in company decision-making, even if it frequently follows a three-way lowest link. Empirical studies have focused on how financial institutions fulfill their social responsibilities to achieve sustainability (Doh et al., 2010). In the contemporary modern period of individualism, cooperation is valued over competition, and society is all about the individuals. In the modern day, measurements of a bank's performance include employee productivity and contentment (Rahman et al., 2016; Mandhachitara and Poolthong, 2011) as well as customer loyalty and satisfaction. However, the social environment in which the banking institutions operate should also be taken into consideration. Although profit maximization is the primary goal of banks, the current environment has modified this goal. The focus shifted from profit maximization to increasing CSR initiatives (Doh et al., 2010).

In support of this endeavor, Indian banks have been advised by the Reserve Bank of India (RBI) (Corporate Social Responsibility, 2007) to adopt nonfinancial reporting pertaining to environmental, social, and economic activities as well as to execute CSR efforts targeted at sustainable development. Various authors, academics, and researchers have defined CSR in different ways. Arora and Puranik (2004) describe CSR as a company's continuous commitment to adhere to ethical standards, foster national economic development, and improve the well-being of its employees and the broader community. In contrast, Hopkins (2003) sees CSR as the moral treatment of stakeholders, both internal and external, with the ultimate goal of improving stakeholders' quality of life while guaranteeing the corporation's financial success.

The banking industry in India places a strong emphasis on CSR projects that address issues such as healthcare, education, rural development, financial literacy, self-employment training, infrastructure development, and environmental preservation. By extending financial services to underserved or unbanked areas of the nation, these initiatives seek to advance financial inclusion (Moharana, 2013; Sahoo and Nayak, 2007; Singh et al., 2013). CSR has become a prominent trend in international corporate operations. It has grown in importance as a subject of conversation among scholars, researchers, government bodies, nongovernmental organizations (NGOs), and business community at large in recent years (Sharma, 2013).

Organizations are increasingly striving to comprehend their roles in social responsibility due to the competitive market landscape. Notably, both government agencies and NGOs believe that CSR is essential for establishing brand communication for corporations (Demir and Budur, 2019). In light of the current competitive environment, many companies across various sectors, including banking, have begun to adopt some form of CSR initiatives (Kaur and Sharma, 2018). This highlights the growing importance of CSR from the perspective of stakeholders, including shareholders, customers, employees, government, and society (Freeman and Velamuri, 2006). Management now recognizes the necessity of transitioning from traditional business concepts to a model focused on responsible business practices (Sharma and Mani, 2013). They acknowledge that CSR activities are crucial for achieving exemplary business practices and effective leadership.

Despite being a prominent and widely discussed topic, CSR has largely escaped rigorous examination within the banking industry, as noted by Lorena (2018). While there is growing interest in exploring the relationship, between CSR initiatives, and their effects on banking reputation and brand image, research in this sector remains limited. Henderson (2009) contended that businesses should focus exclusively on their core competencies such as creating job opportunities and enhancing stakeholder wealth. Furthermore, since the interests of public and private organizations differ, CSR may have little relevance for private entities. Business leaders often lack a voice in societal matters, with only the government holding such authority. Moir (2001) reevaluated the definition and rationale behind CSR by referencing literature, practical examples, and various scenarios, while also examining the theories that elucidate this behavior. The writing effectively differentiates between normative or ethical actions and practical applications. The article concluded by prompting a discussion on when instrumental practices shift from being perceived as social responsibilities to being regarded as business endeavors. Epstein (1987) indicated in his study that the repercussions of corporate operations are the primary focus of CSR. He defined it as "the identification of issues, expectations, and claims on business organisations regarding the effects of policies and behaviour on both internal and external stakeholders."

In their study, Aguilera et al. (2006) developed a multitiered conceptual model to explore the reasons behind the increasing engagement of commercial organizations in CSR activities and to assess whether this involvement can foster positive social change. They stated that there is pressure on commercial organizations to participate in CSR programs and explored the organizational justice model, corporate governance, and different forms of capitalism. Additionally, moral, relational, and instrumental motivations may be used to exert these pressures. According to Sawant and Patil (2017), top firms place a lot of importance on CSR. CSR is used to gain public trust by demonstrating that businesses care about societal needs. The study's main subject was cement companies. According to Chawak and Dutta (2014), CSR is more than just a duty to the community; it also extends beyond the interests of owners or shareholders. The business firm has taken a lot from the community in order to make a profit and accumulate wealth; thus, the viewpoint is to give back to the community. They discovered that the most frequent mistakes made by firms were a lack of vision, nonparticipative management, and a lack of employee and local community participation.

In their research, Graafland and Ven de Ven (2006) emphasized that a high level of professionalism among individuals in the financial sector is essential for a bank's CSR efforts. Consequently, key priorities should include the Code of Ethics, evaluation of employee potential, transparency, stakeholder interests, and support for social institutions. Cheung and Mak (2010) investigated the impact of CSR implementation and its absence on the financial performance of organizations, ultimately concluding that no definitive correlation exists between the two.

In their study, Christopher and Luke (2013) found that banks prioritize social issues and that financial inclusion is the key area of their CSR agenda, but that many of them pay little attention to environmental issues. In their study, Fassin and Gosselin (2011) looked into why large organizations such as Fortis, AIG, and Arthur Andersen & Co. failed despite having a strong CSR culture. In their research, they discovered a discrepancy between the top management's stated policies and actual practices.

Even though CSR has been discussed widely and the banking industry has also gained the attention of globally recognized researchers, there is a need for more extensive research works regarding the CSR approaches in the Indian banking sector, which means that the Indian banking industry can be explored in a more precise manner. So to fulfill this gap, this study attempts to give a detailed analysis of CSR practices running by Indian PSBs and extend to the literature.

8.3 STRUCTURE OF THE INDIAN BANKING INDUSTRY

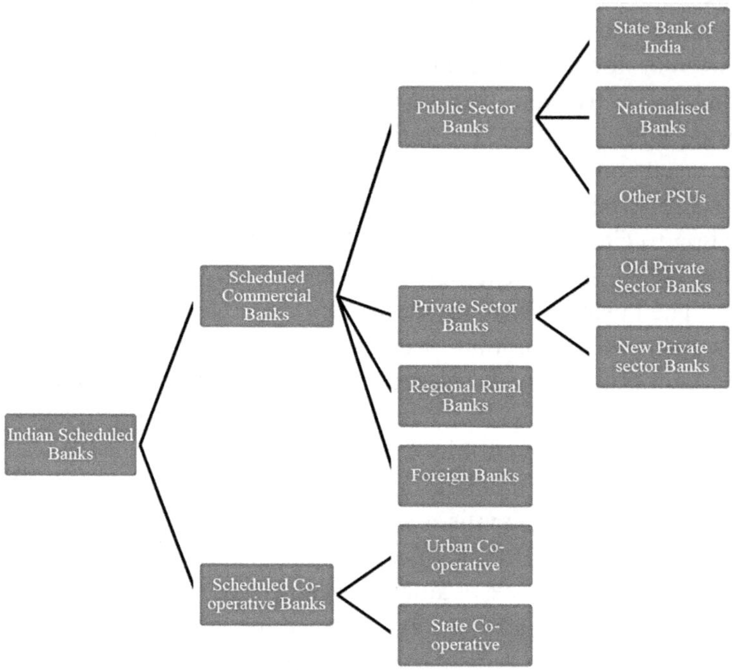

FIGURE 8.2 Banking structure in India.

The Indian banking sector's organizational structure is shown in Figure 8.2. The supreme body overseeing this industry is the RBI. The RBI was founded on April 1, 1935, in accordance with the RBI Act of 1934. It serves as India's central bank (www.rbi.org.in) and is involved in a number of marketing and education campaigns. A commercial bank is a kind of financial organization that serves a wide range of customers and business owners by accepting deposits, lending money for businesses, and other associated services. The main priority for these institutions is profitability. The two main categories of Indian banks are scheduled and nonscheduled banks.

8.3.1 SCHEDULED BANKS

Under the Second Schedule of the RBI Act, 1934, a classification known as scheduled banks has been established by the RBI (www.rbi.org.in). These banks, which fall under Schedule 2 of the RBI Act, benefit from various privileges compared with nonscheduled banks. These privileges include the ability to refinance with the RBI, access to currency vaults, and the right to participate in the Clearing House. Cooperative banks may also be designated as scheduled banks if they satisfy the criteria outlined in the RBI Act of 1934. Scheduled banks comprise both commercial and cooperative banks, with scheduled commercial banks serving as a crucial part of the Indian financial system. This classification includes public, private, foreign, and regional rural banks (www.rbi.org.in), as shown in Table 8.1.

TABLE 8.1 Scheduled Commercial Banks in India (as of December 31, 2018)

Public Sector Banks	Private Sector Banks	Foreign Banks	Regional Rural Banks (RRBs)
1. Allahabad Bank	1. City Union Bank	44 foreign banks	82 RRBs
2. Andhra Bank	2. Karur Vysya Bank		
3. Bank of India	3. Catholic Syrian Bank		
4. Bank of Baroda (Dena Bank + Vijaya Bank)	4. Axis Bank		
	5. DCB Bank		
5. Bank of Maharashtra	6. Yes Bank		
6. Canara Bank	7. ICICI Bank		
7. Central Bank of India	8. HDFC Bank		
8. Corporation Bank	9. Dhanalakshmi Bank		
9. Indian Overseas Bank	10. Federal Bank		
10. Indian Bank	11. Jammu & Kashmir Bank		
11. Oriental Bank of Commerce	12. Kotak Mahindra Bank		
	13. Tamil Nadu Mercantile Bank		
12. Punjab National Bank	14. IndusInd Bank		
13. Punjab & Sind Bank	15. Lakshmi Vilas Bank		
14. State Bank of India	16. Karnataka Bank		
15. Syndicate Bank	17. South Indian Bank		
16. UCO Bank	18. RBL Bank		
17. Union Bank of India	19. IDFC Bank		
18. Union Bank of India (merged with PNB)	20. Bandhan Bank		
	21. IDBI Bank Ltd.		

Source: Adapted from www.rbi.org.in.

On April 1, 2020, six PSBs merged into four major PSBs as part of a significant consolidation involving 10 PSBs, which was announced on August 30, 2019. This merger reduced the number of PSBs to 12. Following the consolidation, Punjab National Bank (PNB) became the second largest PSB, with Canara Bank ranking fourth, Union Bank of India (UBI) fifth, and Indian Bank seventh. However, the government has decided to privatize four of these banks, with plans to privatize two by the end of the fiscal year 2021–2022 (Rai et al., 2021).

8.4 CSR IN THE INDIAN BANKING INDUSTRY

Indian banks should be responsible for overseeing the activities of the nation's financial institutions. The economic system must address the needs of the population and, in turn, reflect those needs. This has led to an increase in the overall socioeconomic value of the output and distribution system. Moreover, it is believed that banks that engage in CSR tend to have higher benefit quality and liquidity ratios, along with improved capital management capabilities (Keffas and Briggs, 2011). A study by Chahoud et al. (2007) indicates that CSR in India largely consists of charitable activities and community development projects. However, Indian businesses and stakeholders are beginning to embrace aspects of the mainstream agenda, such as integrating CSR, into their operational practices and engaging in multistakeholder discussions.

According to the study done by Kennedy (2016), the majority of private banks entered the Indian financial market around the beginning of 1990; several of them were established by prominent business families who advised them on how to engage in charitable endeavors at the time. In India, economic planning began in the latter half of the 20th century. Due to the moral obligation that the banks have to give back to society after receiving support from it year after year, this encouraged the banks to fulfill their social duties. This emphasis was furthered after the nationalization of the banks in 1969 (Cheung et al., 2010). The banks were driven by the banking policy change to support the growth of entrepreneurship and provide job opportunities to the country's poor population. After this reform, the banks have achieved their CSR goals. It was believed that if the local population understood banking practices, participated in them, and started economic activities, this would enhance the degree of capacity building among the populace. Additionally, it was believed that systematic implementation of the particularly created and designed programs would ultimately benefit the recipients (Kennedy, 2016).

Since before the nation's independence, CSR has taken on several forms (Kennedy, 2016). In India before independence, Gandhiji's support and direction for societal improvement and women's emancipation were valued. Prior to the nationalization of the banks, CSR initiatives in the banking sector were primarily concerned with moral and cultural issues in society. The CSR initiatives at the time of the nationalization of the banks were geared toward the environment, society, and employees. Following that, the consolidation phase began, and this time, the banks' attention was on sustainability (Chahoud et al., 2007; Sundar, 2000). The ongoing restructuring of the banking sector has compelled banks to adopt CSR as a fundamental aspect of their investment strategy. During the first phase, known as the Preindependence Period (up to 1947), the CSR focus within the banking industry was directed toward empowering women and promoting societal advancement through Gandhiji's concept of Trusteeship. In the second period, that is, Prenationalization Period (1948–1969), charitable CSR activities, which were focused on social, cultural, and moral and ethical backgrounds were done under the CSR dimensions. In the third period, that is, Nationalization Period (1970–1984), several types of laws, such as the Compensation Act, the Minimum Wages Act, the Labor Law, and the Environmental Law, were introduced by the government, and the main CSR dimensions of the banks were focused toward the development of society and the environment. The Consolidation Period (1985–2000) was the fourth period where companies began putting more emphasis on the well-being of all stakeholders after discovering how it would affect long-term sustainability. The current period is named as Restructuring Period (2001 onward), in which the government, for-profit businesses, and nonprofit organizations all collaborate. A change to the Companies Act makes reporting of CSR projects and investment in CSR initiatives mandatory.

8.5 OBJECTIVE OF THE STUDY

The primary objective of the study is to analyze the CSR practices applied by Indian PSBs.

8.6 RESEARCH METHODOLOGY

The current study is qualitative in character, and content analysis was employed to meet the goal of the investigation. Qualitative content analysis is one of the research methods used to analyze text data (Hsieh and Shannon, 2005). The study examined text data from the fiscal years

2013–2014 through 2021–2022, a span of 9 years. The study's foundation is secondary data that the researchers gathered from the banks' annual reports and business responsibility reports, which are accessible on their websites and in their books. The analysis included all PSBs that were in operation in India before the 2020 merger. There were 18 PSBs in operation in India overall, and they have all been listed.

The bank's CSR initiatives have been thoroughly examined in the yearly reports. The researchers have examined 162 (18*9) bank annual reports in total for the study.

8.7 FINDINGS

8.7.1 CSR PRACTICES OF INDIAN PUBLIC SECTOR BANKS (FY 2013–2014 TO FY 2021–2022)

CSR activities of Indian PSBs are listed in Table 8.2.

TABLE 8.2 CSR Practices Done by Indian Public Sector Banks

No.	Banks	Thematic Areas	Star Initiatives
1.	Allahabad Bank (now merged with Indian Bank)	Financial inclusion Creating rural self-employment training institutes (RSETIs) Financial literacy centers (FLCs) Women welfare Rural Development Trust National Sports Development Fund	1 Youth were given instruction with the aid of RSETIs. 2. Opened FLCs to aid in the promotion of financial literacy. 3. Given scholarships to young women. 4. Electric fans were provided in government schools. 5. Developed the Udayan Care—Udayan Ghar Program to aid orphaned and disadvantaged girls. 6. Made a donation to Swachh Bharat Kosh to build restrooms for girls. 7. Established Farmers' Training Centers and Farmers' Clubs. 8. Contributed money for feminine hygiene. 9. Provided Jagannath Temple in Kidderpore with 100 plastic chairs. 10. Kolkata-based cancer awareness campaign sponsored. 11. Made a donation to the Trust for the Sabarmati Harijan Ashram. 12. Gave sewing machines and water purifiers to schools and ashrams. 13. Made financial contributions to numerous charities. After 2019 due to merger, the final annual reports are in the name of Indian Bank.

TABLE 8.2 *(Continued)*

No.	Banks	Thematic Areas	Star Initiatives
2.	Andhra Bank (merged with Union Bank of India)	Donations to the CM Relief Fund and the National Sports Development Fund for those with physical disabilities.	1. The bank has made significant donations to the CM Relief Fund of many states. 2. Contributed funds for wheelchairs, ambulances, vans, and other necessities to orphanages, local governments, and other charitable trusts. 3. Donated water purifiers, battery-powered automobiles, and school vans. 4. Contributed funds to build senior housing. 5. Made a sizeable donation to help fund the construction of restrooms in schools. 6. In rural locations, donated solar street lighting. 7. Protected water supply in infected areas.
3.	Bank of India	Community welfare Women and children Financial inclusion awareness Swachh Bharat Abhiyan Rural development Education	1. Participated in and supported the Swachh Bharat Abhiyan on a large scale. 2. Contributed to the protection of the environment by offering Carbon Life Gas Fuel conservation products. 3. Provided cancer patients and members of the underprivileged part of society with access to healthcare facilities. 4. Donated computers, laptops, and water coolers to schools. 5. Donated ambulances, tempos, and medical equipment 6. Set up training and skill-development programs for young people in rural areas. 7. Made significant donations for the welfare of blind people, cancer sufferers, and children with AIDS. 8. Made a substantial financial donation to build hostels for the orphanage. 9. Gave victims of a national tragedy financial assistance. 10. Assist in supplying portable water facilities in places vulnerable to drought. 11. Beti Bachao Beti Padhao Abhiyan. 12. Environmental sustainability and ecological balance 13. Health and family welfare including social welfare 14. COVID-19 is geared toward promoting disaster preparedness, cleanliness, and healthcare, especially preventive care.
4.	Bank of Baroda	FLCs RSETIs Education Women welfare	1. Created FLCs to educate rural residents about banking and financial services. 2. Started a number of RSETIs to improve the skills of rural kids and foster entrepreneurial endeavors. 3. Gave an ambulance to the Kerala-based Daya Rehabilitation Trust.

TABLE 8.2 *(Continued)*

No.	Banks	Thematic Areas	Star Initiatives
			4. Made significant donations to socioeconomic programs aimed at helping the less fortunate members of society. 5. Making donations and/or funding of activities as mentioned in Schedule VII of Companies Act, 2013.
5.	Bank of Maharashtra	• Skill development • Primary healthcare • Education	1. Education camps in different areas. 2. Donation to charitable trusts. 3. Donation of stationary items to schools
6.	Canara Bank	• Rural development • Skill development • Primary healthcare • Drinking water • Women empowerment • Financial literacy • Community development • Education	1. Bank has established several training facilities for young people in remote areas. 2. Started a great clinic program in the outlying districts. 3. A retail mobile marketing van was donated in order to support women businesses and self-help groups (SHGs). 4. Install RO plants to provide 217 villages with clean drinking water. 5. Under the Swachh Vidyalaya Scheme, restrooms were built in girl's schools. 6. Started the SC/ST rural female students' Canara Vidya Jyoti Scheme. 7. Started the Canara Gramodaya Scheme to build communities and slums together. 8. Assisted Keralans suffering from endosulfan sickness. 9. Made significant financial contributions to hospitals so they could purchase vans, ambulances, and other equipment to assist those in need. 10. Support for school infrastructure, donated computers, and water 11. Provided financial assistance to the families of the victims of COVID-19. 12. Installed solar-powered "Retail Mobile Marketing Van" to assist women entrepreneurs, SHGs, and artisans to market their products. 13. Distributed food packets, groceries, medical kits, distribution of blankets, ambulance van to the COVID-19 pandemic affected areas.
7.	Central Bank of India	• Education • Health • Natural calamities	1. Because the bank suffered losses in the last fiscal year, so there was no CSR in from 2018 to 2021. 2. No CSR funding for 2016–2017 due to the bank's losses in the preceding fiscal year. 3. As part of the Tribal Welfare Project in Murbad, donated cabinets for the school and library as well as books for the library. 4. Provided financial support to Saunvardhan Pratishthan in accordance with Mumbai's policy on safe drinking water. 5. Donated Mumbai's traffic cops rain gear.

TABLE 8.2 *(Continued)*

No.	Banks	Thematic Areas	Star Initiatives
8.	Corporation Bank (merged with Union Bank of India)	• Corporation Bank Economic Development Foundation • Corporation Bank Self Employment Training Institutes • Grameena Abhyudaya Financial Literacy Trust (GAFLT) • Financial inclusion • Education • Health & hygiene • Art & culture promotion	1. Organizes CSR events with the assistance of Corporation Bank Economic Development Foundation. 2. The bank has given several water purifiers, laptops, benches, tables, cupboards, food cloths, and other items to schools, orphanages, old age homes, etc. through Corp-Kiran, an association of bank executive spouses. 3. Under Corporation Bank Self-Employment Training Institute, numerous training initiatives have been undertaken for rural adolescents in various states. 4. Under the GAFLT, several financial literacy camps were organized at various schools, universities, villages, and institutions. 5. Supported financially a large number of self-help groups. 6. Under the "Swachh Vidyalaya Scheme," more than 150 restrooms at various schools received renovations.
9.	Indian Overseas Bank	• Village Adoption • Women empowerment and women education • RSETI • SHAKTI—trust	1. "Udaan," a unique industry initiative for Jammu and Kashmir's young to receive training and employment, is run by the bank. 2. Set up financial literacy centers to inform the public about financial products and services. 3. Contributed funds to aid flood victims in Chennai. 4. Conducted numerous programs to help women expand their businesses. 5. Several RSETIs have been opened to train SHGs, farmers, and other beneficiaries. 6. Contributed funds to the Pradhan Mantri Awas Yojana, Swachh Bharat Abhiyan, health checkup clinics, provision of ornamental tree samples, and beach cleaning activity. 7. Under the Indian Overseas Bank Sampoorna Project, the bank has started the entire village adoption process. 8. Provided educational assistance to the native people. 9. Established the "Shakti" Indian Overseas Bank Chaitanya Chettyar Memorial Trust to carry out all types of CSRs. 10. No CSR from FY 2016 to FY2021 due to high loses in the last fiscal years.
10.	Indian Bank	• Inclusive growth • Girl child welfare • Healthcare • Village adoption • Clean India movement • Women empowerment	1. A lot of temples, schools, and other establishments now have restrooms thanks to the clean India movement. 2. Contributed funds and funded maintenance projects in Kerala for the well-being of children. 3. The bank has backed numerous government initiatives, including the clean India movement, the Swachh Vidyalaya Scheme, and smart city initiatives.

TABLE 8.2 *(Continued)*

No.	Banks	Thematic Areas	Star Initiatives
			4. Gave deaf school children books, presents, and other educational materials. 5. Incorporated the recently hired trainees into the bank's CSR initiatives. 6. Created a library for female students and set up motivational classes for lady pupils. 7. Hosted several health fairs, eye exams, blood drives, organ donations, etc. 8. Adopted some villages for the inclusive development and advancement of the underprivileged. 9. Bank has provided some war widows with financial assistance. 10. Additionally, the bank has planted trees in numerous locations and views this as a typical practice; to date, the bank has planted more than 1000 trees and seedlings. 11. The bank has added solar-powered systems to many of its branches. 12. The bank has constructed the Dr. A.P.J. Abdul Kalam Skill Development Training Institute in Andhra Pradesh for the purpose of developing and training unskilled labor. 13. Contribution toward PM CARES Fund and CM Relief fund of various States as a humble attempt to fight COVID-19 pandemic. 14. Distribution of PPE kits, hand sanitizers, safety accessories, and other required things to frontline workers in COVID-19 relief camps.
11.	Oriental Bank of Commerce (merged with PNB)	• Green initiatives • Environmental protection • Education • Health • Sports	1. Bank has given the less fortunate individuals school supplies, books, mattresses, pillows, blankets, CCTV cameras, solar lamps, and other items. 2. Installation of water coolers, RO systems, donation of chairs, uniforms, blankets, invertor, computers to various organizations/trusts/institutes, etc. A health awareness camp was held in Gurugram. 3. Gave chairs, water filters, fans, and water coolers to various trusts, schools, and other organizations. 4. Bank adopted, cleaned, and planted Tulsi seedlings, "Lal Ghat" & "Hanuman Garhi Ghat" at Varanasi's River Ganga Ghats. 5. Every year, the bank has constructed more than 100 restrooms in various government schools as part of the Swachh Vidyalaya Campaign. 6. Additional donations to the less fortunate included a mobile e-clinic van, computers, medical supplies, and school buses.

TABLE 8.2 *(Continued)*

No.	Banks	Thematic Areas	Star Initiatives
			7. Made a sizable donation to the Prime Minister Relief Fund. 8. To impart knowledge and develop a top-tier workforce in the field of financial inclusion, the bank established the OBC National Institute of Financial Inclusion.
12.	Punjab National Bank	• Sustainability • Social investments • Health • Green initiatives • Sports promotion	1. PNB established "PNB-Prerna," an association of senior female employees and the wives of senior PNB employees, to promote and aid the bank's CSR programs. 2. Every year, PNB Prerna gives away wheelchairs, audio CDs, fans, mattresses, school dresses, sports equipment, and school bags to children in need. 3. Every year, Ann Daan Seva blood donation camp is organized. 4. Provided pharmaceutical donations to help cancer patients at the cancer rehabilitation center "CanSupport." 5. For the general welfare and development of rural youth, women, and farmers, established the PNB Farmer's Welfare Trust. 6. PNB Centenary Rural Development Trust, which promotes rural development and jobs. 7. Has established a significant number of RSETIs and FLCs to date to educate rural youngsters. 8. PNB-VIKAS is a bank-run village adoption program for general development, including health, education, drinking water, sanitation, and power. 9. The PNB-LADLI program, which aims to educate girls in rural and semiurban areas, has been introduced across India. 10. PNB Asha Kiran, a comprehensive program for rural women's empowerment. 11. As part of the Swachh Vidyalaya Campaign, bathrooms, literature, a library, and sports equipment were provided. 12. PNB-UJALA, a program whereby the bank gives each adopted girl a solar lantern in addition to four solar lights for the villages where she was adopted. 13. To support organic farming, the "PNB Dugdh Vikas Yojana" milk collection centers were established by the agriculture credit bank. 14. Contributed to PM-CARES Fund and CM Relief Fund. 15. Supporting orphanages, underprivileged, and oppressed groups in society. 16. Serving the community by donating books, medicines, etc. as well as the opening and upkeep of libraries and reading rooms. 17. No CSR investment in FY 2017–2018 to 2018–2019 due to losses in the last fiscal years.

TABLE 8.2 *(Continued)*

No.	Banks	Thematic Areas	Star Initiatives
13.	Punjab & Sindh Bank	• Rural development	1. Built a structure for the Rural Self Training Institute in Ludhiana to empower the underprivileged by giving them access to the necessary skills. 2. Providing PPE kits, masks, water dispensers, water coolers, and RO sanitizers as part of the bank's commitment to the PM CARES Fund and the allocation of CSR funding to the Water and Environmental Development Organization. No CSR in 2018–2019 due to losses in the last fiscal years.
14.	State Bank of India	• Education • Healthcare and sanitation • Environmental protection • Assistance in natural calamities • Assistance to underprivileged • Green banking • Protection of national heritage • Empowerment of women, youth, and senior citizens	1. A significant sum of money has been provided by the bank to charitable groups for the purchase of ambulances, medical vans, and other equipment. 2. Every year, the bank organizes health fairs, eye exams, free cancer screenings, fairs for reproductive health, etc. 3. Every year, the bank also gives dustbins, water coolers, sanitary napkins, and water purifiers to needy villages, schools, and other locations. 4. As part of the Swachh Vidyalaya Scheme, the bank constructs restrooms in numerous schools each year. Distributes battery-powered automobiles, solar lamps, and other solar-powered products. 6. Bank RSETIs offer skill-based training to farmers, rural youth, and other groups. 7. SBI youth for India Fellowship Program, which aims to close the gap between young people from rural and urban areas. 8. A sizable sum of money was donated to the Swachh Bharat Mission. 9. Food relief is provided with the help of on-the-ground implementation partners PAN India; PPE kits are distributed PAN India; healthcare infrastructure is strengthened by the purchase of ventilators and other medical equipment for government hospitals PAN India; ECHO India builds the skills of healthcare professionals; and initiatives for students' mental health are supported. 10. Donation toward PM CARES Fund. 11. Supported girls' education as a part of the "Beti Bachao Beti Padhao" campaign of the government. 12. Provided basic infrastructure to various hospitals, NGOs, and trusts working for society under the health sector.
15.	Syndicate Bank (merged with Canara Bank)	• Education • Gender equality and women empowerment community development • Healthcare	1. Donated funds to a blind organization for educational purposes. 2. Gave away toilet chairs, battery-operated garbage trucks, cargo rickshaws, water coolers, school bags, and free nutritional supplements to the less fortunate members of society.

TABLE 8.2 *(Continued)*

No.	Banks	Thematic Areas	Star Initiatives
			3. More than 30 villages were adopted through the SyndSamgra Gramin Yojana. 4. In several cities, financial support was given to various philanthropic organizations, institutions, government hospitals, etc. 5. Made a contribution to the welfare of injured veterans of war. 6. Made a sizable donation to cover the costs of books, equipment, stationary, uniforms, and other study-related fees for tribe youngsters. 7. Built a community utility hall. 8. Made donations to help those affected by natural disasters like the Hudhud Cyclone and the Nepal earthquake. 9. Donated water purifier, LCD projectors, ambulance, water coolers, etc. for required places.
16.	UCO Bank	• RSETI • Women empowerment. • Environmental protection • Village adoption • FLCs.	1. Each year, the bank opens a large number of rural self-training centers in various states. 2. Enacted the "UCO Udaan Scheme" and adopted more than 30 communities. 3. FLCs were established, and information on banking and financial services was distributed. 4. Providing youngsters attending government schools with a free lunch. 5. Encouraged, developed, and supported the women's hockey team. 6. Gave the needy wheelchairs, bicycles, chairs, blankets, etc. 7. Tree planting in various locations across Mathura. 8. In the tribal areas of Odisha, water was delivered using an automated water delivery system called "Jal Doot." 9. A regular magazine called *UCO Tower* has been published by the bank to inform its staff. 10. UCO Samagra Gram Vikas Yojana, which aims to develop the villages under the bank's management on a whole. 11. Made financial contributions to the Clean Ganga Fund and Swachh Bharat Kosh.
17.	Union Bank of India	• Education • Girl child education • Women empowerment • Environmental protection Union Bank Social Foundation Trust	1. Through its "Union Adarsh Gram" village adoption program, the bank has adopted 211 villages to date. 2. Established village knowledge centers in numerous villages to educate farmers on financial services and agriculture. 3. Offered the needy people educational software, ambulances, mobile medical services, health equipment, etc.

TABLE 8.2 *(Continued)*

No.	Banks	Thematic Areas	Star Initiatives
		• Community welfare • Healthcare • Skill development	4. Bathrooms have been built in several schools. 5. Adopted around 300 female students from various villages. 6. Donated a van for cancer patients as part of cancer awareness programs. 7. Offered job-related training to young people without jobs. 8. Supported various schools.
18.	United Bank of India (merged with PNB)	• United Bank Socio Economic Foundation • Women empowerment and social upliftment • Education, social support, and development • Health and sanitation	1. With the assistance of United Bank RSETIs, the bank organized training for young people in rural areas. 2. The bank establishes financial literacy centers each year to educate rural residents about financial services. 3. Donate money to the Swachh Vidyalaya campaign. 4. Supported several initiatives for women's empowerment and gender equality. 5. Made financial contributions to several initiatives including education and research projects for female students. 6. Conduct a range of programs in the healthcare industry 7. Aided in the creation of a healthcare library room, the distribution of e-slates, blankets, and vehicles for social activities, among other public social activities.

Source: Compiled by the authors.

8.8 DISCUSSIONS AND CONCLUSION

CSR encompasses a company's proactive engagement in sustainable development across various sectors such as human resources, customer relations, supplier interactions, corporate governance, environmental stewardship, and contributions to social and community well-being. A key feature of financial institutions, including banks, is their dedication to social responsibility (Bihari and Pradhan, 2011). The forces of globalization have intensified competition among banks and compelled them to adhere to international standards and fair trade practices. As a result, banks are increasingly reallocating resources to engage in more social initiatives (Yin et al., 2020). In the banking sector, the CSR approach primarily focuses on all stakeholders, both internal and external. Internal stakeholders consist of employees and shareholders, while external stakeholders include customers, creditors, suppliers, government entities, and the broader community (Freeman and Velamuri, 2006).

India's current economic growth and development are largely attributable to its robust financial system, which is made up of numerous financial institutions, markets, and instruments. Financial markets and institutions both play a key role in the financial system by offering a variety of financial services to the Indian populace (Bihari and Pradhan, 2011). CSR has gained importance in business, particularly the banking sector. This country's economic growth is greatly influenced by the banking sector, which is a key component of the economy. As a result, the banking industry has played a vital role in driving the economy. Corporate governance and social responsibility have become more significant in recent years on a global basis.

RBI has also directed the banks in 2007 to perform CSR activities for sustainable development and to initiate nonfinancial reporting related to the investments in social, economic, and environmental areas in their annual reports. These kinds of social initiatives are done by the banks with the help of the banks' employees on a regular basis and this helps the society. Following the 2013 modification to the Companies Act, the Indian banking industry has given the notion of CSR a lot of attention. The promotion of environmentally friendly banking and lending services has begun by banks. The RBI instructed Indian banks to take the necessary steps for social advancement and environmental sustainability in 2007 (www.rbi.org.in). The total CSR expenditure should be reported for each project or activity conducted under CSR in the business responsibility report that banks are obligated to publish, either individually or as part of their annual report.

The various CSR activities undertaken by the Indian banking sector are shown in Figure 8.3.

By addressing issues such as financial literacy, financial inclusion, career training, education, healthcare, sports and culture endorsement, culture, eradicating poverty, malnutrition abolishment, infrastructural facilities, protection of the environment, community engagement, youth engagement, developing skills, livelihood creation, building capacity, girl child care, and women's rights, the Indian PSBs aims to provide the overall social and environmental development. Different banks have different stances on their CSR initiatives. Indian PSBs primarily focus their CSR efforts on key areas such as rural self-employment training institutions, financial literacy, women's welfare, education, environmental protection, and community development. With the aid of rural self-employment training institutes (RSETIs), banks offer job training to the unemployed youths, improving their employability and professionalism.

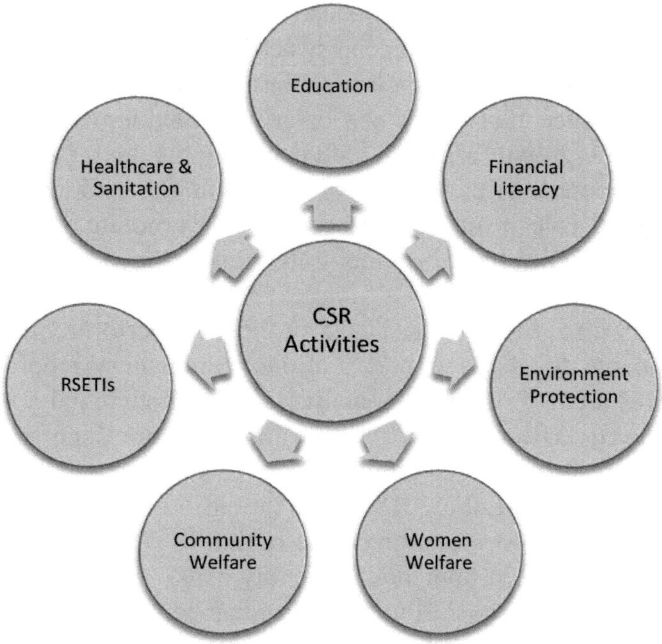

FIGURE 8.3 CSR activities by Indian public sector banks.

Source: Compiled by the author.

Every year, banks open a variety of RSETIs in various locations throughout several states to assist the underprivileged young. Banking institutions have helped millions of people by using these RSETIs.

Almost all Indian banks, both public and private, invest substantial funds in CSR activities related to financial literacy centers. These organizations aim to enhance the lives of millions of uneducated rural residents by increasing their awareness of financial and banking services and products in those areas. Rural youth, farmers, senior persons, self-help groups, and women from rural areas can access essential knowledge and information about financial services at little to no cost through these learning centers. Women's welfare or women's empowerment is another key area of CSR operations. Banks place a high value on empowering women from all backgrounds, and caring for young girls is a component of this effort. Under this heading, banks offer financial support to underprivileged or socially disadvantaged women and girls in need for things such as education, health, sanitation, and maintaining hygiene. This category includes the general development of women and girls, and banks are heavily funding it.

Any nation's success is largely dependent on its level of education. Any nation's entire economic and social progress depends heavily on education. Banks are placing more of an emphasis on education as part of their CSR initiatives. Banks are spending a significant amount of money on educating the less privileged and weaker members of society. The education of girls is included in this section as well. Banks are giving school supplies such as books, notebooks, stationary, and uniforms to municipal and other schools to support the underprivileged students. Environmental protection is yet another key area. Banks are focusing on actions related to environmental protection such as reducing trash, recycling garbage, planting trees and saplings, reforestation, and the use of solar energy. Community welfare is the last but certainly not the least significant element of CSR efforts. Researchers determined from the yearly reports that banks are giving more attention to general social and community development. Investments are made by banking institutions in the areas of healthcare, sports, culture, and the arts, village adoption, environmental efforts, and any other CSR-related activities.

8.9 LIMITATION AND FUTURE SCOPE

The current study has certain limitations. Since it focuses solely on Indian PSBs, the findings cannot be generalized to the global banking industry. Additionally, the data is confined to the financial years 2014–2022. Future research may consider other parameters beyond those discussed. It would also be valuable to examine how these CSR investments impact various factors. Furthermore, a comparative analysis of CSR investments between Indian PSBs and private sector banks may provide additional insights.

KEYWORDS

- **corporate social responsibility**
- **banking institutions**
- **CSR practices**
- **sustainability**

REFERENCES

Aguilera, R. V., Williams, C. A., Conley, J. M., Rupp, D. E. (2006). Corporate governance and social responsibility: A comparative analysis of the UK and the US. *Corporate Governance: An International Review, 14*(3), 147–158.

Arora, B., Puranik, R. (2004). A review of corporate social responsibility in India. *Development, 47*(3), 93–100.

Bihari, S. C., Pradhan, S. (2011). CSR and performance: The story of banks in India. *Journal of Transnational Management, 16*(1), 20–35.

Chahoud, T., Emmerling, J., Kolb, D., Kubina, I., Repinski, G., Schläger, C. (2007). *Corporate Social and Environmental Responsibility in India—Assessing the UN Global Compact's Role* (Vol. 26, p. 127). DEU.

Chaudhury, S. K., Das, S. K., Sahoo, P. K. (2012). Practices of corporate social responsibility (CSR) in banking sector in India: An assessment. *Research Journal of Economics, Business and ICT, 4*.

Chawak, S., Dutta, D. (2014). Corporate social responsibility: Trends and challenges. *Abhinav-International Monthly Refereed Journal of Research In Management & Technology (Online ISSN 2320-0073), 3*(4), 32–39.

Chen, Y. C., Mingyi, H., Wang, Y. (2018). The effect of mandatory CSR disclosure on firm profitability and social externalities: Evidence from China. *Journal of Accounting and Economics, 65*(4), 169–190.

Cheung, P., Mak, W. (2010). *The Relation between Corporate Social Responsibility Disclosure and Financial Performance: Evidence from the Commercial Banking Industry* (Doctoral dissertation, Beedie School of Business-Segal Graduate School).

Cheung, Y. L., Tan, W., Ahn, H. J., Zhang, Z. (2010). Does corporate social responsibility matter in Asian emerging markets. *Journal of Business Ethics, 92*(3), 401–413.

Christopher, E., Luke, M. (2013). The pivotal role of corporate social responsibility perception on consumer behavior. *Journal of Business Management Research, 2*, 47–55.

Corporate Social Responsibility, Sustainable Development and Non-Financial Reporting —Role of Banks, (2007) https://www.rbi.org.in/scripts/NotificationUser.aspx?Mode=O&Id=3987

Demir, A., Budur, T. (2019). Roles of leadership styles in corporate social responsibility to non-governmental organizations (NGOs). *International Journal of Social Sciences & Educational Studies, 5*(4), 174–183.

Doh, J. P., Howton, S. D., Howton, S. W., Siegel, D. S. (2010). Does the market respond to an endorsement of social responsibility? The role of institutions, information, and legitimacy. *Journal of Management, 36*(6), 1461–1485.

Epstein, E. M. (1987). The corporate social policy process: Beyond business ethics, corporate social responsibility, and corporate social responsiveness. *California Management Review, 29*(3), 99–114.

Fassin, Y., Gosselin, D. (2011). The collapse of a European bank in the financial crisis: An analysis from stakeholder and ethical perspectives. *Journal of Business Ethics, 102*(2), 169–191.

Fatma, M., Ruiz, A. P., Khan, I., Rahman, Z. (2020). The effect of CSR engagement on eWOM on social media. *International Journal of Organizational Analysis, 28*(4), 941–956.

Freeman, R. E., Velamuri, S. R. (2006). A new approach to CSR: Company stakeholder responsibility. In: *Corporate Social Responsibility* (pp. 9–23). Palgrave Macmillan: London.

Galan, J. I. (2006). Corporate social responsibility and strategic management. *Journal of Management Studies*, *43*, 1629–1641.

Gauba, R. (2012). The Indian banking industry: Evolution, transformation & the road ahead. *Pacific Business Review International*, *5*(1), 85–97.

Graafland, J., Van de Ven, B. (2006). Strategic and moral motivation for corporate social responsibility. *Journal of Corporate Citizenship*, (22), 111–123.

Henderson, D. (2009). Misguided corporate virtue: The case against CSR, and the true role of business today 1. *Economic Affairs*, *29*(4), 11–15.

Hopkins, M. (2003). *The Planetary Bargain: Corporate Social Responsibility Matters*. Earthscan: UK.

Hsieh, H. F., Shannon, S. E. (2005). Three approaches to qualitative content analysis. *Qualitative Health Research*, *15*(9), 1277–1288.

Kaur, A., Sharma, P. C. (2018). Social sustainability in supply chain decisions: Indian manufacturers. *Environment, Development and Sustainability*, *20*(4), 1707–1721.

Keffas, G., Olulu-Briggs, O. V. (2011). Corporate social responsibility: How does it affect the financial performance of banks? Empirical evidence from US, UK and Japan. *Journal of Management and Corporate Governance*, *3*(1), 8–26.

Kennedy, A. S. (2016). *A Comparative Study on Corporate Social Responsibility between Public sector and Private Sector Banks*. Annamalai University. Retrieved from http://hdl.handle.net/10603/206001

Limbore, N. V., Mane, B. S. (2014). A study of banking sector in India and overview of performance of Indian banks with reference to net interest margin and market capitalization of banks. *Review of Research*, *3*(6), 1–11.

Lorena, A. (2018). The relation between corporate social responsibility and bank reputation: A review and roadmap. *European Journal of Economics and Business Studies*, *4*(2), 7–21.

Mandhachitara, R., Poolthong, Y. (2011). A model of customer loyalty and corporate social responsibility. *Journal of Services Marketing*, *25*(2), 122–133.

Mansoor, K. (2016). CSR standards and Islamic banking practice. *The Journal of Development areas, 50*(5), 295–306.

Moharana, S. (2013). Corporate social responsibility: A study of selected public sector banks in India. *IOSR Journal of Business and Management*, *15*(4), 1–9.

Moir, L. (2001). What do we mean by corporate social responsibility. *Corporate Governance: The International Journal of Business in Society*, *1*(2), 16–22.

Rahman, S., Haski-Leventhal, D., Pournader, M. (2016). The effect of employee CSR attitudes on job satisfaction and organizational commitment: Evidence from the Bangladeshi banking industry. *Social Responsibility Journal*, *12*, 228–246.

Rai, A. K., Yadav, K. P., Malik, A., Gupta, P. (2021). Impacts of bank mergers on shareholder's wealth: An event study on Indian public sector banks. *International Journal of Accounting, Business and Finance*, *1*(1), 8–14.

Rani, P., Khan, M. S. (2014). Corporate social responsibility (CSR) rules and regulation under Companies Act, 2013. *International Journal of Multidisciplinary Research and Development*, *1*(7), 528–261.

Roy, A. (05 11, 2015). Commercial banking in India, NCFM publication. www.nseindia.com/content/ncfm/ncfm_CBBM_workbook.pdf.

Sahoo, P., Nayak, B. P. (2007). Green banking in India. *The Indian Economic Journal, 55*(3), 82–98.

Sawant, P. D., Patil, M. R. (2017). Corporate social responsibility performance of select manufacturing companies in India: An empirical study. *International Journal of Engineering and Management Research (IJEMR)*, *7*(4), 216–223.

Sawant, S. (2020). Corporate social responsibility audit in the context of India. *International Journal of Management Research and Reviews, 10*(8), 1–10.

Sharma, S. (2011). Corporate social responsibility in India. *Indian Journal of Industrial Relation, 46*(4), 637–649.

Sharma, S. K. (2013). Relationship of corporate social responsibility with consumer buying behavior: An Indian perspective. *Romanian Economic Journal, 16*(50), 101–130.

Sharma, E., Mani, M. (2013). Corporate social responsibility: An analysis of Indian commercial banks. *AIMA journal of Management & Research, 7*(1/4), 0974–497.

Singh, N., Srivastava, R., Rastogi, R. (2013). CSR practices & CSR reporting in Indian Banking Sector. *International Journal of Scientific and Research Publications, 3*(12), 472–479.

Singh, A., Verma, P. (2017). How CSR affects brand equity of Indian firms. *Global Business Review, 18*(3_suppl), S52–S69.

Skare, M., Golja, T. (2014). The impact of government CSR supporting policies on economic growth. *Journal of Policy Modeling, 36*(3), 562–577.

Sundar, P. (2000). *Beyond Business: From Merchant Charity to Corporate Citizenship.* Tata McGraw Hill: New Delhi.

Tran, Y. T. H. (2014). CSR in banking sector a literature review and new research directions.

Yin, H., Yang, J., Lu, X. (2020). Bank globalization and efficiency: Host-and home-country effects. *Research in International Business and Finance*, *54*, 101305.

Yuan, W., Bao, Y., Verbeke, A. (2011). Integrating CSR initiatives in business: An organizing framework. *Journal of Business Ethics, 101*(1), 75–92

CHAPTER 9

Farmer Connect Mobile App and Agricultural Equipment E-Business

ASMITA SHIVAJI DEORE, VAISHNAVI SANDEEP TILEKAR, JITENDRA MUSALE, SAYALI SANTOSH MAHADIK, SAMBHAJI NAWALE, AND ARVIND SANJAY MUSALE*

Department of Computer Engineering, ABMSP's Anantrao Pawar College of Engineering and Research, Pune, Maharashtra, India

*Corresponding author

ABSTRACT

This chapter aims to assist farmers by providing a platform where they can rent modern agricultural machinery and equipment, as well as secure storage facilities for their crops based on their needs. The application can be integrated with government systems to monitor crop records, fertilizer usage, government subsidies, and other relevant data. To implement this, a database will be maintained to track the availability of equipment and storage spaces. Indians are highly dependent on agriculture for livelihood, yet many farmers remain disconnected from technological advancements in farming. Most lack knowledge of market rates for their crops and usually sell their produce at cheaper rates. Currently, farmers rely on newspapers and television for information, but they rarely receive real-time updates about local markets. As a result, they are forced to sell their goods at unfavorable rates. Our platform, designed for smartphone users, provides real-time updates on vegetable and fruit prices across markets in India, enabling farmers to sell at fair rates. Additionally, the app includes government notifications related to farming and informs farmers about various schemes. It also features market condition-based predictions for

Industry 5.0: Innovation and Adoption for Sustainability.
Tilottama Singh, Pradeep Suri, and Zulkarin Jahangir (Eds.)

DOI: 10.1201/9781779641878-9

crop prices and offers weather updates, allowing farmers to plan their activities for the next few days.

9.1 INTRODUCTION

India is known as agricultural nation, with approximately 70% of the population depends on farming for their livelihood. Agriculture also contributes significantly to the national economy, accounting for nearly one-third of the country's income. The economic growth of India is closely tied to the advancement of its agricultural sector. In the coming years, the agriculture industry is expected to undergo significant changes. However, many Indian farmers, particularly small-scale producers, face challenges in accessing the information and technology that could help them improve crop yields and secure better prices for their produce. While farming-related data is available from various sources such as print media, radio, television, internet, and mobile platforms, these sources often present information in different formats, making it difficult for farmers to navigate and comprehend. Furthermore, converting this information from one format to another often involves complex manual processes.

Many farmers in India remain isolated from modern advancements in farming techniques and market information. They often lack awareness of current market prices for their crops and, as a result, end up selling their produce at any available price, sometimes far below its true value. Currently, farmers rely on newspapers and television for updates, but many lack the time to read or watch broadcasts consistently. This lack of timely information leaves them uninformed about government farming schemes and market prices, ultimately forcing them to sell their goods at cheaper rates. Due to the low income, many farmers are compelled to take out loans, often at high interest, further worsening their financial situation.

The first part in here introduces—this new concept of mobile innovation not only disrupts the conventional business models of the mobile industry but also opens fresh opportunities in the mobile market. The rapid growth of app development is a relatively recent phenomenon, driven by the widespread popularity and use of smartphones among consumers. The mobile app market is intensely competitive. Smartphones and apps offer significant advantages to users, such as portability, location-based services, and constant accessibility. As smartphone computing power continues to advance and mobile applications increasingly dominate digital interaction, apps have emerged as a new frontier for advancing field experimentation methods.

The second part: Introduction of basic models and methods: This mobile application allows farmers to directly connect with equipment suppliers, storage facility providers, and buyers. The platform provides a way for farmers to share their location and receive personalized services, such as finding nearby storage options or equipment rentals. It also allows farmers to communicate directly with suppliers, local buyers, and other farmers through a built-in chat feature. Additionally, farmers can sign up to receive real-time updates on market prices, weather conditions, and government schemes, ensuring they make informed decisions about their crops and produce.

Training and texting model: In developing the mobile app, machine learning models are employed to provide accurate predictions and personalized recommendations to farmers. They can make informed decisions regarding crop sales, planting schedules, and resource management, which increase the productivity and profitability.

Agricultural applications can assist farmers in optimizing farm management by offering real-time information on weather, soil health, and crop management strategies. This is one example of an attention mechanism. Farmers may increase productivity and efficiency by using this data to make educated decisions about when to sow and harvest their crops.

Datasets, evaluation, and other parameters: These help to support vector machines (SVMs) work by taking data and projecting it into an N-dimensional feature space, and they are capable of classifying your data even when your data points are not perfectly linearly separable. SVM recognizes a separation line between different categories and translates it into a hyperplane in the new data.

Open machine learning: This method is more appropriate when we want to predict something such as crop yield based on soil quality and weather; the farmers can make farming more data driven.

Goals: You can access all the agricultural content you require in one convenient location. Whether you are looking for information on poultry, pigs, dairy, feed, or precision farming, this app has you covered. The articles are designed for easy reading on your mobile device, featuring a user-friendly reading view.

9.2 IMPORTANCE OF TECHNOLOGY

Android-based applications will help the farmers to connect with the rental organizations and provide extra advantage to the renowned agricultural sector.

- **Improved connectivity:** Android apps can help farmers connect with one another for sharing their knowledge, resources, and experiences. It could also facilitate best practices and information about farming in a community setting.
- **Better access to machinery:** An android app allows instant availability of agricultural machine rental services and this can be important for improving productivity and efficiency. Less expensive machinery can be particularly beneficial to smallholders, as it opens access to cost-prohibitive or specialty equipment.
- **Higher working efficiency:** Due to the interconnectivity and availability of machinery for rent, we are able to run our farming operations in a relatively efficient manner. With the reduction of time and effort spent on searching for and renting machinery, farmers will be able to use their time and other resources wisely, resulting in increased productivity of output as well as better results.

  This can allow farmers to know more about what others are using and why and likewise share their experience or plans for implementing new technology, such as artificial intelligence (AI), in farming. The knowledge sharing is also an aspect that helps farmers in adopting modern agricultural practices resulting in better quality and quantity of yield.
- **Farmers can cut overall operational cost:** Android apps reduce the cost by making it accessible for farmers to share the resources. Shared machinery can free farm operators from the need to invest in costly equipment, allowing them to devote their resources more flexibly.
- **Market access:** Farmers are likely to be able to get their produce to the market by using a technology-driven strategy. To enhance market integration and boost profits, farmers may make informed decisions about what and when to produce by using Android applications to gather data on demand, prices, and market trends.

9.3 LITERATURE SURVEY

In nations that rely heavily on agriculture, crop fields are the source of economic growth, and farmers rely increasingly on them for their livelihood. The project's main goal is to boost crop yield and production by developing a system that uses temperature and soil moisture sensors to stack sensors and monitor the fields' moisture and temperature. Water is then delivered to the plants based on the data from the sensors. Geofencing for agricultural regions is another benefit of radiofrequency identification (RFID) tagging that keeps animals and unauthorized people out (Dahiya, 2023). An alarm will sound to alert the farmer to any trespassing.

Smart agriculture requires sophisticated hardware and software tools to achieve it quality objective of higher operational efficiency and productivity (Jararweh et al., 2023). SWAMP: Scalable tools to maintain asset water management via Internet of Things (IoT) and machine learning platform: It provides up-to-date and historical moisture information from the fields, in both map and chart formats available, to improve farmers' water-level situational awareness.

The mobile internet speeds up the realization of modern agriculture, encourages the agricultural supply side to reform, improves the innovative power to build high-level agricultural information construction, and penetrates the foundation of a prosperous society in all aspects. In light of this, this study develops a model of agriculture app information adoption behavior based on the theoretical model of Unified Theory of Acceptance and Use of Technology (UTAUT) and tests the relationship between the influencing elements of agriculture app information adoption behavior using an empirical questionnaire and structural equation (Williams et al., 2015).

The Philippines' agricultural sector is crucial to the country's economic growth. In addition to providing the organization with plant goods, it creates a sizable number of jobs in rural areas. However, due to the significant imports from nearby Asian countries, farmers have recently been facing difficulties due to the low price of products grown locally. The majority of farmers are in the network's periphery and are occasionally accorded far less weight when it comes to government assistance. As a result, as the population grows, so does the need for food products.

Smartphones is considered as one of the greatest innovations of the century and have become indispensable devices that fit in anyone's pocket. Because cellphones facilitate verbal communication, gaming, training, entertainment, technical assistance, information, daily news, and business, their use is spreading more quickly. Software applications are created specifically for smartphones, and other devices are known as mobile programs. With innovative features and content in educational and advisory offers, there is a need to develop and grow a mobile application for farmers, college students, and agricultural professionals.

Mobile-based services aimed at smallholder farmers can be taken advantage of because mobile technology is widely available to vast parts of the population. In an effort to target smallholder farmers involved in the production of crops and livestock, these platforms have been implemented throughout the nation.

Nowadays, most farmers own a smartphone. One of the most fundamental human tools, mobile phones, is essential to day-to-day living. But with traditional farming methods, the advancement is more gradual. Despite the existence of current farming-enhancing instruments and technology, their application in agriculture is very limited. The most recent advancements in farming and the most recent prices for their crops and goods are still unknown to many farmers.

Agriculture and Indian economy are the backbone of the country, and most of the population in India is agrarian based (Majumdar, 2020). Agriculture since antiquity is said to be least predictable profession because of various reasons. Other than natural phenomenon—worship, the biggest reason for this is the absence of knowledge in agriculture science.

9.4 EXISTING SYSTEM

- Market research: Conduct extensive market research to learn what your target market wants and needs. Look for the pain points that exist in the current system and discover how your app can offer an alternative to these.
- Ideate and plan: First, you must define the main features and functions of your application. It can have user registration to create profile, machinery is listed, for users—location-based services will be provided, communication channels have also been included, and secure payment gateways and many more features are provided.
- Development: Right technology stack is taken, and framework is decided to build your application.

9.4.1 IMPORTANT FUNCTIONALITIES

- **Registration and authentication:** A mechanism for users to create accounts and verify their identities.
- **Profile creation:** Allow users to set up full profiles outlining their farming demographics and equipment.
- **Listing machines:** Let farmers list their own equipment for rent, including its specifications, availability, and cost. To assist other farmers in locating the ideal agribusiness, include search and filtering capabilities. You may also search by geography, machinery availability, and other criteria.
- **Communication:** Add a messaging system where users can communicate with each other, talk about terms of the rental, farming tips, and everything else related to hemp.
- **Rating:** Establish a system of reviews and ratings to assist users in making well decisions regarding renting or collaborating with other farms.
- **Payments:** To provide clients with a secure method of paying for services such as renting equipment and set up a payment gateway.
- **Testing:** Test the software thoroughly to ensure that it works correctly and is error free. To gather feedback and improvements from a selected user base, do beta and alpha testing.
- **Launch:** Release the app on the Google Play Store, which is marketing strategy to reach the target demographic.

9.5 ARCHITECTURAL DIAGRAM

9.5.1 LAYER 1: EQUIPMENT FOR BUSINESS

9.5.1.1 MODERN FARMING: THE ROLE OF AGRICULTURAL EQUIPMENT TRACTORS

These are versatile machines used for plowing, planting, and hauling materials. Balers are used to efficiently crush crops that have already been harvested, for instance hay or straw, into a compact shape suitable for storage or to be transported. Unlike traditional harvesting implements, combines cut, thresh, and clean grains in a single operation. Plows (both past and present) are used to prepare soil by turning it, breaking it up, and incorporating nutrients. Mowers are responsible for maintaining fields by cutting grass or crops. Planters disperse seeds in the Earth, which markets the seeds strait apart from every one other for crop health. Because of sprayers, this pest and fertilizer exposure become easier, so that they help enhance crop yield and quality. In spite of new methods and technology, the axe, which is an old hand tool, is still used for many farm jobs such as land cleaning, tree trimming, and wood processing. Together with this apparatus, it establishes practice and constitutes efficient, sustainable, and environmentally friendly agriculture, supporting national and world food security, as well as national economy.

9.5.2 LAYER 2: SECURITY

9.5.2.1 FIREBASE AUTHENTICATION

Firebase Authentication provides the backend services that allow you to authenticate users with email and password, phone numbers, or social media accounts and also make it easier than ever to get your users signed in while also improving the onboarding experience for your users. You can use it to create an email and password account, phone authentication, Google, Twitter, Facebook, and so on.

An electronic message, also known as an email sent via the internet, is referred to by an email ID. An address for both the sender and the recipient is required for emails, just like it is for any physical correspondence.

Password: In your large G hormonal exertion, sugar chloride is your password. Go to passwords. Google for accounts that have passwords saved. Use Chrome to view your passwords. In order to access passwords, you must log in again (Figure 9.1).

9.5.3 LAYER 3: DATA MANAGEMENT

Rental equipment: Generally, rental equipment is recorded as part of an operating expense since it is used in the ordinary course of business.

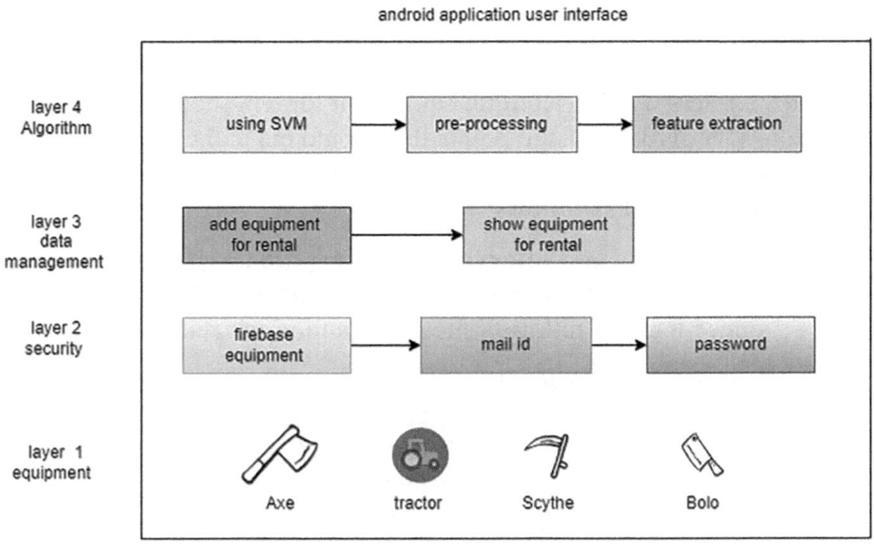

FIGURE 9.1 Layer 2—security.

Equipment rental is the most conventional expense category, and it is usually the easiest to classify.

Model equipment for rent: Equipment rental, or plant hire, is a service industry that rents machinery, equipment, and tools to other businesses and individuals.

9.5.4 LAYER 4: ALGORITHM

SVM is an illustration of a supervised algorithm. Although SVM is a highly strong supervised algorithm, it performs best on small, complex datasets. Although the SVM can be used for both classification and regression tasks, it generally did well in classification tasks.

Preprocessing includes:

1. Word tokenization and deleting all blank rows from the data.
2. Eliminate stop words.
3. Eliminate any nonalpha text.
4. Lemmatization of words.

9.6 RESEARCH METHODOLOGIES

For example, the research processes needed for creating an Android app that links farmers with agricultural equipment for rent should be extensive enough to identify key design points that would make the app functional, useful, and usable. A few methods that can be applied during the development phase are given as follows:

- **User surveys and interviews:** Surveys are taken to gain insights into their specific requirements, challenges, and preferences for connecting with fellow farmers and renting agricultural machinery.
- **Feature extraction:** This is the initial stage in creating an effective SVM forecaster. Principal component analysis (PCA), independent component analysis based on SVM, and kernel PCA are used in this study to extract features. To find out about farmers' requirements, preferences, and difficulties with renting farm equipment and interacting with other farmers, do user surveys and interviews. This will enable you to determine which features will actually inspire people to use them more and accomplish more.
- **Analysis of competitors:** Examine the competition, features, and areas for improvement in the agricultural app and platform ecosystem. By carrying out this research, you will be able to comprehend the market environment, determine your unique selling proposition (USP), and take advantage of distinctive features in your app.
- **Focus groups:** Hold focus groups with farmers and other agricultural stakeholders to get their opinions on the app's potential features, user interface, and prototype. You do get a thorough understanding of how the app functions and is usable from the end user's point of view from these sessions.
- **Pilot studies:** Assessing the app's usability, functionality, and capacity to meet the needs of farmers in particular farming communities or geographic regions is the aim of pilot studies. Get customer feedback, so that any necessary changes can be made before the app becomes widely used. Feedback to make any necessary corrections.

- **Data analysis:** Based on agricultural data, market trends, and reports, maintain awareness of all the relevant issues that are being faced by farmers, machinery demand, or the prevalent farm practices of any particular region. This analysis will assist in identifying features and services that need to be developed to meet the needs of different types of farms.
- **Technology scouting:** Keep track of latest technology trends within the agriculture domain such as IoT devices, precision agriculture, and robotic automation in agriculture.
- **Regulation compliance research:** Examine agricultural regulations and policies, including those pertaining to tractor rentals, data privacy, and user safety. For this reason, ensure that the app complies with all applicable rules and regulations to avoid potential legal issues.

9.7 ALGORITHM

1. Loading libraries.
2. Importing the dataset, and extraction of the *X* variables and *Y* separately is done.
3. Dataset is being divided into train and test.
4. Initializing SVM classifier.
5. Fitting the SVM classifier model.
6. Predictions are done.
7. End

It has been proposed that an Android App should be developed to bring together farmers and through an Android App, agricultural machinery can be rented. So, the flow start from the registration of farmers and the listing of machineries. Farmers are then able to follow-up with each other or rent equipment. Rental transactions are open, and farmers can look back on their rental history. The app encourages sustainable agricultural practices and cooperation between farmers.

9.8 FLOW DIAGRAM OF PROPOSED WORK (FIGURE 9.2)

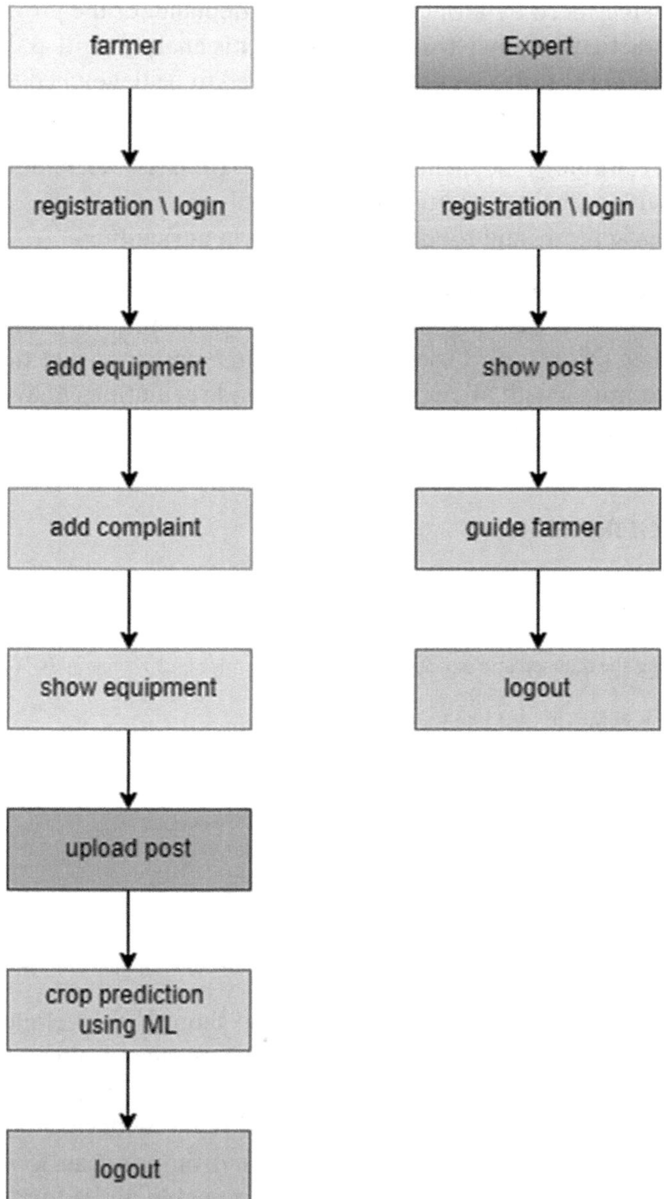

FIGURE 9.2 Flow diagram of the proposed system.

9.9 ADVANTAGES OF THE PROPOSED MODEL

- **Increased efficiency:** With the help of this app, the agricultural operations can be done with less effort. The farmers will easily find nearby farmers who are providing specific service or have required machinery; consequently, there will be minimum downtime or delay.
- **Resource optimization:** This in turn allows farmers to share agricultural machinery and equipment, thus saving the capital invested in machinery. We can channel it better and save on costs too.
- **Access to modern equipment:** Rentals allow farmers who are not be able to afford expensive machinery to take advantage of modern farming methodologies.
- **Cost reduction:** Instead of buying and maintaining equipment you seldom use, it may prove to be more cost-effective to rent machinery as needed. This can be very helpful for tasks that arise seasonally or are one-time events.
- **Income generation:** Farmers with machinery can earn money from renting it out for the time when they are not using this equipment.
- **Community building:** Using the app helps foster a real sense of community between farmers. This promotes the spirit of cooperation and collaboration and networking among various stakeholders in agriculture.
- **Enhanced productivity of agriculture:** Having machines provides the farmers' productivity, so that they will be able to produce greater bequests of crops and, thus, greater income.
- **Data and insights:** The apps can extract key data about the availability, demand, and utilization of equipment. The data that is produced will assist with future planning in order to drive the agricultural sector in a better direction.
- **Reduced environmental impact:** This also helps lower the environmental impact and creates a more sustainable farming practice,

since fewer farmers needs to purchase and operate the same type of machinery.

- **Market expansion:** For those into renting out machines, the app helps you gain a wider market and grow your clientele.
- **Streamlined communication:** The app serves as a communication and coordination hub for farmers, simplifying the process of hiring rentals or services.
- **Personalization and grading system:** Users are allowed to set their search criteria (location, type of equipment, price, etc.) on the app. Moreover, it can include a rating and review mechanism to ensure credibility and reliability.
- **For rural communities:** The app can help both economic development and better living standards in a largely rural and agricultural part of the world.
- **Time saving:** The app help farmers find and rent machines or services within a short time; if they want to rent machine, they save time without go look for out one by one it has machine they will take hours or days to find rent machine. The proposed app has so many changes it can make like it can bridge the gap between farmers and people willing to rent out their machinery for a certain amount of time and modernize farming. It can also improve local resource use and finally improve lives among the farming community. It proposes a pragmatic, cheap, and crowd-sourced way of doing agriculture.

9.10 CONCLUSION

When farmers will be able to access the price of their produce from different local and far-off markets, it will enable them to make smart choices decision regarding where to sell their agriculture produce overcoming the issue within the farming sector, thus getting an increased benefit. This would help to guard the crops in another kind of climate that the cadre is mainly for procurement and sale of the crops, which supports the farmer to sell his commodity and increases a good profit rate more especially than the market. Transport feature assists the farmer to get selected with others who are following the similar route. This dividend is shared out from the compensation, leaving profits for the farmers. We utilized Android Studio and Firebase, some of the current tools and platforms we have. In the course of development, we have realized how much of project development and management is done on an individual level and how much is done in a group context that is alongside a team or among them. Our improved communication through presentations of our project at several seminars and exhibitions and demonstration of professional ethics paved the road for creating a culture of lifelong learning.

9.11 FUTURE SCOPE

1. Marketplace expansion: Slowly open the platform to a larger area. You can begin at a local scale and then scale it nationally and internationally.
2. Services by crop: Have features that are specific to a certain crop or type of agriculture. For instance, details and resources relevant to crops such as rice, wheat, cotton, or any staple crop in your area.
3. Weather forecast integration: Providing access to live weather data and forecasts can assist farmers in planning their activities more effectively. This can be very important when planting and harvesting crops.
4. Agricultural expertise: Provide access to agricultural experts or consultants to guide crop management, pest control, and other key farming practices.
5. Mobile payment integration: Mobile payment and rent machinery will be available as in-app mobile payments.
6. Marketplaces for agricultural equipment reviews: Users using this app can rate as well as review agricultural equipment and services through our app, which will help others make informed decisions.
7. Build a community: Make forum, chat, and user-generated content inside the app. This would assist farmers in exchanging ideas and experiences.
8. Offline functionality: Internet connectivity in many rural parts may not be ubiquitous. Make your app offline-friendly. Make sure that users can look up important details even when their internet connection is far from stable.

9. Subscription models: Charge users for more advanced features or insights they would want to have access just so the app can get some money.
10. Regional support: If the audience for your app speaks mainly local or regional languages, you need to add multilingual support to the app for better accessibility.
11. Integrate information on government subsidies: Weave in information on government agricultural subsidies and support programs to enable farmers to access financial assistance.
12. Sustainable farming practices: Educate and provide tools to support sustainable farming practices, which promote organic farming, no-till farming, or precision agriculture.
13. Collaborate with agricultural experts and other people in the agricultural field to expand network and add value to your app; this will leverage the farmer help.
14. Commodity price prediction using machine learning: Use machine learning to predict commodity prices as this will help farmers to decide on when to sell their commodities.
15. Connect farmers to local markets, wholesalers, and retailers to integrate the supply chain and reduce waste while improving profits.
16. Insurance integration: Collaborate with insurance providers and incorporate crop insurances in the app to give farmers financial security.
17. Sustainability metrics: Provide farmers with easy-to-use tools to help them measure metrics related to sustainability.

KEYWORDS

- **Android**
- **MySQL**
- **smartphone**
- **weather forecasting**
- **market rates**
- **vegetables**

REFERENCES

Dahiya, P. (2023). Internet of drone things. In *Machine Intelligence, Big Data Analytics, and IoT in Image Processing* (pp. 269–304). Wiley. https://doi.org/10.1002/9781119865513.ch11

Jararweh, Y., Fatima, S., Jarrah, M., & AlZu'bi, S. (2023). Smart and sustainable agriculture: Fundamentals, enabling technologies, and future directions. *Computers and Electrical Engineering, 110*, 108799. https://doi.org/10.1016/j.compeleceng.2023.108799

Majumdar, K. (2020). Rural transformation in India: Deagrarianization and the transition from a farming to non-farming economy. *Journal of Developing Societies*, *36*(2), 182–205. https://doi.org/10.1177/0169796X20912631

Williams, M. D., Rana, N. P., & Dwivedi, Y. K. (2015). The unified theory of acceptance and use of technology (UTAUT): A literature review. *Journal of Enterprise Information Management*, *28*(3), 443–488. https://doi.org/10.1108/JEIM-09-2014-0088

CHAPTER 10

Adoption of Blockchain Technology in the Banking Sector

SATYAM PRAKASH SRIVASTAVA,[1*] RUPA KHANNA MALHOTRA,[2] and PRIYANSHU CHAUHAN[1]

[1]*Graphic Era Deemed to be University, Dehradun, Uttarakhand, India*

[2]*Department of Commerce, Graphic Era Deemed to be University, Dehradun, Uttarakhand, India*

*Corresponding author

ABSTRACT

The blockchain technology in the banking sector is a decentralized ledger system and has been more commonly known by it being the basis of cryptocurrencies, such as Bitcoin. However, blockchain is to Bitcoin, as email is to internet; the possible areas of examination with the use of blockchain technology are gigantic. One of such areas is the banking sector, which has certain ambiguities, which needs to be spoken about proximately, such as the lack of distinct accountability, disorganization in deliverance on time, deficiency of transparency, and "too big to operate effectively" kind of attitude. This research work is based on the careful exploration, examination, and evaluation of the possible methods of applying blockchain technology in the banking sector.

10.1 INTRODUCTION

Blockchain entails the creation, use, and utilization of several systems, technologies, tools, and procedures to resolve issues, boost productivity, and improve human existence. There are many other disciplines that fall

Industry 5.0: Innovation and Adoption for Sustainability.
Tilottama Singh, Pradeep Suri, and Zulkarin Jahangir (Eds.)

DOI: 10.1201/9781779641878-10

under the umbrella of technology, including computer science, electronics, telecommunications, engineering, and more. It has had a significant impact on human society and changed how we interact with one another in terms of how we live, work, and communicate.

From decades, technology has been one key factor, which has played a major role in the enhancement and the advancement of every possible nation. Due to the introduction of concept of liberalization, privatization, and globalization, technological boundaries were shattered, and it led to the integration of technology and innovation on a worldwide level. Technology has led to the upgradation of standard of living and has given ease to the modern civilization. The term technology itself means a series of methods, processes, and practices that are developed in order to make the human life easier. Technology usually covers smartphones, software, appliances, and so on. Technology is used in every aspect of our lives, be it the manufacturing sector, healthcare sector, communication sector, logistics sector, agriculture sector, information technology sector, tour and travel sector, banking sector, and so on. Overall, technology is a crucial part of the modern world and plays an imperative role in shaping our regular activities.

Modern definitions of "technology" frequently include digital innovations, such as computers, smartphones, internet, and artificial intelligence that have fundamentally altered how people communicate and access information. Technology, however, also includes nondigital developments, such as manufacturing procedures, medical equipment, and transportation infrastructure.

10.2 DEVELOPMENT OF TECHNOLOGY

The evolution of technology has seen remarkable growth over the decades. During the prehistorical era, the humans used basic tools carved out of stone and woods to carry out their daily activities, which later led to the development of wheels, carts, and weapons to hunt and fulfill the basic requirements of that particular era. It was the 18th and the 19th centuries that turned out to be a turning point in terms of technology. During this century, industrial revolution came into existence, which led to the development of machines. Machines made the task easier and more feasible.

Over the last few decades, advances in computing power, data processing, and communication technologies have accelerated technological development. The following are some significant technological advancements:

1. Computing power: As a result of computers' exponentially growing computing power, more sophisticated software applications, data analytics, and artificial intelligence systems have been created.
2. Connectivity: With the development of the internet, people can now connect with one another and access information from all over the globe. People may now stay online constantly because to the rise of mobile devices, which has further improved connectivity.
3. Data processing: It is now essential for both businesses and governments to be able to process and analyze massive amounts of data. Technologies for data analytics have advanced to make it possible to gather, analyze, and analyze huge datasets in real time.
4. Automation: Human-preformed tasks are progressively being automated using automation technologies, such as robotics, machine learning, and artificial intelligence. In many industries, this has enhanced productivity and efficiency.
5. Advances in augmented and virtual reality technology have made it possible to provide users with immersive experiences. It is adopted in number of areas of education, health care, games, and so on.
6. Renewable energy: It is now possible to produce electricity without making use of fossil fuel. This is feasible through renewable energies in form of solar and wind mitigating greenhouse gas emissions and their impact on climate change.
7. The way we live, work, and communicate has been completely transformed by technology, and it will only continue to advance and change the globe in the years to come.

10.3 BLOCKCHAIN TECHNOLOGY

Blockchain technology is frequently utilized in crypto market and mostly like Bitcoin. It is utilized for the purpose of maintaining database that is used by network of computers globally to distribute and validate scattered transactions. They are controlled and regulated by large community instead of central bank. The past record of transactions cannot be misused for any purpose or security breach. Usually, banking systems are working-specific servers, whereas blockchain is divided between software users. While making use of this technology since everyone on network has access to everyone else entries, the technology makes it difficult for central authority to control the network. The transactions are based on certain set of algorithms and in this way once block of indifferent user information is occupied, it is converted in form of secured chain called blockchain. This technology is based on decentralized network, which is also known as peer-to-peer network.

Blockchain is described as "a digital, immutable, distributed ledger that chronologically records transactions in close to real time." The block-chain-based system uses the internet to carry out value trades devoid of the need for a middleman. Three benefits of using blockchain technology include (1) process integration, (2) quick processing, and (3) decreased operational expenses. When creating business apps for banks and other companies with a financial relationship, blockchain is a technology that is used. Many financial institutions with a focus on cash, such as UBS, Deutsche Bank, JP Morgan, the Bank of America, and other nations, are experimenting with blockchain technology. Due to the blockchain, the financial sector could see an exciting development. It is likely to alter the regular flow of events while advancing preceding theories. In order to cut mitigate time, labor, and expense involved in interbank transactions, some institutions have created and shared a verified, circulating database of client information. Blockchain technology, a decentralized and secure ledger system, enables transactions between users that are transparent and impervious to manipulation without the involvement of middlemen, such as banks or centralized authorities. This distributed database is protected from fraud, hacking, and censorship since it is built on the concept of nodes that verify financial dealings.

Figure 10.1 shows ranking for payment options in various category of financial transactions. It involves cash pooling, person to person, interbank cross border, corporate payments, cross-border remittances, and intrabank cross border. Ranking for intrabank cross border is highest compared to other payments followed by cross-border remittances, incorporate payments, and interbank cross border.

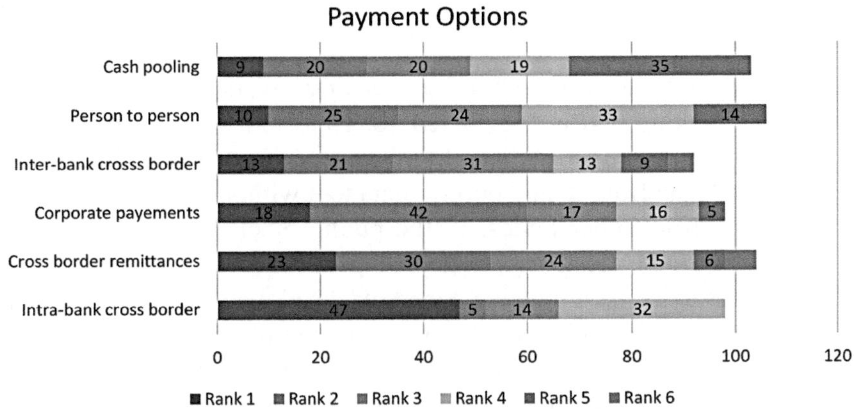

FIGURE 10.1 Ranking for payment options in different transactions.

Figure 10.2 reflects the anticipated advantages of blockchain technology in multiple ways. It involves minimizing costs, administration costs, reducing clearance duration, minimize mistakes, and minimum cost of capital.

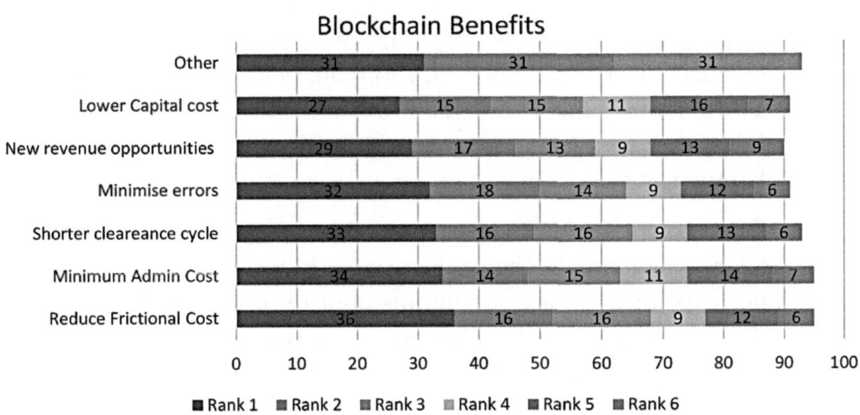

FIGURE 10.2 Major anticipated blockchain benefits.

10.3.1 CLASSIFICATION OF BLOCKCHAIN TECHNOLOGY

Number of nodes in this technology might range from a few to millions, all spread out around the globe. Layers of blockchain are segregated into three ways: (1) public, (2) private, and (3) consortium. These three types are depending on how the network layer is set up.

1. Anybody can join a public blockchain network as a node for mining and taking part in the procedure. This is known as a permissionless or public blockchain. In this case, nodes can join the network, read data from it, and publish data to it without requesting permission from other nodes. In the public blockchain, any data may be written using cryptographic principles, and miners validate transactions. No outside parties or developers have any influence on the public blockchain, which is completely decentralized. In this blockchain, transaction speed is, however, slower. Among public blockchain networks, Bitcoin is a well-known one.
2. Blockchain networks that are closed and have authorization to operate are known as private or permissioned blockchains. This form of blockchain network is typically built by businesses that wish to exchange their data with other businesses, and reading and writing are, thus, considered to be a part of an organization. In a private blockchain, decisions are made by the network owner, and the consensus process is not frequently employed. The central authority of a private blockchain network also predetermines all the rules and regulations surrounding who receives what. A contract that each user signs when they join the blockchain network serves as the management tool for it. As a result, a private blockchain is quicker than a public one. The private blockchain known as Ripple is a typical example.
3. Blockchain network, known as the consortium, was created to address the drawbacks of the private blockchain. The blockchain network in this case is administered by a group of users rather than a single central authority. In this case, a preselected group of nodes, who are in charge of making decisions for a consortium blockchain network, manage the consensus mechanism and the process. Therefore, rather than having just one entity in command, there are numerous. For organizations, such as government agencies, ports, banks, and other enterprises, the option of bringing in neutral partners or entities is provided via consortiums. Using a consortium of 15 financial institutions as an illustration, each of which runs a node, and at least 10 financial institutions.

Blockchains are usually associated with virtual currencies, such as Bitcoin because they use the same technology. However, it also has applications outside of the world of cryptocurrencies, including supply chain management, voting systems, and identity verification. One of the key benefits of blockchain technology is its immutability, which precludes additions to or deletions of transactions once they have been added to the ledger. This makes blockchains especially well suited for use cases where accountability and openness are crucial.

Even though it has revolutionized many industries, this technology still relatively new and rapidly growing area. Only a few of the issues still need to be resolved, including scalability, interoperability, and energy usage. However, blockchain technology has a lot of promise, and it is anticipated that it will become more prominent in the digital economy in the years to come.

10.3.2 BLOCKCHAIN TECHNOLOGY IN BANKING SECTOR

The term "banking sector" includes all those financial institutions that provide range of different services, such as disbursing funds, managing investments, and rendering other financial goods and services. The banking companies are relevant among all sectors as furnishes credit to all the individuals and companies for their specific purposes. Depending on their roles and duties, banks can be commercial, investment, or central banks. For instance, commercial banks offer services to people and corporations, such as savings accounts, mortgages, and personal and corporate loans. On the other side, investment banks offer services, such as mergers and acquisitions, the underwriting of new securities, and the trading of securities. The control of a nation's monetary policy rests with its central banks.

One of India's oldest and most effective financial intermediaries is the banking sector. Since then, a few significant adjustments have been made to the way the financial division operates. From "traditional banking" to "accommodation banking," banks in India have undergone a radical transformation. The Reserve Bank of India (RBI) established a computerization commission in 1988, with Dr. C. Rangarajan as its chairman. Technological innovations, dynamism in robotics, and digital platforms will bring transformation in the next 5 years. The blockchain application innovation, which is the most inventive and is envisaged major cause of disruptions, is being used by the banks to advance digitalization. The Fourth Modern Upheaval around the globe will begin with the blockchain innovation.

Foreign transactions, financial market, insurance sector, business regulatory framework, Smurf, banking, and financial industries have got many uses for blockchain technology. International payments are made possible by the tamper-proof record that blockchain technology provides for sensitive information or data, which enables its usage in cross-border settlements and fund transfers. In April 2018, Banco Santander, for instance, launched Santander One Pay FX, the first blockchain-based money transfer service. To enable customers to send money internationally the same day or the following day, this service makes advantage of Ripple's xCurrent. Santander One Pay FX minimizes the number of middlemen necessary in these kinds of foreign transactions by automating every step of the financial settlement process through blockchain network. It helps to minimize the aggregate overall cost of transactions done globally.

10.3.2.1 CAPITAL MARKETS

Blockchain-based systems have the potential to be better financial markets. Blockchain technology can have an influence on a consolidated audit trail, quicker clearance and settlement, and routine upgrades in the financial markets. Axoni is a firm built on blockchain technology that was created specifically for the capital markets in 2013. The establishment of a distributed ledger network recently Axoni is used to oversee equity swap agreements.

10.3.2.2 TRADE FINANCING

The lengthy processing periods of conventional trade financing techniques, which may force the entire operation to halt and make maintaining liquidity challenging, are a serious problem. Trade finance transactions may become more logical, thanks to blockchain technology. It enables businesses to effortlessly transact with one another across geographical borders.

Figure 10.3 shows adoption of this technology in banking. Among trending conditions, it shows involvement in POCs with other companies is highest compared to other factors.

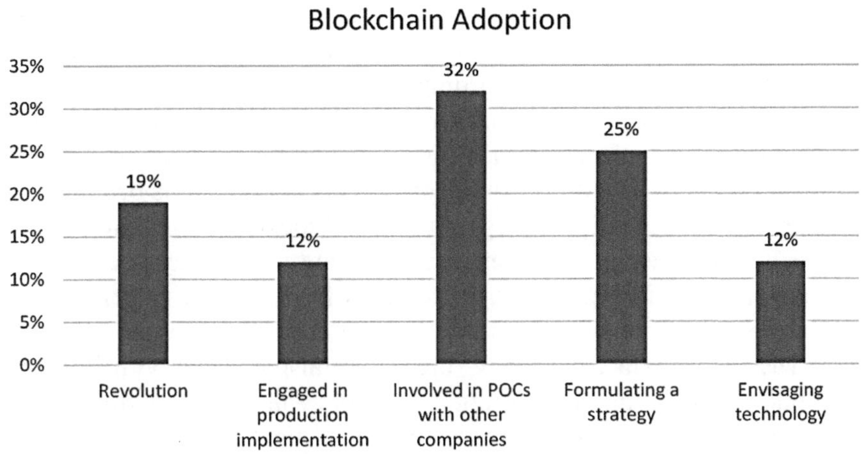

FIGURE 10.3 Banks' current stage in blockchain adoption.

10.4 LITERATURE REVIEW

Being involved in the project has really aided in learning about blockchain technology in the banking industry. The bank uses blockchain technology, which is highly important to its efficient functioning. Since there are many facets to this topic, it would be impossible to discuss them all in a limited amount of time. The data kept on the blockchain, however, is tamper proof and cannot be altered because every effort has been taken to cover most of the critical features that employ encryption and digital signatures. In the end, it was decided that the banking sector would implement blockchain technology. One of India's oldest and largest financial intermediaries is the banking sector (Abou Jaoude and Saade, 2019). The development of blockchain technology may have a significant impact on how transactions are managed, resources are improved, money is controlled, and a variety of other processes.

The future of employment might be changed by blockchain technology. In a nutshell, it can be felt that it has the potential for digitization while also assisting with the work of the auditor and other tasks and areas. It has a track record of adapting to meet the demands of an economic environment that is changing quickly. Akcora et al. (2018) stated that blockchain is, thus, an evolution rather than a revolution in the age of technology.

Technological upgradations, such as Bitcoins, cryptocurrencies, artificial intelligence, and robotics (Böhme et al., 2015) have advanced the working in financial sector.. Everyone is talking about blockchain technology since it is so well known. Bitcoin and cryptocurrency are the parents of this technology. According to a study by Chen and Bellavitis (2020), banking system's financial services will undergo change because of this shared, decentralized ledger, which only participants may use without the intervention of a centralized authority. This innovation is beneficial.

The decentralized digital ledger known as blockchain prevents hackers from achieving their goals. Therefore, adopting technology is crucial for Sri Lankan financial sectors, such as banks from a security perspective. Furthermore, this contributes to the banking sector's increased effectiveness. The blockchain technology offers several prospects with immense potential. By permitting the simplicity of money in the globe, this offers a special technique to build cryptographic transactions.

Blockchain also gives customers improved privacy in addition to more secure transactions (Levy and Ellis 2006). Incorrect manual data input hazards are decreased by reducing the quantity of human labor required (Silva and Junior, 2016). Since a "third party" is not necessary, transactions may be made considerably more affordably. Furthermore, because the entire system is transparent, transactions are safe overall. Even though transactions are typically straightforward, they cannot be free since they require a significant amount of computer time, dependable hardware, and a substantial amount of electricity.

10.5 RESEARCH OBJECTIVES

1. To study the role of blockchain in banking industry.
2. To study how banking companies are benefited.
3. To study the level of security blockchain provides for the customer credential.

10.6 RESEARCH METHODOLOGY

Uses and the intricacies of using this technology in Indian banking industry are examined in this chapter. Some regions, for which there are few research, have been illuminated using a qualitative method. The study's chosen topic is rather technical. Considering this, only experts or groups with a specific level of knowledge in the financial and technology fields would provide reliable and representative replies. The breadth of the study and analysis would have been constrained by the difficulty of locating the appropriate respondents. Because of this, the study only includes data that already exists, and information important to our research was obtained from secondary sources of data.

A thorough assessment of the results of earlier research on the chosen topic is the primary research methodology used. This evaluation covered books, scholarly articles, research papers, and other pertinent materials. To achieve the predetermined objectives, this method's goal was to explain, summarize, and make inferences from the available information. In addition to offering support for the study's assumptions, they also sparked thoughts about potential advancements and the field's potential future. To maintain consistency and impartiality in literature reviews, the following framework with many processes was developed.

The quantity of already published material was first calculated. It was decided how many papers would be evaluated to specify the review's scope and choose the best strategy.

To find relevant material for review, an approach was then developed. To guarantee that the search keywords and phrases were pertinent to the main subject, they were chosen. As an illustration, consider blockchain in India, blockchain in banking and finance, Indian fintech innovation, and so on. To restrict the search results to high-quality, pertinent results, these phrases were chosen.

The strategy for conducting the actual review was also chosen. To verify the quality of the material and its coherence with the topic of study, the abstract of each article was to be examined first. Following the selection of articles based on the abstract, the full-text article was read and evaluated. To find more pertinent sources and data, the publications' references were further looked through. The research includes the following objectives and methodology adopted:

- Sources of data: Primary and secondary
- Sample: 243
- Method of data collection: Questionnaire
- Sampling technique: Simple random sampling
- Research design: Descriptive

Secondary sources:

1. Books—textbooks and reference works written by academics and scholars.
2. Reports—trade- and business-related reports released by organizations and associations, such as Deloitte, McKinsey, and ASSOCHAM. These include quantitative information that makes it easier to do more thorough study.
3. Publications by organizations—Institutions, such as the World Bank and the World Economic Forum occasionally carry out and publish surveys and case studies.
4. Government data—The state or federal government may make data available or form committees to look at a certain problem or area. For instance, reports are published by Niti Aayog, National Payments Corporation of India (NPCI), and so on.
5. Research papers and academic publications—Academics, experts, and scientists in the fields of banking, finance, computer science, and so on, publish their research on websites.

10.7 LIMITATIONS OF STUDY

There were several limitations of this study, including:

1. Total dependence on secondary data—the study's data came only from secondary sources. As a result, the information supplied might now be gathered or independently confirmed.
2. Resources were difficult to get; several research and reports that were pertinent to the paper's main results were unavailable. This was because some websites only let subscribers to view them, some required institutional affiliations, and some charged excessively expensive one-time access fees. However, to have a comprehensive understanding, articles utilizing data from the sources were read.
3. Longitudinal impacts—there was not enough time to thoroughly examine and study the area of investigation.

10.8 CONCLUSION

Although the banking industry is thought to be the beating heart of a strong economy, the state of this sector right now is far from ideal. The absence of effective regulatory management and control has led to several issues in the sector. Additionally, the industry is suffering because hackers from all around the world are exploiting this weak infrastructure. However, blockchain technology is also developing quickly and is proving to be quite powerful in the financial industry. The implementation of blockchain technology in India became vital in recent years as the banking system underwent updates and blockchain's global influence grew.

The industries of banking, financial services, and insurance in India have embraced blockchain technology the most. In the public sector, this technology has also been actively employed for use cases, including farm insurance, electronic health record management, vehicle life cycle management, and land title registration.

The finance minister of India has said that there are always open doors for individuals to do experiments with blockchain, bitcoins, or other cryptocurrencies. She claimed that because fintech relies on such trials, that time will be open. According to the government, India has a significant competitive edge in the blockchain technology sector because so many fintech businesses have achieved outstanding advancements in this field. State government is busy with a lot of work. In Gandhinagar's International Financial Services Centre (IFSC) or GIFT City, there are also plans to take it seriously.

The early barriers for India caused people to have a negative attitude toward blockchain technology due to a lack of awareness and the absence of a regulating agency, among other factors that limited the expansion of the technology. According to recent research from PwC, blockchain's total economic impact on India is expected to increase to $62.2 billion by 2030. Nevertheless, this figure is less than one-sixth of what it will be in the United States and China. In light of this, India is speeding up the application of this technology and striving to update rules, particularly in the financial sector, to keep up with current global market trends.

In a groundbreaking move, the Indian Banks' Blockchain Infrastructure Co Pvt. Ltd. organization would permit 15 banks to join forces toward the end of 2021 to build a new business that would execute inland letters of credit using blockchain technology. According to bankers, the new system would speed up transactions and reduce the risk of fraud by making use of bills for GST tax and electronic invoices to validate data. Ten of these banks are private, four are public, and one is foreign.

These banks are all public sector lenders. This demonstrates the growing investment in blockchain technology by Indian banks, which will significantly mitigating the risk of fraudulent practices and enhance the current fintech environment.

KEYWORDS

- **blockchain technology**
- **banking sector**
- **usage**
- **limitations**

REFERENCES

Abou Jaoude, J., Saade, R. G. (2019). Blockchain applications—Usage in different domains. *IEEE Access*, *7*, 45360–45381.

Akcora, C. G., Dixon, M. F., Gel, Y. R., Kantarcioglu, M. (2018). Bitcoin risk modeling with blockchain graphs. *Economics Letters*, *173*, 138–142.

Albayati, H., Kim, S. K., Rho, J. J. (2020). Accepting financial transactions using blockchain technology and cryptocurrency: A customer perspective approach. *Technology in Society*, *62*, 101320.

Allen, F., McAndrews, J., Strahan, P. (2002). E-finance: An introduction. *Journal of financial Services Research*, *22*(1), 5–27.

Andolfatto, D. (2018). Blockchain: What it is, what it does, and why you probably don't need one. *Federal Reserve Bank of St. Louis Review*, *100*(2), 87–95.

Ante, L. (2020). Smart contracts on the blockchain—A bibliometric analysis and review. *Telematics and Informatics*, *57*, 101519.

Ante, L., Fiedler, I. (2021). Market reaction to large transfers on the Bitcoin blockchain—Do size and motive matter. *Finance Research Letters*, *39*, 101619.

Asongu, S., Nnanna, J. (2018). ICT in reducing information asymmetry for financial sector competition. *DBN Journal of Economics and Sustainable Growth (Forthcoming)*, AGDI Working Paper No. WP/18/035.

Bai, S., Yang, G., Rong, C., Liu, G., Dai, H. (2020). QHSE: Anecient privacy-preserving scheme for blockchain-based transactions. *Future Generation Computer Systems*, *112*, 930–944.

Böhme, R., Christin, N., Edelman, B., Moore, T. (2015). Bitcoin: Economics, technology, and governance. *Journal of Economic Perspectives*, *29*(2), 213–238.

Breidbach, C. F., Tana, S. (2021). Betting on Bitcoin: How social collectives shape cryptocurrency markets. *Journal of Business Research*, *122*, 311–320.

Brereton, P., Kitchenham, B. A., Budgen, D., Turner, M., Khalil, M. (2007). Lessons from applying the systematic literature review process within the software engineering domain. *Journal of Systems and Software*, *80*(4), 571–583.

Chang, V., Baudier, P., Zhang, H., Xu, Q., Zhang, J., Arami, M. (2020). How Blockchain can impact financial services—The overview, challenges and recommendations from expert interviewees. *The Logical Forecasting and Social Change*, *158*, 120166.

Chen, Y., Bellavitis, C. (2020). Blockchain disruption and decentralized finance: The rise of decentralized business models. *Journal of Business Venturing Insights*, *13*, e00151.

Cioroianu, I., Corbet, S., Larkin, C. (2021). The differential impact of corporate blockchain-development as conditioned by sentiment and financial desperation. *Journal of Corporate Finance*, *66*, 101814.

Corbet, S., Larkin, C., Lucey, B., Meegan, A., Yarovaya, L. (2020). Crypto-currency reaction to fomc announcements: Evidence of heterogeneity based on blockchain stack position. *Journal of Financial Stability*, *46*, 100706.

Cronin, P., Ryan, F., Coughlan, M. (2008). Undertaking a literature review: A step-by-step approach. *British Journal of Nursing*, *17*(1), 38–43.

Ertz, M., Boily, É. (2019). Rise of the digital economy: Thoughts on blockchain technology and cryptocurrencies for the collaborative economy. *International Journal of Innovation Studies*, *3*(4), 84–93.

Fanning, K., Centers, D. P. (2016). Blockchain and its coming impact on financial services. *Journal of Corporate Accounting & Finance*, *27*(5), 53–57.

Frizzo-Barker, J., Chow-White, P. A., Adams, P. R., Mentanko, J., Ha, D., Green, S. (2020). Blockchain as a disruptive technology for business: A systematic review. *International Journal of Information Management*, *51*, 102029.

Gao, F., Zhu, L., Shen, M., Sharif, K., Wan, Z., Ren, K. (2018). A blockchain-based privacy-preserving payment mechanism for vehicle-to-grid networks. *IEEE Network*, *32*(6), 184–192

Hadidi, R. (2003). The status of E-finance in developing countries. *Electronic Journal of Information Systems in Developing Countries*, *11*(1), 1–5.

Harwick, C., Caton, J. (2020). What's holding back blockchain technology? On the possibility of decentralized autonomous finance. *Quarterly Review of Economics and Finance*, *84*, 420–429. https://doi.org/10.1016/j. qref.2020.09.006.

Hassani, H., Huang, X., Silva, E. (2018). Banking with blockchain big data. *Journal of Management Analytics*, *5*(4), 256–275.

Kar, A. K., Navin, L. (2020). Diffusion of blockchain in insurance industry: An analysis through the review of academic and trade literature. *Telematics and Informatics*, *58*, 101532.

Kastner, M., Tricco, A. C., Soobiah, C., Lillie, E., Perrier, L., Horsley, T., Welch, V., Cogo, E., Antony, J., Straus, S. E. (2012). What is the most appropriate knowledge synthesis method to conduct a review? Protocol for a scoping review. *BMC Medical Research Methodology*, *12*, 114.

Levy, Y., Ellis, T. J. (2006). A systems approach to conduct an effective literature review in support of information systems research. *Forming Science*, *9*.

Lewis, R., McPartland, J., Ranjan, R. (2017). Blockchain and financial market innovation. *Economic Perspectives*, *41*(7), 1–17.

McLean, S., Deane-Johns, S. (2016). Demystifying blockchain and distributed ledger technology—Hype or hero. *Computer Law Review International, 17*(4).

Meegan, A., Corbet, S., Larkin, C., Lucey, B. (2021). Does cryptocurrency pricing response to regulatory intervention depend on underlying blockchain architecture. *Journal of International Financial Markets, Institutions and Money, 70, 101280 BMC Medical Research Methodology, 12*(1), 1–10.

Naughton, J. (2016). Is Blockchain the most important IT invention of our age. *Guardian, 24.*

Nawari, N. O., Ravindran, S. (2019). Blockchain and the built environment: Potentials and limitations. *Journal of Building Engineering, 25*, 100832

Silva, M., Junior, M. (2016). A secure protocol for exchanging cards in P2P trading card games based on transferable e-cash. *Journal of Information Security and Cryptography (Enigma), 3*(1), 26. https://doi.org/10.17648/enig.v3i1.57

Taylor, P. J., Dargahi, T., Dehghantanha, A., Parizi, R. M., Choo, K. K. R. (2020). A systematic literature review of blockchain cyber security. *Digital Communications and Networks, 6*(2), 147–156

Wątorek, M., Drożdż, S., Kwapień, J., Minati, L., Oświęcimka, P., Stanuszek, M. (2021). Multiscale characteristics of the emerging global cryptocurrency market. *Physics Reports, 901*, 1–8.

Yli-Huumo, J., Ko, D., Choi, S., Park, S., Smolander, K. (2016). Where is current research on blockchain technology? A systematic review. *PLoS One, 11*(10), e0163477.

Zhong, L., Wu, Q., Xie, J., Guan, Z., Qin, B. (2019). A secure large-scale instant payment system based on blockchain. *Computers & Security, 84*, 349–364. https://www.datasciencecentral.com/the-benefits-and-use-cases-for-blockchain-technology-in-banking/

CHAPTER 11

Blockchain and Six Sigma: A New Paradigm for Supply Chain Quality Assurance

RIKTESH SRIVASTAV

City University, Ajman, United Arab Emirates

ABSTRACT

The study looks at the integration of blockchain technology and Six Sigma methodology to improve supply chain quality assurance. The project intends to provide a solid platform for solving challenges such as transparency and procedural inefficiencies by merging the decentralized and secure characteristics of blockchain with Six Sigma. The research investigates the synergies between blockchain and Six Sigma using a quantitative approach and a survey of 49 participants, incorporating variables such as blockchain integration, Six Sigma implementation, supply chain challenges, quality assurance framework, and supply chain quality assurance. The chapter addresses five research issues while identifying a research gap in the explicit link between quality assurance frameworks and blockchain integration. Convergent and discriminant validity are established, and hypothesis testing shows substantial connections, emphasizing the benefits of integration and the crucial role of a well-implemented quality assurance system in enhancing overall supply chain quality.

Industry 5.0: Innovation and Adoption for Sustainability.
Tilottama Singh, Pradeep Suri, and Zulkarin Jahangir (Eds.)

DOI: 10.1201/9781779641878-11

11.1 INTRODUCTION

The merging of blockchain technology with Six Sigma techniques has emerged as a viable area in the pursuit of supply chain quality assurance in today's rapidly growing global marketplace. Supply chains, which comprise the foundation of many companies, suffer a range of problems, including but not confined to transparency issues, counterfeit products, knowledge asymmetries, and inefficiencies in procedures. To handle these multifaceted difficulties, it is necessary to first know the underlying notions of blockchain, Six Sigma, and supply chain management.

Blockchain, which is frequently lauded as a game-changing distributed ledger technology, offers a decentralized and tamper-proof platform for recording and authenticating transactions across a network of users. Six Sigma, on the other hand, is a data-driven method aiming at minimizing defects and variations in processes, thereby increasing quality and efficiency. Supply chains, on the other hand, are intricate networks with multiple stakeholders, making them susceptible. Traditional supply chain management has failed to appropriately address these challenges.

Many of these challenges may be alleviated by introducing blockchain technology into supply chain management. Blockchain enhances traceability and accountability, decreases the threat of counterfeit items, and optimizes information flow by providing an immutable and transparent record. However, it is vital to assure the quality and usefulness of this integration. This is where Six Sigma comes in, with its tried-and-true Define, Measure, Analyze, Improve, Control (DMAIC) methodology, which gives a systematic approach to process optimization and continuous improvement.

This research will look at how blockchain technology may be utilized to tackle existing supply chain difficulties and, when integrated with Six Sigma ideas, establish a robust foundation for supply chain quality assurance. This study attempts to establish a thorough understanding of how these two paradigms may act in harmony to achieve excellence in quality assurance by employing the DMAIC processes of Six Sigma to the adoption and management of blockchain in the supply chain.

The relevance of this work arises from its potential to increase supply chain efficiency, transparency, and reliability. As organizations rely increasingly on global supply networks, rigorous quality assurance methods are more critical than ever. This research intends to deliver important insights and recommendations for enterprises and governments alike by utilizing the synergy between blockchain and Six Sigma, opening the way for a new age of supply chain excellence and reliability.

The chapter follows a structured organization comprising distinct sections. In Section 11.2, a comprehensive review of pertinent literature unfolds systematically, encompassing themes such as blockchain integration in supply chains, the role of Six Sigma in quality management, and the challenges faced by supply chains. Section 11.3 delves into the research methodology, elucidating the chosen research design, sampling methodology, and the specified sample size. It outlines the variables identified through an extensive literature review, details the formulation of research questions, and expounds on the development of the conceptual framework and hypotheses. In the concluding sections, specifically Sections 11.4 and 11.5, the chapter undertakes a thorough data analysis, presenting a comprehensive examination of the study's outcomes and detailed analyses.

11.2 LITERATURE REVIEW

The review of the literature in this study offers a complete investigation of major issues linked to the usage and benefits of employing blockchain for supply chain management. The literature study has been categorized into three distinct sections, namely (1) blockchain integration in supply chain, (2) Six Sigma for quality improvement, and (3) supply chain challenges. The literature highlights the aspects that plays an essential part in integration of blockchain with supply chain, as well as also discusses about the research gap that companies encounter via the integration.

11.2.1 BLOCKCHAIN INTEGRATION IN SUPPLY CHAIN

Blockchain technology has attracted substantial interest in recent years as a possible solution to promote transparency and traceability inside supply chains (Risso et al., 2023). Scholars have examined the use of blockchain in many supply chain areas, highlighting its potential to establish secure, irreversible records of transactions and commodities movement (Alazab et al., 2021; Blossey et al., 2019; Cole et al., 2019; Gurtu and Johny, 2019; Kopyto et al., 2020). Research by Helo and Hao (2019), Kshetri (2021), Montecchi et al. (2019), and Musamih et al. (2021) demonstrated how blockchain might alleviate concerns connected to counterfeit items by creating a tamper-proof record. Additionally, Cui et al. (2023), Gerakoudi-Ventouri (2022), Kim and Shin (2019), Rejeb et al. (2021), Schmidt and Wagner (2019), Tiwari et al. (2023), Wan et al. (2020), Wang et al. (2018), and Xue et al. (2020) indicated that blockchain may eliminate information asymmetry among supply chain stakeholders, leading to increased cooperation and decision-making.

11.2.2 SIX SIGMA IMPLEMENTATION FOR QUALITY IMPROVEMENT

The Six Sigma technique, noted for its data-driven approach to process improvement, has been a cornerstone in quality management for decades. Recent study has explored its flexibility and efficacy in the setting of supply chains (Fragapane et al., 2023; Javaid et al., 2021; Laureani and Antony, 2019; Rowlands and Milligan, 2020; Saporito et al., 2023; Thekkoote, 2021; Zhang et al., 2023). Ahmed et al. (2019, 2022), Belhadi et al. (2023), Krishnan et al. (2022), Ramadan et al., (2022), Sharma et al. (2021), and Trehan et al. (2019) studied about the effective use of Six Sigma techniques to minimize defects and enhance efficiency in a supply chain scenario. Furthermore, the significance of Six Sigma in attaining operational excellence within supply chain operations has been stressed (Rossini et al., 2022; Tsarouhas, 2020).

11.2.3 SUPPLY CHAIN CHALLENGES

The discipline of supply chain management confronts a multitude of issues necessary for organizational resilience and competitiveness (Emrouznejad et al., 2023; Ramezani and Camarinha-Matos, 2020). Globalization adds complications in managing operations across multiple cultures and geographies (Cumbers and Andrew, 2018; Mangan and Lalwani, 2016; Saberi et al., 2019). The fast integration of technologies such as blockchain and AI provides both benefits and concerns, necessitating a careful balance between innovation and security (Dutta et al., 2020; Fosso Wamba et al., 2022; Gerakoudi-Ventouri, 2022; Paliwal et al., 2020; Queiroz et al., 2019; Sodhi et al., 2022; Wong et al., 2020). The intricate challenges in modern supply chain management necessitate adaptable strategies to effectively navigate the multidimensional nature of globalization, technological advancements, sustainability imperatives, and risk management complexities.

There is a clear research gap regarding the explicit relationship between quality assurance frameworks and the integration of blockchain in supply chains, as well as the larger context of supply chain quality assurance, despite the literature review's thorough analysis of blockchain integration, Six Sigma for quality improvement, and general supply chain challenges. The potential of blockchain integration to improve transparency, traceability, and remove information asymmetry is the main emphasis of the literature on supply chain integration (Alazab et al., 2021; Helo and Hao, 2019; Kopyto et al., 2020). The alignment of supply chain management's use of blockchains with well-established quality assurance frameworks, however, has not received much attention. Comparably, the literature on Six Sigma for supply chain quality improvement highlights the methodology's efficacy in reducing errors and boosting productivity (Javaid et al., 2021; Zhang et al., 2023). Therefore, modern supply chain management may result from the integration of Six Sigma methods with blockchain technology and the identification of its influence on supply chain quality assurance with the mediating effect of quality assurance framework.

11.3 RESEARCH METHODOLOGY

The research methodology in this study is delineated into six distinct steps. First, it involves the selection of a research design, followed by the choice of an appropriate sampling strategy. The identification of variables, substantiated by a comprehensive review of the literature, is a crucial step, wherein the rationale for variable selection is thoroughly supported. Subsequently, a conceptual framework is constructed to provide a structured basis for the study. The process then entails the formulation of research questions and the development of hypothesis statements, contributing to a methodical and systematic approach to research design and inquiry.

11.3.1 RESEARCH DESIGN

This study's research approach is quantitative in nature and is based primarily on a survey given to a specific sample. This approach guarantees the gathering of data that is quantifiable, enabling statistical analysis and offering a systematic comprehension of the combination of blockchain technology with Six Sigma for supply chain quality assurance.

11.3.2 SAMPLING

The research employs a probability sampling approach to guarantee the selection of participants is both representative and unbiased. Within this methodology, the chosen approach is stratified sampling, and a purposeful selection process is designed to focus on participants with specific and pertinent experience in both utilizing blockchain technology and engaging in supply chain activities within a company. This targeted selection aligns with the study's objectives. Despite the small sample size of 49 participants, the approach is exhaustive in meeting the study's specific requirements, considering the unique conditions imposed on the respondents.

11.3.3 VARIABLES

The comprehensive literature review sought to identify a range of variables crucial for constructing the conceptual framework. Throughout this exploration, a multitude of variables were observed in use by authors across diverse settings. In this study, the identified independent variables include blockchain integration, Six Sigma integration, and supply chain challenges, while the mediating variable is the quality assurance framework, and the dependent variable is supply chain quality assurance. An interesting observation emerged, revealing a relative scarcity of studies specifically focusing on the application of these variables within the context of the integration of blockchain with Six Sigma and its implications for supply chain dynamics. These variables are intricately woven into the construction of the conceptual framework, reflecting their significance in understanding the complex interplay within the studied phenomenon.

11.3.4 CONCEPTUAL FRAMEWORK

The conceptual framework that resulted from our thorough literature study is shown in Figure 11.1. The methodology identifies supply chain challenges, Six Sigma implementation, and blockchain integration as independent variables. Supply chain quality assurance is the dependent variable, while the quality assurance framework is the mediating variable. The preliminary investigation of the three independent variables' impact on the quality assurance framework and the subsequent analysis of its implications for supply chain quality assurance constitute the research's distinguishable contribution to present. This finding adds an important new dimension to the body of information and opens up the possibilities for further study on the effects of combining Six Sigma and blockchain on supply chain management.

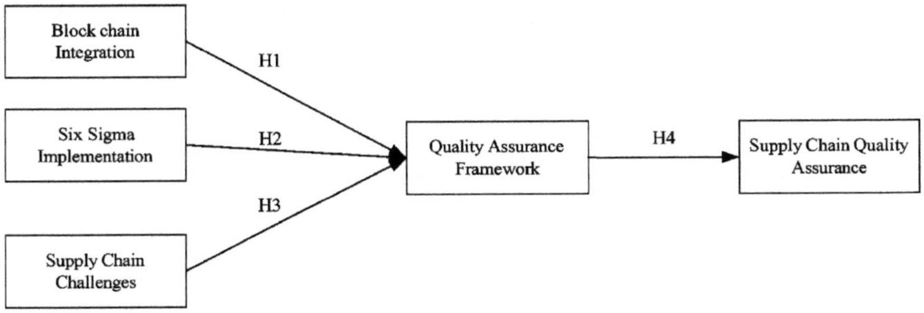

FIGURE 11.1 Proposed conceptual framework.

Source: Author.

11.3.5 RESEARCH QUESTIONS

Based on the suggested conceptual framework (Figure 11.1), five research questions (RQs) have been developed as a result of the analysis. These research questions are carefully crafted to explore different aspects of the conceptual framework, allowing for a comprehensive investigation of the relationships and dynamics between its fundamental components. The methodical direction of the examination of every element in the framework is aided by the thoughtful selection of these research topics. This methodology advances a full comprehension of the research goals by encouraging an in-depth investigation of the complex linkages outlined in the conceptual framework.

1. *RQ1 (blockchain integration): To what extent has blockchain technology been integrated into your organization's supply chain management processes?*

2. *RQ2 (Six Sigma implementation): How extensively are Six Sigma methodologies applied to improve quality and efficiency within your organization's supply chain?*
3. *RQ3 (supply chain challenges): To what degree do the following challenges affect your organization's supply chain operations?*
4. *RQ4 (quality assurance framework): To what extent has the implementation of the quality assurance framework resulting from the integration of blockchain technology and Six Sigma methodologies contributed to improving quality assurance within your organization's supply chain?*
5. *RQ5 (supply chain quality assurance): How satisfied are you with the overall effectiveness and reliability of quality assurance processes within your organization's supply chain, considering the implementation of the quality assurance framework?*

11.3.6 HYPOTHESIS STATEMENTS

The section explains the hypothesis statements that correlate to the four research questions. These hypotheses will be thoroughly examined and scrutinized in the section, where their validity and consequences will be rigorously investigated and researched.

H1: Blockchain integration and quality assurance framework

H0: There is no significant relationship between the extent of blockchain technology integration and the implementation of the quality assurance framework.

H1: There is a positive relationship between the extent of blockchain technology integration and the implementation of the quality assurance framework.

H2: Six Sigma implementation and quality assurance framework

H0: The extent of Six Sigma methodology implementation is not significantly associated with the implementation of the quality assurance framework.

H2: The extent of Six Sigma methodology implementation is positively associated with the implementation of the quality assurance framework.

H3: Supply chain challenges and quality assurance framework

H0: The degree to which supply chain challenges affect the organization's supply chain operations is not significantly related to the implementation of the quality assurance framework.

H3: The degree to which supply chain challenges affect the organization's supply chain operations is negatively related to the implementation of the quality assurance framework.

H4: Quality assurance framework and supply chain quality assurance

H0: The implementation of the quality assurance framework does not significantly influence the satisfaction with the overall effectiveness and reliability of quality assurance processes within the organization's supply chain.

H4: The implementation of the quality assurance framework positively influences the satisfaction with the overall effectiveness and reliability of quality assurance processes within the organization's supply chain.

11.4 DATA ANALYSIS

Three parts make up the comprehensive data analysis for the suggested conceptual framework. The first part begins with a discussion of convergent validity, explaining the relationships between the variables used in the conceptual framework. The second part then explores discriminant validity, looking at how unique each variable is and making sure survey questions are not related to each other. The third and final step is hypothesis testing, whereby the significance of the proposed links within the conceptual framework is assessed using a two-tailed *t*-test.

11.4.1 CONVERGENT VALIDITY

Convergent validity is achieved when all items in a test reflect the underlying notion consistently. The composite reliability (CR) value for all variables exceeds 0.70, and the average variance extracted (AVE) value exceeds 0.5. Table 11.1 contains the specifics, displaying the CR and AVE values for each variable within the suggested conceptual framework. The findings confirm that all of the investigated structures are dependable and may be successfully used for future investigation.

TABLE 11.1 Convergent Validity Using CR and AVE

Variables	CR	AVE
Blockchain integration	0.731	0.561
Six Sigma implementation	0.743	0.593
Supply chain challenges	0.702	0.541
Quality assurance framework	0.814	0.665
Supply chain quality assurance	0.841	0.708

11.4.2 DISCRIMINANT VALIDITY

The HTMT criteria are used to assess discriminant validity. If all HTMT values are less than 1, it indicates that the constructs have discriminant validity. The values of discriminant validity for the five variables employed in the proposed conceptual framework are shown in Table 11.2.

TABLE 11.2 Discriminant Validity using HTMT

Variables	Blockchain Integration	Six Sigma Implementation	Supply Chain Challenges	Quality Assurance Framework	Supply Chain Quality Assurance
Blockchain integration	1.000	0.722	0.631	0.501	0.596
Six Sigma implementation	0.722	1.000	0.618	0.583	0.609
Supply chain challenges	0.631	0.618	1.000	0.560	0.567
Quality assurance framework	0.501	0.583	0.560	1.000	0.773
Supply chain quality assurance	0.596	0.609	0.567	0.773	1.000

The discriminant validity for all HTMT constructs is less than 1, indicating that they are justified.

11.4.3 HYPOTHESIS TESTING

The findings of the inner model, including the *r*-squared output, parameter coefficient, and *t*-statistic, are employed to evaluate hypotheses. In the test, the *t*-statistic > 1.96 was adopted, with a *p*-value of 0.05 (5%) significant level and a positive beta coefficient. The findings are shown in Table 11.3.

TABLE 11.3 Hypothesis Testing

Hypothesis	Relationship Between Construct	β-Coefficients	T-Values	P-Values	Finding
H1	Blockchain integration > Quality assurance framework	0.238	3.456	0.001	Significant
H2	Six Sigma implementation > Quality assurance framework	0.196	2.879	0.005	Significant
H3	Supply chain challenges > Quality assurance framework	−0.152	−2.125	0.038	Significant
H4	Quality assurance framework > Supply chain quality assurance	0.342	4.786	0.014	Supported

Note: Significant at 0.05 (two-tailed).

The positive beta coefficients for blockchain integration, Six Sigma implementation, and quality assurance framework indicate a positive association with quality assurance framework, while the negative beta coefficient for supply chain challenges indicates a negative association. The hypothesis test also signifies that based on the *P*-values, all the variables indicate significant relationships within the proposed conceptual framework.

11.5 CONCLUSION

The comprehensive review of the suggested conceptual framework provides useful insights into the juxtaposition of blockchain and Six Sigma for supply chain optimization. This research may be used as a practical reference for businesses looking to simplify their supply chain operations using blockchain technology while maintaining quality requirements. The following are the important findings from this thorough research:

1. ***Positive integration contributions*****:** The positive associations discovered between blockchain integration, Six Sigma implementation, and the quality assurance framework highlight their collective contribution to the establishment and effective implementation of quality assurance measures throughout the supply chain. This implies that combining blockchain with Six Sigma has a favorable influence on quality assurance.

2. ***Quality assurance challenges***: The discovered negative relationship between supply chain challenges and the quality assurance framework reveals possible barriers to providing quality assurance within the context of supply chain operations. This conclusion highlights the need of taking proactive steps to address and overcome the obstacles of maintaining consistent quality standards.
3. ***Impact on supply chain quality assurance***: The discovered positive relationship between the quality assurance framework and supply chain quality assurance supports the hypothesis that a well-implemented quality assurance framework is critical in improving the overall effectiveness and reliability of quality assurance processes throughout the supply chain. This stresses the need of a strong quality assurance system in guaranteeing supply chain quality integrity.

These insights help to gain an in-depth understanding of the complex dynamics involved in integrating blockchain and Six Sigma for successful supply chain management.

KEYWORDS

- **blockchain**
- **Six Sigma**
- **supply chain**
- **blockchain integration**
- **Six Sigma implementation**
- **supply chain challenges**
- **quality assurance framework**
- **supply chain quality assurance**

REFERENCES

Ahmed, A., Olsen, J., Page, J. (2022). Integration of Six Sigma and simulations in real production factory to improve performance—A case study analysis. *International Journal of Lean Six Sigma, 14*(2), 451–482. https://doi.org/10.1108/IJLSS-06-2021-0104

Ahmed, A., Page, J., Olsen, J. (2019). Enhancing Six Sigma methodology using simulation techniques: Literature review and implications for future research. *International Journal of Lean Six Sigma, 11*(1), 211–232. https://doi.org/10.1108/IJLSS-03-2018-0033

Alazab, M., Alhyari, S., Awajan, A., Abdallah, A. B. (2021). Blockchain technology in supply chain management: An empirical study of the factors affecting user adoption/acceptance. *Cluster Computing, 24*(1), 83–101. https://doi.org/10.1007/s10586-020-03200-4

Belhadi, A., Kamble, S. S., Gunasekaran, A., Zkik, K., M., D. K., Touriki, F. E. (2023). A Big Data Analytics-driven Lean Six Sigma framework for enhanced green performance: A case study of chemical company. *Production Planning and Control, 34*(9), 767–790. https://doi.org/10.1080/09537287.2021.1964868

Blossey, G., Eisenhardt, J., Hahn, G. (2019). Blockchain technology in supply chain management: An application perspective. In: *Hawaii International Conference on System Sciences 2019 (HICSS-52)*. https://aisel.aisnet.org/hicss-52/os/impact_of_blockchain/6

Cole, R., Stevenson, M., Aitken, J. (2019). Blockchain technology: Implications for operations and supply chain management. *Supply Chain Management: An International Journal, 24*(4), 469–483. https://doi.org/10.1108/SCM-09-2018-0309

Cui, Y., Gaur, V., Liu, J. (2023). Supply chain transparency and blockchain design. *Management Science, 70*(5). https://doi.org/10.1287/mnsc.2023.4851

Cumbers, D. M., Andrew. (2018). *An Introduction to Economic Geography: Globalisation, Uneven Development and Place*, 3rd edn., Routledge. https://doi.org/10.4324/9781315684284

Dutta, P., Choi, T.-M., Somani, S., Butala, R. (2020). Blockchain technology in supply chain operations: Applications, challenges and research opportunities. *Transportation Research Part E: Logistics and Transportation Review, 142*, 102067. https://doi.org/10.1016/j.tre.2020.102067

Emrouznejad, A., Abbasi, S., Sıcakyüz, Ç. (2023). Supply chain risk management: A content analysis-based review of existing and emerging topics. *Supply Chain Analytics, 3*, 100031. https://doi.org/10.1016/j.sca.2023.100031

Fosso Wamba, S., Queiroz, M. M., Guthrie, C., Braganza, A. (2022). Industry experiences of artificial intelligence (AI): Benefits and challenges in operations and supply chain management. *Production Planning and Control, 33*(16), 1493–1497. https://doi.org/10.1080/09537287.2021.1882695

Fragapane, G., Eleftheriadis, R., Powell, D., Antony, J. (2023). A global survey on the current state of practice in Zero Defect Manufacturing and its impact on production performance. *Computers in Industry, 148*, 103879. https://doi.org/10.1016/j.compind.2023.103879

Gerakoudi-Ventouri, K. (2022). Review of studies of blockchain technology effects on the shipping industry. *Journal of Shipping and Trade, 7*(1), 2. https://doi.org/10.1186/s41072- 021-00105-2

Gurtu, A., Johny, J. (2019). Potential of blockchain technology in supply chain management: A literature review. *International Journal of Physical Distribution and Logistics Management, 49*(9), 881–900. https://doi.org/10.1108/IJPDLM-11-2018-0371

Helo, P., Hao, Y. (2019). Blockchains in operations and supply chains: A model and reference implementation. *Computers and Industrial Engineering, 136*, 242–251. https://doi.org/10.1016/j.cie.2019.07.023

Javaid, M., Haleem, A., Pratap Singh, R., Suman, R. (2021). Significance of quality 4.0 towards comprehensive enhancement in manufacturing sector. *Sensors International, 2*, 100109. https://doi.org/10.1016/j.sintl.2021.100109

Kim, J.-S., Shin, N. (2019). The Impact of Blockchain Technology Application on Supply Chain Partnership and Performance. *Sustainability, 11*(21), 21. https://doi.org/10.3390/su11216181

Kopyto, M., Lechler, S., von der Gracht, H. A., Hartmann, E. (2020). Potentials of blockchain technology in supply chain management: Long-term judgments of an international expert panel. *Technological Forecasting and Social Change, 161*, 120330. https://doi.org/10.1016/j.techfore.2020.120330

Krishnan, S., Mathiyazhagan, K., Sreedharan, V. R. (2022). Developing a hybrid approach for lean six sigma project management: A case application in the reamer manufacturing industry. *IEEE Transactions on Engineering Management, 69*(6), 2897–2914. https://doi.org/10.1109/TEM.2020.3013695

Kshetri, N. (2021). Blockchain and sustainable supply chain management in developing countries. *International Journal of Information Management, 60*, 102376. https://doi.org/10.1016/j.ijinfomgt.2021.102376

Laureani, A., Antony, J. (2019). Leadership and Lean Six Sigma: A systematic literature review. *Total Quality Management and Business Excellence, 30*(1–2), 53–81. https://doi.org/10.1080/14783363.2017.1288565

Mangan, J., Lalwani, C. (2016). *Global Logistics and Supply Chain Management*. John Wiley and Sons.

Montecchi, M., Plangger, K., Etter, M. (2019). It's real, trust me! Establishing supply chain provenance using blockchain. *Business Horizons, 62*(3), 283–293. https://doi.org/10.1016/j.bushor.2019.01.008

Musamih, A., Salah, K., Jayaraman, R., Arshad, J., Debe, M., Al-Hammadi, Y., Ellahham, S. (2021). A blockchain-based approach for drug traceability in healthcare supply chain. *IEEE Access, 9*, 9728–9743. https://doi.org/10.1109/ACCESS.2021.3049920

Paliwal, V., Chandra, S., Sharma, S. (2020). Blockchain technology for sustainable supply chain management: A systematic literature review and a classification framework. *Sustainability, 12*(18), 18. https://doi.org/10.3390/su12187638

Queiroz, M. M., Telles, R., Bonilla, S. H. (2019). Blockchain and supply chain management integration: A systematic review of the literature. *Supply Chain Management: An International Journal, 25*(2), 241–254. https://doi.org/10.1108/SCM-03-2018-0143

Ramadan, M. A., Al Dhaheri, M. K., Maalouf, M., Antony, J., Bhat, S., Gijo, E. V. (2022). Application of Six Sigma methodology to enhance the productivity and performance of a hotel in the UAE. *The TQM Journal, 35*(2), 554–576. https://doi.org/10.1108/TQM-11- 2021-0325

Ramezani, J., Camarinha-Matos, L. M. (2020). Approaches for resilience and antifragility in collaborative business ecosystems. *Technological Forecasting and Social Change, 151*, 119846. https://doi.org/10.1016/j.techfore.2019.119846

Rejeb, A., Keogh, J. G., Simske, S. J., Stafford, T., Treiblmaier, H. (2021). Potentials of blockchain technologies for supply chain collaboration: A conceptual framework. *The International Journal of Logistics Management, 32*(3), 973–994. https://doi.org/10.1108/IJLM-02-2020-0098

Risso, L. A., Ganga, G. M. D., Godinho Filho, M., Santa-Eulalia, L. A. de, Chikhi, T., Mosconi, E. (2023). Present and future perspectives of blockchain in supply chain management: A review of reviews and research agenda. *Computers and Industrial Engineering*, *179*, 109195. https://doi.org/10.1016/j.cie.2023.109195

Rossini, M., Powell, D. J., Kundu, K. (2022). Lean supply chain management and Industry 4.0: A systematic literature review. *International Journal of Lean Six Sigma*, *14*(2), 253–276. https://doi.org/10.1108/IJLSS-05-2021-0092

Rowlands, H., Milligan, S. (2020). Quality-driven industry 4.0. In: *Key Challenges and Opportunities for Quality, Sustainability and Innovation in the Fourth Industrial Revolution* (pp. 3–30). World Scientific. https://doi.org/10.1142/9789811230356_0001

Saberi, S., Kouhizadeh, M., Sarkis, J., Shen, L. (2019). Blockchain technology and its relationships to sustainable supply chain management. *International Journal of Production Research*, *57*(7), 2117–2135. https://doi.org/10.1080/00207543.2018.1533261

Saporito, A., Tassone, C., Di Iorio, A., Barbieri Saraceno, M., Bressan, A., Pini, R., Mongelli, F., La Regina, D. (2023). Six Sigma can significantly reduce costs of poor quality of the surgical instruments sterilization process and improve surgeon and operating room personnel satisfaction. *Scientific Reports*, *13*(1), 1. https://doi.org/10.1038/s41598- 023-41393-x

Schmidt, C. G., Wagner, S. M. (2019). Blockchain and supply chain relations: A transaction cost theory perspective. *Journal of Purchasing and Supply Management*, *25*(4), 100552. https://doi.org/10.1016/j.pursup.2019.100552

Sharma, A., Bhanot, N., Gupta, A., Trehan, R. (2021). Application of Lean Six Sigma framework for improving manufacturing efficiency: A case study in Indian context. *International Journal of Productivity and Performance Management*, *71*(5), 1561–1589. https://doi.org/10.1108/IJPPM-05-2020-0223

Sodhi, M. S., Seyedghorban, Z., Tahernejad, H., Samson, D. (2022). Why emerging supply chain technologies initially disappoint: Blockchain, IoT, and AI. *Production and Operations Management*, *31*(6), 2517–2537. https://doi.org/10.1111/poms.13694

Thekkoote, R. (2021). Understanding big data-driven supply chain and performance measures for customer satisfaction. *Benchmarking: An International Journal*, *29*(8), 2359–2377. https://doi.org/10.1108/BIJ-01-2021-0034

Tiwari, S., Sharma, P., Choi, T.-M., Lim, A. (2023). Blockchain and third-party logistics for global supply chain operations: Stakeholders' perspectives and decision roadmap. *Transportation Research Part E: Logistics and Transportation Review*, *170*, 103012. https://doi.org/10.1016/j.tre.2022.103012

Trehan, R., Gupta, A., Handa, M. (2019). Implementation of Lean Six Sigma framework in a large scale industry: A case study. *International Journal of Six Sigma and Competitive Advantage*, *11*(1), 23–41. https://doi.org/10.1504/IJSSCA.2019.098710

Tsarouhas, P. (2020). Reliability, availability and maintainability analysis of a bag production industry based on the six sigma DMAIC approach. *International Journal of Lean Six Sigma*, *12*(2), 237–263. https://doi.org/10.1108/IJLSS-09-2019-0101

Wan, P. K., Huang, L., Holtskog, H. (2020). Blockchain-enabled information sharing within a supply chain: A systematic literature review. *IEEE Access*, *8*, 49645–49656. https://doi.org/10.1109/ACCESS.2020.2980142

Wang, Y., Han, J. H., Beynon-Davies, P. (2018). Understanding blockchain technology for future supply chains: A systematic literature review and research agenda. *Supply Chain Management: An International Journal*, *24*(1), 62–84. https://doi.org/10.1108/SCM-03- 2018-0148

Wong, L.-W., Leong, L.-Y., Hew, J.-J., Tan, G. W.-H., Ooi, K.-B. (2020). Time to seize the digital evolution: Adoption of blockchain in operations and supply chain management among Malaysian SMEs. *International Journal of Information Management*, *52*, 101997. https://doi.org/10.1016/j.ijinfomgt.2019.08.005

Xue, X., Dou, J., Shang, Y. (2020). Blockchain-driven supply chain decentralized operations—Information sharing perspective. *Business Process Management Journal*, *27*(1), 184–203. https://doi.org/10.1108/BPMJ-12-2019-0518

Zhang, G., Yang, Y., Yang, G. (2023). Smart supply chain management in Industry 4.0: The review, research agenda and strategies in North America. *Annals of Operations Research*, *322*(2), 1075–1117. https://doi.org/10.1007/s10479-022-04689-1.

CHAPTER 12

Augmenting Human Resources: Exploring the Integration of Artificial Intelligence in Human Resource Management for Enhanced Efficiency and Decision-Making

TANUJA TOMER

School of Business Studies, Jigyasa University, Dehradun, India

ABSTRACT

This chapter examines how artificial intelligence (AI) can be integrated with Human Resource Management (HRM) to improve organizational decision-making and operational efficiency. As technology advances, HRM stands at the forefront of transformative changes, with AI presenting unprecedented opportunities for optimization. The study discovers the various HRM AI applications, with talent attainment, employee engagement, performance management, and predictive analytics. Through an analysis of case studies, industry best practices, and emerging trends, this study seeks to shed light on the possible advantages and problems of implementing AI in HRM. It delves into how AI-driven tools and algorithms can streamline routine HR responsibilities, allowing human resources professionals to focus on strategic initiatives and complex decision-making. Furthermore, the research investigates the impact of AI on employee experiences and organizational dynamics, considering ethical considerations and the need for transparent and fair AI practices in HRM. The results of

Industry 5.0: Innovation and Adoption for Sustainability.
Tilottama Singh, Pradeep Suri, and Zulkarin Jahangir (Eds.)

DOI: 10.1201/9781779641878-12

this study add to the growing body of knowledge regarding the mutually beneficial link between AI and HRM by highlighting the significance of a well-rounded strategy that incorporates both human and technology know-how. The chapter concludes with recommendations for organizations seeking to leverage AI in HRM, highlighting key considerations for successful implementation and ongoing management. Ultimately, the research aims to guide HR professionals, policymakers, and researchers in navigating the evolving landscape of HRM in the era of AI.

12.1 INTRODUCTION

In the ever-evolving landscape of the modern workplace, the intersection of human resources and technology stands as a pivotal arena for innovation, and artificial intelligence (AI) in Human Resource Management (HRM) transformation is becoming a strategic necessity as organizations struggle with the challenges of managing human capital in a time of fast technological innovation. This research explores the profound implications of augmenting human resources through the exploration of AI integration, with a specific focus on enhancing efficiency and decision-making processes within HRM.

12.1.1 BACKGROUND AND CONTEXT

Historically, HRM has been at the forefront of adapting to societal, economic, and technological changes. The advent of AI brings forth a paradigm shift, promising not only to streamline routine HR tasks but also to revolutionize the very nature of HRM practices. AI integration in HRM encompasses a wide spectrum of applications, ranging from talent acquisition and employee engagement to predictive analytics that can forecast workforce trends.

12.1.2 SIGNIFICANCE OF ARTIFICIAL INTELLIGENCE INTEGRATION WITH HUMAN RESOURCE MANAGEMENT

The significance of AI in HRM lies in its potential to address long-standing challenges faced by HR professionals. By automating repetitive tasks, AI frees up valuable time, and HR teams have more focus on strategic initiatives and human-centric aspects of workforce management. Moreover, the predictive capabilities of AI have the power to transform decision-making processes, offering data-driven insights that can shape talent development strategies, improve employee experiences, and contribute to the overall success of an organization.

12.2 RESEARCH OBJECTIVES

The study explores comprehensively the integration of AI in HRM, shedding light on both the promises and pitfalls of this technological evolution. Key objectives include:

- **Investigating AI applications in HRM**: Analyzing how AI is presently being applied in various facets of HRM, including recruitment, employee engagement, performance management, and predictive analytics.
- **Assessing success stories and challenges**: Examining real-world examples of AI implementation in HRM, identifying success stories, and addressing the challenges faced by organizations in the process.
- **Understanding decision support in HRM**: Exploring how AI contributes to strategic decision-making within HR, offering insights into workforce planning, talent development, and organizational culture enhancement.
- **Evaluating human–AI collaboration**: Investigating the evolving roles of HR professionals in the era of AI, emphasizing the need for upskilling and fostering a harmonious collaboration between human expertise and AI capabilities.

Through these objectives, this research seeks to contribute to the ongoing dialogue surrounding the augmentation of human resources through AI, providing valuable insights for HR professionals, organizational leaders, and researchers navigating the dynamic landscape of HRM in the digital age.

12.3 LITERATURE REVIEW

Kapoor (2010) observed the function and application of business intelligence in HRM. This study examined the business intelligence and data analytics elements integrated into HRM modules by looking into the top business intelligence vendors.

Jain (2018) identified the function of AI in HRM. According to the report, the majority of businesses have been implementing contemporary technology in a variety of HR procedures, which includes hiring, performance reviews, and cloud-based HR systems.

Dirican (2015), in the paper titled "The Impact of Robotics, Artificial Intelligence on Business and Economics," described use of robotics and AI in business may hurt an organization's overall functions, including production, performance management, sales, strategic planning, customer relationship management, banking system, coaching, training, and taxes.

Buzko et al. (2016), in their research paper "Artificial Intelligence Technologies in Human Resource Development," examined the challenges posed by AI in the ground of human resources and pointed out that AI is particularly good at determining the ROI of training expenditures. The authors of the research article observed that AI technologies help humans analyze data more quickly.

R and D (2018), in "Recruitment through Artificial Intelligence: A Conceptual Study," narrated the AI's role in recruitment where AI plays a significant role in the recruitment process. AI supports employee relations, interview scheduling, candidate screening, and automatically created messaging to candidates.

Jarrahi (2018), in the research article titled "Artificial Intelligence and the Future of Work: Human–AI Symbiosis in Organizational Decision-Making," discussed how AI has proven valuable in assisting humans with decision-making, particularly in handling uncertainty and ambiguity within organizations. However, the paper emphasizes that human involvement remains crucial in industries, as technology still relies on human intuition and subconscious judgment when evaluating decisions.

Merlin and Jayam (2018) draw attention to the AI's function in human resources in the study titled "Artificial Intelligence in Human Resource Management." According to the author, AI is helpful in the workplace and may assist HR authorities in understanding how they operate and seeing issues and trends ahead of time.

12.4 RESEARCH METHODOLOGY

A descriptive research design was used in this study. The researcher used secondary data in the investigation. Through published journals, survey reports, online publications, and blogs, the secondary data was collected.

12.4.1 ARTIFICIAL INTELLIGENCE IN HUMAN RESOURCE MANAGEMENT

1. **Talent acquisition and recruitment:** AI-driven algorithms and machine learning are applied to streamline the recruitment process. Automated screening of resumes, predictive analytics for candidate suitability, and chatbots for initial candidate engagement contribute to more efficient and data-driven talent acquisition. This expedites the hiring process and increases the probability of finding applicants who meet organizational needs.
2. **Employee engagement and retention:** AI tools are increasingly employed to gauge and improve employee engagement. "Natural Language Processing" can investigate the employee's sentiments from surveys, feedback, and social platforms, providing insights into the overall workplace atmosphere. Predictive analytics models can identify factors contributing to employee turnover, enabling proactive measures for retention.
3. **Performance management:** AI assists in the development of fair and objective performance evaluation systems. Algorithms can analyze performance data, identify patterns, and provide real-time feedback. AI-driven performance management systems contribute to personalized development plans, fostering continuous improvement and aligning individual goals with organizational objectives.
4. **Predictive analytics in HR:** Predictive analytics powered by AI enhances decision-making in HR. By analyzing historical data, AI models can forecast trends related to workforce planning, talent development, and employee turnover. HR managers may plan for future organizational needs, strategically allocate resources, and make educated decisions thanks to this data-driven strategy.
5. **HR analytics for strategic decision-making:** AI-driven analytics tools enable HR professionals to derive actionable insights from large datasets. Whether it is identifying skill gaps, predicting training needs, or assessing the impact of HR policies, AI contributes to strategic decision-making. The AI integration in HR analytics enhances the precision and depth of analysis, guiding HR strategies aligned with organizational goals.
6. **Chatbots and virtual assistants:** AI-powered chatbots and virtual assistants are used in HR for a variety of purposes. They can

handle routine queries from employees, provide information about HR policies, and assist in benefits administration. This not only improves efficiency by reducing the workload on HR staff but also ensures consistent and instant responses to employee inquiries.

7. **Learning and development:** AI applications in learning and development facilitate personalized training programs. AI algorithms are used by adaptive learning platforms to evaluate user learning styles, provide pertinent information, and customize training courses. This will enhance the workforce's overall skill development in addition to enhancing the educational experience.
8. **Bias mitigation and fairness:** AI applications in HR also address the challenge of bias in decision-making processes. Algorithms are designed to recognize and ease biases in recruitment, performance evaluation, and other HR functions. This ensures fair treatment and opportunities for all employees, contributing to a more inclusive and diverse workplace.
9. **Robotic process automation (RPA):** It uses software robots to systematize monotonous and rule-based processes. Tasks in HR that can benefit from RPA adoption include payroll processing, benefits administration, and data input. This reduces the likelihood of errors and frees up HR professionals to work on more important and strategic projects.
10. **Employee well-being and mental health support:** AI applications are increasingly integrated into employee well-being programs. Chatbots and AI-driven tools can monitor employee stress levels, provide mental health resources, and offer support in real time. This proactive approach contributes to a healthier and more supportive work environment.

12.5 CHALLENGES AND ETHICAL CONSIDERATION

12.5.1 CHALLENGES

1. **Data privacy and security:**

   Challenge: AI systems often require access to large amounts of sensitive employee data, raising concerns about privacy and security.

Mitigation: Implement robust data encryption, access controls, and compliance with data protection regulations.

2. **Bias and fairness:**

   Challenge: AI algorithms may perpetuate biases present in historical HR data, leading to unfair treatment of certain groups.

   Mitigation: Regularly audit and adjust algorithms to ensure fairness, transparency, and reduce bias in decision-making processes.

3. **Employee acceptance and trust:**

   Challenge: Resistance from employees who may feel uncomfortable with AI-driven decision-making in HR processes.

   Mitigation: Communicate transparently about the purpose and limitations of AI, involve employees in the process, and provide training to build trust.

4. **Integration with existing systems:**

   Challenge: Integrating AI into existing HR systems may be complex and costly.

   Mitigation: Plan for a phased implementation, ensure compatibility with existing systems, and provide sufficient training for HR staff.

5. **Lack of understanding:**

   Challenge: HR professionals and employees may lack understanding of how AI works, leading to misconceptions and mistrust.

   Mitigation: Offer training programs to enhance understanding, promote awareness of AI capabilities and limitations, and address concerns proactively.

12.5.2 ETHICAL CONSIDERATIONS

1. **Transparency:**

   Ethical consideration: Ensuring transparency in AI decision-making processes to avoid "black box" scenarios.

Action: Clearly communicate how AI is used in HR processes, what data it analyzes, and how decisions are made.

2. **Accountability:**

   Ethical consideration: Establishing accountability for AI-driven decisions and actions.

   Action: Clearly define roles and responsibilities and ensure that there are mechanisms for addressing any negative consequences arising from AI decisions.

3. **Informed consent:**

   Ethical Consideration: Obtaining informed consent from employees regarding the use of AI in HR processes.

   Action: Communicate openly about AI implementation, its purpose, and the impact on employees, allowing them to make informed decisions.

4. **Equity and inclusion:**

   Ethical consideration: Ensuring that AI applications promote equity and inclusion.

   Action: Regularly assess and address any biases in AI algorithms and actively seek to enhance diversity and inclusion in HR processes.

5. **Continuous monitoring:**

   Ethical consideration: Implementing continuous monitoring to identify and rectify any ethical issues arising from AI use.

   Action: Regularly audit AI systems, solicit feedback from employees, and make necessary adjustments to ensure ethical use.

12.6 CONCLUSION

In conclusion, the AI integration in HRM has tremendous potential for enhancing efficiency and decision-making within organizations. The exploration of this dynamic integration has revealed numerous benefits, from streamlined recruitment processes to data-driven insights that empower

HR professionals to make informed decisions. But it is crucial to recognize and deal with the difficulties and moral dilemmas this life-changing experience presents. As organizations navigate this evolving landscape, it is essential to view AI as an augmentative tool that complements human expertise rather than a replacement. A collaborative approach, wherein AI empowers HR professionals to focus on strategic decision-making and interpersonal aspects, can foster a harmonious balance between technology and human touch.

In conclusion, the integration of AI in HRM is not merely a technological upgrade but a transformative journey that demands an all-inclusive understanding of its implications on organizational culture, employee relations, and societal values. By addressing challenges and embracing ethical considerations, organizations can unlock the full potential of AI in augmenting human resources, ushering in an era of enhanced efficiency, innovation, and equitable HR practices.

KEYWORDS

- **artificial intelligence**
- **human resource management**
- **efficiency**
- **decision-making**

REFERENCES

Buzko, I., Dyachenko, Y., Petrova, M., Nenkov, N., Tuleninova, D., Koeva, K. (2016). Artificial intelligence technologies in human resource development. *Computer Modelling & Newtechnologies*, *20*(2), 26–29.

Dirican, C. (2015). The Impacts of robotics, artificial intelligence on business and economics. *Procedia: Social and Behavioral Sciences*, *195*, 564–573.

Jain, D. S. (2018). Human resource management and artificial intelligence. *International Journal of Management and Social Sciences Research (IJMSSR)*, *7*(3), 56–59.

Jarrahi, M. H. (2018). Artificial intelligence and the future of work: Human-AI symbiosis in organizational decision making. *Business Horizons*, *61*(4), 1–10. https://doi.org/10.1016/j.bushor.2018.03.007

Kapoor, B. (2010). Business intelligence and its use for human resource management. *The Journal of Human Resource and Adult Learning*, *6*(2), 21–30.

Merlin, P. R., Jayam, R. (2018). Artificial intelligence in human resource management. *International Journal of Pure and Applied Mathematics*, *119*(14), 1891–1895. Retrieved fromhttp://www.acadpubl.eu/hub/

Geetha, R., BhanuShree Reddy, D. (2018). Recruitment through artificial ntelligence: A conceptual study. *International Journal of Mechanical Engineering and Technology (IJMET)*, *9*(7), 63–70. Retrieved fromhttp://www.iaeme.com/ijmet/issues.asp?JType=IJMET&VType=9&IType=7

CHAPTER 13

Beyond Conventional Banking: A Conceptual Framework for Artificial Intelligence-Enhanced Digital Banking Services and Environmental Sustainability

MOHAMMED ARIF HUSSAIN* and SUDHA VEMARAJU

GITAM School of Business, GITAM University, Hyderabad, Telangana, India

*Corresponding author

ABSTRACT

The banking sector in this tech era provides digital services enabled by artificial intelligence (AI) technology to gain a competitive advantage and offer customized and personalized banking services. Focusing on sustainable development and leveraging new technologies are essential to digitalize banking services. This chapter attempts to show how customer loyalty and satisfaction in digital banking are substantially impacted by AI technology for long-term growth and environmental sustainability. By utilizing the theory of the technology acceptance model, the chapter intends to develop a conceptual model that connects these themes through a theoretical framework with the modified service quality dimensions and the modified AI technology acceptance elements. The chapter gathered 278 valid responses from banking customers using a purposive sampling strategy and a structured questionnaire. Software tools such as SMARTPLS4 and SPSS29 are used to measure the model. The findings

Industry 5.0: Innovation and Adoption for Sustainability.
Tilottama Singh, Pradeep Suri, and Zulkarin Jahangir (Eds.)

DOI: 10.1201/9781779641878-13

indicate that the altered AI technology factors and service quality aspects are fulfilling the threshold limits of respective statistical tests conducted in the chapter, thus resulting in improved user satisfaction and encouraging users to keep using AI-enabled digital banking. This research explores future directions; develops model and theory; and offers helpful advice to researchers, practitioners, and environmentalists to enhance digital banking research and environmental sustainability.

13.1 INTRODUCTION

Artificial intelligence (AI) is a sophisticated computing technology that assists in executing complex tasks that necessitate human intellect. AI and sustainability are the modern game-changing trends that will shortly shape financial services and the advancement of predominant sectors. Sustainable finance deals with environmental, social, and governance (ESG) factors during the investment and choice of underwriting. In most cases, these factors are mutually quantitative and qualitative; incorporating these facts into a complicated decision-making framework and best practices are still unfolding. Accumulating sustainable financial data and winding the data gaps will be crucial in evidence-based, efficient policymaking, financial stability, fiscal matters analysis, and risk and reserve management. Assessing the pertinent protection level against sustainability risks is an objective for central banks. A confiding solution would be to opt for just one distinct data source for each of the ESG disclosures across probable assets and depend entirely on this estimation. AI computing platforms will enable the use of AI, big data processing, efficient data filtering, and data science visualization to implement sustainable finance, which are critical technological tools in this context.

13.2 INCREASING SIGNIFICANCE OF ARTIFICIAL INTELLIGENCE CONCERNING SUSTAINABILITY

According to PwC UK study by 2020, applying AI for ecological purposes offers revolutionary opportunities aimed at economic growth and emissions reduction of CO_2; as digitalization and decarbonization gain momentum worldwide, the discharge of greenhouse gases might shrink by 1.5–4.0% by 2030 analogous to that of the combined anticipated perpetual releases of

Canada, Australia, and Japan and boost the global economy by 3.1–4.4% (Gillham et al., 2020). Applications of AI can produce US$3.6–5.2 trillion in benefits over the four key industries: (1) agriculture, (2) water, (3) transportation, and (4) energy. The primary sources of these benefits would be enhanced output productivity, better input utilization, and automated processes of repetitive laborious activities (Gillham et al., 2020).

As a potent tool, conscious AI can assist businesses in reconsidering the design, implementation, upkeep, and application of algorithms while empowering wealth managers to offer more knowledgeable investment advice. Thus, the future of the financial and technology sectors depends on responsible AI and sustainable digital finance.

To consider opportunities and risks related to ESG, investors can use AI to gather and analyze previously impossible data. Sentiment analysis algorithms are a significant source of AI's potential in ESG investing. Therefore, increased cooperation between financial institutions and regulatory bodies is necessary for sustainable AI in finance. For sustainable AI usage to be established across multiple nations and jurisdictions, macro-level norms, integrated regulation, and ongoing international connections are required (Williams et al., 2018).

13.3 ARTIFICIAL INTELLIGENCE AND SUSTAINABLE BANKING

Financial institutions are undergoing a rapid digital shift in the era of the Fintech revolution. Lloyds Banking Group Inc. has made strategic investments of £3 billion to develop products for digital banking and update its workforce for the future of banking. Moreover, as per Williams et al. (2018), JP Morgan allocates 16% of its whole budget to digital services, while Deutsche Bank spends over 4 billion dollars yearly in technological advancements.

Financial services, which deal with large amounts of data and demand organized problem-solving, are ideally suited for AI technology. When financial institutions adopt AI, automation is undoubtedly the most noticeable aspect. To enhance customer interactions and assist staff in resolving complex issues, progressive banks actively create client-facing chatbots, also known as "virtual assistants." By utilizing more than 220 robotic devices for repetitive operations such as error correction and data processing, BNY Mellon made significant advancements in AI. An 88% improvement in overall processing time was made possible by this AI initiative at BNY Mellon (Williams et al., 2018).

Sustainability (ESG) factors are forecast to be significant trends in the digitalization of banking operations over the next 10 years. The State Bank of Vietnam has also made decisions and directives to advance the nation's Sustainable Development Goals to actively promote the National Green Growth Action Plan. In addition to a plan of action for the banking sector to adopt the National Strategy on Green Growth, they involve promoting green credit and managing social and environmental risks in credit-granting activities. Approval of a Green Banking Development project is another aspect of it. Its goals include fighting climate change, directing credit flows toward environmentally friendly finance projects, and raising public awareness of the banking system and environmental protection.

A guidebook on issuing green, social, and sustainable bond guidelines was released in April 2021 by the Staff Selection Commission in association with multiple international financial institutions. The guidebook aims to raise capital for eco-projects from the private sector.

In August, 13 Vietnamese banks were surveyed by EY Consulting Vietnam JSC. By 2023, over half of them will have ESG portfolios and likely invest more in these areas. This allows them to issue green bonds to raise capital.

To encourage more initiatives to combat climate change and to fund projects that adhere to ESG standards, one of the largest banks in Vietnam is implementing a comprehensive digital transformation and making several significant agreements with prestigious financial institutions to cooperate.

A leading local bank also reported in its annual report for 2021 that it had successfully implemented the policy of allocating credit capital to meet socioeconomic development requirements, that it was encouraging the growth of green credit and green banking, and that it was increasing the percentage of bank credit allocated to clean and renewable energy projects. A typical instance of how ESG was used to design the bank's digital transformation system is its Microsoft Office electronic office system, which streamlines operations while cutting down on paper waste.

13.3.1 PROBLEM STATEMENT

Integrating AI and big data suggests that businesses possess a great deal of customer data (Wilson and Daugherty, 2018), increasing their privacy risk. However, severe business and social welfare risks are associated with relying too much on AI. The information asymmetry between the public and the industries that design AI causes risks and dangers (Katyal, 2019). A crucial aspect of marketing success is understanding customers' intentions toward loyalty, one of the most critical marketing issues.

13.3.2 OBJECTIVES OF THE STUDY

The primary aim of this chapter is to find out how AI technology affects service quality. The influence of service quality factors on bank customers' loyalty and satisfaction is also evaluated in this study. Finally, the chapter examines how technology acceptance factors [technology acceptance model (TAM)] influence customers' satisfaction, how the perceived risks affect this relationship, and how these factors together shape their long-term loyalty to banks.

13.4 BACKGROUND AND DEVELOPMENT OF HYPOTHESES OF THE PRESENT PROPOSAL

13.4.1 THEORETICAL BACKGROUND

Digital banking is a technology-driven banking service that has transformed conventional banking activities into a digital environment (Sarma, 2017). As a result, customers do not have to visit banks, and also banks do not have to attend their customers physically for any transaction, eventually making digital banking a service-oriented one as well (Van Looy et al., 1998).

While there is a considerable growth in the usage of digital banking, an advanced study has yet to trace all the customer-related issues that may be limited due to partial findings and methods of study (Hoehle et al., 2012). Hence, this research presents a more comprehensive study in India.

13.4.2 RELATIONSHIP BETWEEN ARTIFICIAL INTELLIGENCE TECHNOLOGY FACTORS AND SERVICE QUALITY DIMENSIONS

13.4.2.1 ARTIFICIAL INTELLIGENCE

John McCarthy introduced the "AI" concept in 1957 at a Dartmouth workshop. He described AI as being "the science and engineering of creating intelligent machines" (McCarthy, 2007). The "General Problem Solver" computer program, which mimics human problem-solving skills and essentially resembles human thinking capacity, is where AI first emerged (Newell et al., 1958). In order to determine whether or not a computer could think and act like a human, Alan Turing proposed the Turing test in 1950 (Turing, 1950). Russell and Norvig (1995) proposed the four dimensions of AI, which include thinking and acting like human brains, reasoning, and applying analytical thinking to complex problems. AI in banking must "do the right thing." For instance, a check deposited into the appropriate account is an example of rational AI behavior.

13.4.2.2 THEORY OF TECHNOLOGY ACCEPTANCE MODEL

Davis (1989) designed TAM on the way to theorize the computer technology behavioral usage. Perceived usefulness and perceived ease of use are TAM indicators that constitute user acceptance (Rana et al., 2017). Over the last 20 years, TAM has been improved to comprehend the intention to use technology.

13.4.2.3 FACTORS OF MODIFIED TECHNOLOGY ACCEPTANCE MODEL

- *Perceived usefulness*: It is the degree of belief of a person that digital technology could enhance the banking performance (Davis, 1989). Therefore, if the users perceive AI-based banking services as more valuable, then their usage has a positive impact on their satisfaction.
- *Perceived ease of use*: It is the decisive outcome of a person's belief that a certain system will enhance achievement and favorably stimulate users' perspective concerning modern technology that constitutes satisfaction and loyalty (Davis, 1989).
- *Perceived risk*: Consumer perception of perceived risk refers to how uncertain and unfavorable they believe purchasing a good or service will be (Yang et al., 2015). Customers should be conscious of the risks involved, so strategies should be developed with that awareness. It has been discovered that there is a discrepancy

between the opinions of users and the technology's actual functionality. A customer's likelihood of having a bad experience with banking services increases with the level of risk they perceive.

- Numerous studies have indicated that a significant psychological factor crucial to banking is customer emotional behavior. The psychological variables related to individual-level experiences and cognitive processes, such as a person's thoughts, feelings, and beliefs, are perceived trust and perceived benefit, which influence a customer's satisfaction and loyalty (Khalid et al., 2015).
- *Perceived trust*: It ascertains the degree of self-confidence linked to every financial operation that corresponds directly with the individual's satisfaction. Additionally, it supports safeguarding the transactional association between the customer and the bank (Upton, 2013). Thus, the perceived trust has been considered as one of the predominant factors for evaluating customer loyalty in digital banking in the proposed study.
- *Perceived benefit*: It is the privilege that consumers would become comfortable with digital banking. Online consumers have observed many benefits such as cost savings, time savings, increased convenience, and service variety in comparison with the conventional mode of banking (Peha and Khamitov, 2004). Therefore, more the consumers perceive benefits, more likely they are to perform online transactions, leading to satisfaction and loyalty.

13.4.2.4 SERVICE QUALITY

An intangible act or performance rendered to another party is known as a service without granting ownership of anything (Kotler and Keller, 2009). In contrast, *quality* assesses the integrality of a product or service's features and characteristics and whether they can meet any stated needs (Kotler et al., 2002). The SERVQUAL model is a highly valuable tool for measuring customer service quality dimensions.

The model developed by SERVQUAL has proven five elements of service quality that together are referred to as **RATER** (Parasuraman et al., 1988).

1. Reliability: The potential to execute the ensured services accurately and dependably.

2. Assurance: The awareness and generosity of employees along with their capability to motivate the customers' confidence and trust.
3. Tangibility: The personnel accomplishments and appearances.
4. Empathy: The firm's benevolent and singularized attention toward its clients.
5. Responsiveness: The eagerness to help and provide prompt customer services.

Although numerous attempts have been made to use the SERVQUAL model in conventional banking globally, there has been limited attention given to assessing the service quality, satisfaction, and loyalty in the context of digital banking customers.

Hence, a five-dimensional modified SERVQUAL model called **RACER** (***R**eliability, **A**ssurance, **C**ustomization, **E**mpathy, and **R**esponsiveness*) and modified technology acceptance factors (**P**erceived **U**sefulness, **P**erceived **E**ase of **U**se, **P**erceived **R**isk, **P**erceived **T**rust, and **P**erceived **B**enefits) that influences customer's satisfaction and loyalty has been proposed by using a conceptual framework in the present research through the following hypotheses.

13.4.2.4.1 Reliability

Reliability has been suggested as the appropriate technical functioning of a self-service technology and accurate service delivery (Weijters et al., 2005). Many reported letters have identified reliability as a significant dimension in determining service quality (Bagozzi, 1990).

13.4.2.4.2 Assurance

Significant affirmation in an intangible service indicates the consumable services communicated to the related customers effectively, thereby strengthening the positive outcome of the services encountered. Further, it appropriately increases the trust of the users and decreases risks while performing any transaction (Parasuraman et al., 1991).

13.4.2.4.3 Customization

Customization has been well-defined as the build-to-order tactic for offering a service that exceeds the wishes of the customer (Kaplan and Haenlein, 2007). Firms could build a powerful customer bond through customization, leading to loyalty or higher customer retention (Lovelock and Wirtz, 2004).

13.4.2.4.4 Empathy

The conveyance of the feelings that the customer is unique and special is the basic idea of empathy (Zeithaml et al., 2002). A SERVQUAL model describes a quantitative investigation that recognizes dimensions of service quality, which include security, access, and credibility, to measure empathy that affects satisfaction.

13.4.2.4.5 Responsiveness

The promptness with which banking operators respond to customers is called responsiveness (Sheng and Liu, 2010). Research indicates that responsiveness is critical in determining the eminence of banking services digitally and how it impacts the satisfaction and loyalty of customers (Suleman et al., 2012).

H_1: AI technology factors positively influence service quality dimensions.

13.4.2.5 CUSTOMER SATISFACTION

Oliver (1999) has defined satisfaction as the accomplished response of a consumer. Margherio (1998) stated that satisfaction is a postservice activity estimated index that estimates the mindset of the customer's feelings about past services and their experiences. Customers' decisions are greatly influenced by the experienced services that measure the degree of satisfaction.

13.4.2.6 CUSTOMER LOYALTY

It relates to the commitment that is deeply associated with repatronize or repeat purchase of the preferred services consistently in the future, thus initiating a similar set of brands to purchase repeatedly regardless of the marketing efforts and situational impacts that have the ability to cause a shift in behavior (Oliver, 1997).

The dimensions on which customer loyalty is based are purchase intentions (Oliver, 1999), word-of-mouth chat (Boulding et al., 1993), and commitment (Gremler et al., 2001). Purchase intention is the propensity to make a future purchase of a good or service (Moorman et al., 1992). Noncommercial oral communication between people about a brand, product, or service is known as word of mouth (Ranaweera et al., 2003). Strong relationships and an intention to maintain them are components of customer commitment (Arndt, 1967).

13.4.3 INTERRELATIONSHIP AMONG SERVICE QUALITY, CUSTOMER SATISFACTION, AND CUSTOMER LOYALTY

Today's competitive business environment, with the rapid market entry of new service innovations, requires a rigorous understanding of service quality, customer satisfaction, and loyalty, which are considered significant factors for success, survival, and the basis of marketing strategy (Bansal et al., 2003). Parasuraman et al. (1996) introduced a conceptual model on service quality, customer satisfaction, and service loyalty. The findings of Bloemer et al. (1998) determined that the service quality possesses an indirect effect through satisfaction on loyalty, and eventually satisfaction possesses a direct effect on loyalty (Zeithmal et al., 2009).

H2: Service quality dimensions positively influences customer satisfaction.

H3: Service quality dimensions positively influences customer loyalty.

13.4.4 INTERRELATIONSHIP BETWEEN ARTIFICIAL INTELLIGENCE TECHNOLOGY FACTORS, CUSTOMER SATISFACTION, AND CUSTOMER LOYALTY

According to Bloemer et al. (1998), the TAM has two main dimensions: (1) perceived usefulness and (2) perceived ease of use.

The chapter aims to add perceived risk to basic TAM with the inclusion of perceived trust and perceived benefits in order to examine by what means these factors influence the banking customers' satisfaction and loyalty. Hence, the following hypotheses are proposed to guide this research:

H_4: *AI technology factors have no influence on customer satisfaction.*

H_5: *AI technology factors have no influence on customer loyalty.*

13.4.5 RELATIONSHIP BETWEEN CUSTOMER SATISFACTION AND LOYALTY

The studies conducted by Douglas et al. reveal a positive relationship between customer satisfaction and customer loyalty (Davis, 1989). Banks are recommended to achieve customer loyalty by influencing customer satisfaction positively. Customers would have lower brand switching behavior with high satisfaction (Douglas et al., 2017).

H_6: *Customer satisfaction positively influences customer loyalty.*

13.4.6 CONCEPTUAL FRAMEWORK (FIGURE 13.1)

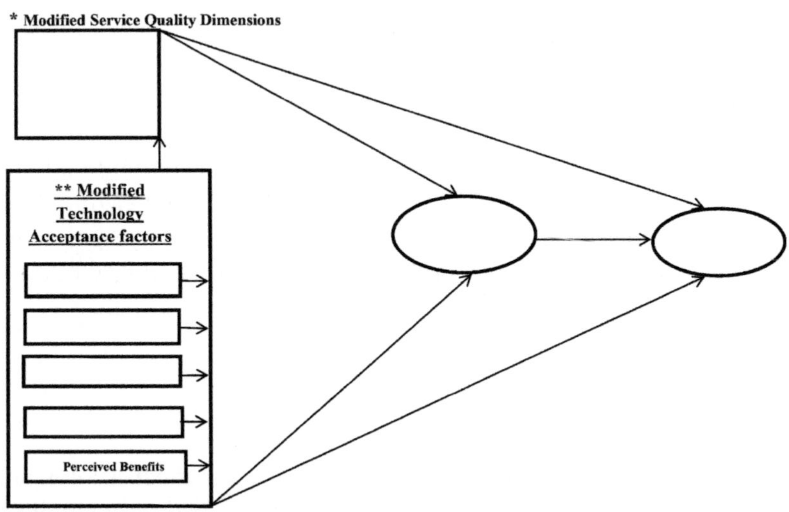

FIGURE 13.1 A conceptual model.

Source: *Modified service quality dimensions. Adapted from Zeithaml et al. (2002).

***Modified TAM factors. Adapted from Davis (1989).*

13.5 RESEARCH METHODOLOGY

Of the 300 banking customers who participated in the survey, 278 valid responses were selected for further investigation. Data were gathered from Secunderabad and Hyderabad in Telangana, India. A purposive sampling technique is employed to collect data. The primary method used to collect data is a structured questionnaire via the Likert scale, where 1 represents disagree strongly and 5 represents agree strongly. Income, age, gender, qualification, and occupation are the five demographic elements included in the questionnaire. Thirty-six items were defined, three related to each of perceived usefulness, perceived ease of use, perceived risk, perceived trust, perceived benefit, customer satisfaction, and customer loyalty. Each three items is related to reliability, assurance, customization, empathy, and responsiveness.

The chapter's objectives and hypothesis testing were measured by extracting empirical results. To ensure the loaded items performed well for their assigned constructs with high reliability, exploratory factor analysis (EFA) was conducted to assess the model's efficacy on the data generated (Hair et al., 2010).

13.6 EMPIRICAL RESULTS AND DISCUSSIONS

13.6.1 AN ANALYSIS OF DESCRIPTIVE STATISTICS

As shown in Table 13.1, 65.4% of the samples are males, while 34.5% are females. Most respondents (35.6%) are 40–59 years old, while 32% are between 20 and 39 years old. Besides, 30.5% of the respondents are postgraduates. Finally, 96 respondents had 34.5% income, while 84 respondents had 30.2% income.

13.6.1.1 DESCRIPTIVE ANALYSIS

Using descriptive statistics, value deviations from their mean scores are displayed. Table 13.2 lists each variable's descriptive statistics that include for both independent and dependent variables. A lower degree of variability and spread in the variables suggests that most results are near their means.

TABLE 13.1 Respondents Demographic Characteristics

		N	%
	Female	96	34.5
Gender	Male	182	65.4
	Total	**278**	**100.0**
	Below 20	73	26.2
	20–39	89	32.0
Age (years)	40–59	99	35.6
	60 and above	17	6.1
	Total	**278**	**100.0**
	Under graduation	23	29.9
	Graduation	85	8.2
Qualification	Postgraduation	98	30.5
	Above postgraduation	72	25.8
	Total	**278**	**100.0**
	Student	65	22.3
	Service	84	28.7
	Business	25	8.9
Occupation	Professional	94	33.8
	Retired	12	4.3
	Homemaker	2	1.7
	Total	**278**	**100.0**
	Below 20,000	64	23.0
	20,000–39,999	84	30.2
Income	40,000–59,999	96	34.5
	60,000 and above	34	12.2
	Total	**278**	**100.0**

13.6.2 INFERENTIAL STATISTICAL ANALYSIS

13.6.2.1 CORRELATION STUDY

The statistical technique known as correlation is used to determine the relationship between two variables. The range of values for the coefficient of correlation is −1 to +1. The relationship directions are indicated by the coefficient sign; a variable has a negative correlation if its correlation coefficient is close to −1 and a positive correlation if it is close to +1.

TABLE 13.2 Measurement Scales and Descriptive Statistics

Construct	Final Item Code	Mean	Standard Deviation	*N*
Reliability (RL)	RL1	3.97	1.118	278
	RL2	3.98	1.118	278
	RL3	3.96	1.077	278
Assurance (ASS)	ASS1	3.92	1.123	278
	ASS2	3.94	1.051	278
	ASS3	3.96	1.101	278
Customization (CUS)	CUS1	3.97	1.092	278
	CUS2	3.98	1.111	278
	CUS3	3.92	1.045	278
Empathy (EM)	EM1	3.88	1.137	278
	EM2	3.79	1.139	278
	EM3	3.81	1.149	278
Responsiveness (RE)	RE1	3.8	1.142	278
	RE2	3.92	1.101	278
	RE3	3.94	1.086	278
Perceived usefulness (PU)	PU1	3.93	1.027	278
	PU2	3.93	1.03	278
	PU3	3.98	1.096	278
Perceived ease of use (PEU)	PEU1	3.88	1.117	278
	PEU2	3.88	1.123	278
	PEU3	3.84	1.118	278
Perceived risk (PR)	PR1	3.68	1.139	278
	PR2	3.53	1.12	278
	PR3	3.6	1.12	278
Perceived trust (PT)	PT1	3.77	1.096	278
	PT2	3.88	1.115	278
	PT3	3.87	1.12	278
Perceived benefit (PB)	PB1	3.96	1.058	278
	PB2	3.95	1.062	278
	PB3	3.91	1.081	278
Customer satisfaction (CS)	CS1	3.93	1.096	278
	CS2	3.91	1.104	278
	CS3	3.9	1.079	278
Customer loyalty (CL)	CL1	3.93	1.089	278
	CL2	3.85	1.125	278
	CL3	3.92	1.071	278

The present investigation demonstrates that CL has a moderately strong correlation with PEU, PR, PT, RL, RE, PU, and CS but a weak correlation with ASS, CUS, and EM. It also possesses a strong correlation with PB and CS. There is a positive correlation found between all of the variables (refer Table 13.3).

TABLE 13.3 Correlation Matrix

	RL	ASS	CUS	EM	RE	PU	PEU	PR	PT	PB	CS	CL
RL	1											
ASS	0.774**	1										
CUS	0.730**	0.878**	1									
EM	0.751**	0.911**	0.918**	1								
RE	0.846**	0.861**	0.788**	0.837**	1							
PU	0.867**	0.775**	0.682**	0.728**	0.848**	1						
PEU	0.873**	0.802**	0.723**	0.762**	0.893**	0.924**	1					
PR	0.860**	0.677**	0.645**	0.648**	0.759**	0.836**	0.829**	1				
PT	0.859**	0.666**	0.640**	0.650**	0.752**	0.836**	0.818**	0.964**	1			
PB	0.896**	0.727**	0.658**	0.672**	0.812**	0.874**	0.886**	0.926**	0.926**	1		
CS	0.903**	0.703**	0.663**	0.668**	0.786**	0.868**	0.857**	0.929**	0.932**	0.959**	1	
CL	0.899**	0.753**	0.694**	0.721**	0.807**	0.870**	0.878**	0.875**	0.878**	0.924**	0.929**	1

**(Two-tailed) significant Pearson correlation exists at the 0.01 level.

13.6.2.2 ANALYZING REGRESSION

Regression analysis is a crucial method for figuring out how various variables that are recognized as variables that are independent have an impact on the variable that is dependent.

A multivariate regression technique is used as this chapter has multiple independent variables and a single dependent variable.

Table 13.4 shows the influence of PU, PEU, PR, PT, PB, RL, ASS, CUS, EM, and RE on the CS and CL's acceptance of AI in banking. Thus, at $P < 0.001$, the predictive model is significant enough, as indicated by the *F*-statistics. In other words, if the model is valid, the outcomes produced by applying it are likewise trustworthy and broadly applicable. R^2 and adjusted R^2 convey the explanatory power of the independent variables, which shows the level to which the independent variable describes the dependent variable.

According to H_1, AI technology positively impacts banking service quality dimensions. The findings for H_1 demonstrate statistically significant $P < 0.001$, a significant and positive correlation between AI technology factors and service quality dimensions. H_2 states that a positive correlation exists between customer satisfaction and service quality dimensions in banking. H_2 finding was significant and positive, with a value of $P < 0.001$. Therefore, in the banking sector, the service quality dimensions positively impact customer satisfaction. Similarly, H_4 holds that customer satisfaction in banking is unaffected by AI technology factors. With $P < 0.001$, the AI technology factors did not impact customer satisfaction. In banking, customer loyalty is positively impacted by the service quality dimensions, according to H_3. The results of H_3 were also very significant and positive, indicated by $P < 0.001$.

In banking, customer loyalty is not impacted by AI technology, according to H_5. A vital difference ($P < 0.001$) was observed between the AI technology factors and customer loyalty. Customers' satisfaction positively impacts on loyalty to banking. With $P < 0.001$, the results are positive and highly significant in accordance with H_6.

Tables 13.4 and 13.5 show that the chosen influencing factors result in $R = 0.977\%$ of the variations in customer loyalty. There is 0.955% adjusted R^2. The coefficient of determination is 0.846; therefore, the selected variables explain about 84.60% of the variation in the satisfaction data. The regression equation is beneficial for making recommendations as the value of R^2 is closest to 1.

TABLE 13.4 Regression Matrix

Variables Entered/Removed[a]			
Model	**Variables Entered**	**Variables Removed**	**Method**
1	PB, PR, EM, PT, PEU, RL, PU, RE, CUS, and ASS[b]		Enter

[a]Dependent variable: CL.

[b]All requested variables entered.

TABLE 13.5 Descriptive Analysis

Model Summary[b]				
Model	***R***	***R*²**	**Adjusted *R*²**	**Standard Error of the Estimate**
1	0.977[a]	0.955	0.953	0.6959

[a]Predictors: (constant), technology acceptance aspects/factors (PU, PEU, PR, PT, and PB). Service quality dimensions/elements (RL, ASS CUS, EM, and RE).

[b]Dependent variable: CL.

The chosen influencing variables account for $R = 0.977\%$ of the variations in customer loyalty. Table 13.6 shows the highly significant F-test result, giving us an indication that the model does fit the data in this case.

TABLE 13.6 ANOVA

ANOVA[a]						
Model		**Sum of Squares**	**df**	**Mean Square**	**F**	**Sig.**
1	Regression	2450.74	10	245.07	506.06	<0.001[b]
	Residual	114.774	237	0.484		
	Total	2565.51	247			

[a]Dependent variable: CL.

[b]Predictors: (constant), PB, PR, EM, PT, PEU, RL, PU, RE, CUS, and ASS.

It is well known that the β value quantifies the degree to which each predictor variable affects the criterion variable. In addition, standard deviation units are used to measure the β. As shown in Table 13.7, perceived benefit has the highest β value (0.635), meaning that a change in perceived benefit will shift customer loyalty to public and private banks in Hyderabad and Secunderabad by 0.635 standard deviation. Therefore, the higher the β value, the more significant the predictor variable's influence on the criterion variable (Table 13.8).

TABLE 13.7 Inferential Analysis

Coefficients[a]							
Model	**Unstandardized Coefficients**		**Standardized Coefficients**	***t***	**Sig.**	**Collinearity Statistics**	**Collinearity Statistics**
	B	**Standard Error**	***β***			**Tolerance**	**Variance Inflation Factor**
1 (Constant)	0.006	0.182		0.034	0.973		
RL	0.202	0.074	0.203	2.713	0.007	0.034	2.531
ASS	−0.21	0.092	−0.202	−2.247	0.026	0.023	3.857
CUS	−0.04	0.078	−0.035	−0.46	0.646	0.033	3.524
EM	0.111	0.047	0.115	2.332	0.021	0.078	2.828
RE	0.174	0.064	0.172	2.73	0.007	0.047	1.145
PU	−0.15	0.062	−0.145	−2.436	0.016	0.053	1.781
PEU	0.247	0.045	0.251	5.513	<.001	0.091	1.02
PR	−0.07	0.026	−0.071	−2.656	0.008	0.268	3.733
PT	0.076	0.046	0.076	1.66	0.098	0.091	1.028
PB	0.647	0.06	0.635	10.722	<.001	0.054	1.562

TABLE 13.8 Regression Analysis Results Summarized for the Hypotheses Testing

Hypotheses	Remarks—Supported
H_1	Yes
H_2	Yes
H_3	Yes
H_4	Yes
H_5	Yes
H_6	Yes

13.6.2.3 An Analysis of Variance Test

To ascertain whether age and gender have an impact on a customer's loyalty to banking services, an analysis of variance (ANOVA) test was conducted. Table 13.9 displays the F-statistics using customer loyalty in banking for the variables age and gender.

There is no age difference in terms of banking customer loyalty as per the significant F-value, as shown in Table 13.9.

TABLE 13.9 ANOVA Test Customer Loyalty in Banking with Respect to Age

Descriptives

CL

	N	Mean	Standard Deviation	Standard Error	95% Confidence Interval for Mean	
					Lower Bound	Upper Bound
Below 20	70	3.628	1.02939	0.139	3.3494	3.9059
20–39	89	4.079	0.84507	0.091	3.8978	4.2602
40–59	99	4.013	0.97832	0.103	3.8078	4.2176
60 and above	20	3.899	0.96429	0.234	3.4033	4.3949
Total	278	3.943	0.95552	0.061	3.823	4.062

Tests of Homogeneity of Variances

	Levene Statistic	df1	df2
CL	1.206	11	267
	0.621	11	267
	0.621	11	266
	1.629	11	267

ANOVA

CL

	df	*F*	Sig.
Between groups	11	2.81	0.04
Within groups	267		
Total	278		

There is no gender difference in terms of banking customer loyalty as per the significant F-value, as shown in Table 13.10.

TABLE 13.10 ANOVA Analysis of Customer Loyalty in Banking with Respect to Gender

	N	Mean	Standard Deviation	Standard Error	95% Confidence Interval for Mean	
					Lower Bound	Upper Bound
Male	182	4.018	0.90743	0.071	3.877	4.1585
Female	96	3.801	1.0307	0.111	3.5798	4.0218
Total	278	3.943	0.95552	0.061	3.823	4.062

Tests of Homogeneity of Variances

		Levene Statistic	df1	df2	Sig.
CL	Based on mean	5.203	1	276	0.023
	Based on median	3.009	1	276	0.084
	Based on median and with adjusted df	3.009	1	272.368	0.084
	Based on the trimmed mean	5.672	1	276	0.018

ANOVA

CL

	Sum of Squares	df	Mean Square	*F*	Sig.
Between groups	2.644	1	2.644	2.919	0.089
Within groups	272.869	276	0.906		
Total	275.513	277			

13.6.3 MEASUREMENT MODEL

13.6.3.1 KAISER–MEYER–OLKIN AND BARTLETT'S TEST

TABLE 13.11 Sampling Adequacy Measurement

KMO and Bartlett's Test		
Kaiser–Meyer–Olkin Measure of Sampling Adequacy		**0.963**
	Approximately χ^2	22,203.455
Bartlett's test of sphericity	df	630
	Sig.	<0.001

Table 13.11 presents the findings of the KMO test and Bartlett's test of factors based on data collected on particular variables. When values are high (0.50–1.0), factor analysis is appropriate. Factor analysis might only be suitable if the value is at least 0.50. KMO is a tool used to assess the suitability of factor analysis by measuring the sphericity of sampling adequacy. KMO indicates that the sampling adequacy in this instance is (value of $0.50 < \text{KMO} < 1.0$), meaning that 0.963 for a selected group of customers using bank facilities is appropriate. The statistical test known as Bartlett's test of sphericity is employed to assess the hypothesis of correlation between the variables in the population. Every variable in the population correlation matrix ($r = 1$) has a perfect correlation with itself but not with any of the other variables ($r = 0$). This matrix is known as an identity matrix. In this chapter, as per Bartlett's test of sphericity, the approximated χ^2 value is <0.001 and the significance level is 22203.455 with 630 (df). Factor analysis is, therefore, a suitable method.

13.6.3.2 FACTOR LOADING, CRONBACH'S ALPHA, COMPOSITE RELIABILITY, AND AVERAGE VARIANCE EXTRACTED

As per Hair et al. (2013), the standard level of Cronbach's alpha (α) and composite reliability (CR) is 0.70. Each construct's AVE threshold value must be >0.5 (Hair et al., 2013). The output of Cronbach's α, CR, and AVE for each construct that satisfies the reliability and validity criteria is also shown in Table 13.12. For all constructs that meet convergent validity, the factor loading results are shown in Table 13.12, as each predictor/indicator has a loading factor value >0.70. Thus, the findings of this chapter in Smart PLS4 have passed the outer model test.

TABLE 13.12 Results Related to Items, Factor Loading, Cronbach's α, CR, and AVE

Items	Factor Loadings	Cronbach's α	CR	AVE
RL1	0.977			
RL2	0.974	0.974	0.983	0.96
RL3	0.974			
ASS1	0.953			
ASS2	0.961	0.962	0.976	0.93
ASS3	0.979			

TABLE 13.12 *(Continued)*

Items	Factor Loadings	Cronbach's α	CR	AVE
CUS1	0.978			
CUS2	0.969	0.97	0.98	0.943
CUS3	0.967			
EM1	0.969			
EM2	0.98	0.975	0.983	0.951
EM3	0.976			
RE1	0.936			
RE2	0.974	0.958	0.972	0.939
RE3	0.968			
PU1	0.987			
PU2	0.987	0.981	0.987	0.963
PU3	0.97			
PEU1	0.985			
PEU2	0.968	0.974	0.983	0.951
PEU3	0.972			
PR1	0.963			
PR2	0.962	0.967	0.977	0.935
PR3	0.975			
PT1	0.947			
PT2	0.962	0.962	0.974	0.925
PT3	0.977			
PB1	0.984			
PB2	0.991	0.987	0.992	0.975
PB3	0.987			
CS1	0.985			
CS2	0.99	0.988	0.992	0.977
CS3	0.99			
CL1	0.991			
CL2	0.964	0.981	0.987	0.97
CL3	0.989			

The communalities in Table 13.13 show that the amount of deviation in the variable can be adequately explained by the factors that were identified. For example, the variance/difference in customer satisfaction is demonstrated by approximately 94%, while the variance in customer loyalty to banks is presented by over 95%.

TABLE 13.13 Result of Communalities

Items		Initial	Extraction
RL1	My bank accomplishes AI ……. right the first time	1.00	0.918
RL2	I consider my bank offers AI ….. promise to do so	1.00	0.907
RL3	I see accuracy ………. AI technology	1.00	0.913
ASS1	My bank provides clear …… digital banking services	1.00	0.908
ASS2	AI technology enables me …… more quickly	1.00	0.922
ASS3	I consider my bank is committed ….. banking services	1.00	0.909
CUS1	My bank greets customers ……services	1.00	0.924
CUS2	Bank communicates ……banking transactions	1.00	0.903
CUS3	AI technology fulfills …… customers	1.00	0.908
EM1	Bank staff serves well ………services	1.00	0.908
EM2	My bank pays individual ….services queries	1.00	0.862
EM3	Employees of my bank …their customers	1.00	0.848
RE1	I get in time information ….. banking services	1.00	0.854
RE2	My bank's AI…… speed is high	1.00	0.896
RE3	Overall the cooperation ….its customers	1.00	0.923
PU1	AI technology is very much useful …. transactions	1.00	0.928
PU2	AI technology improves… performance	1.00	0.9
PU3	AI enabled digital banking services ….anytime	1.00	0.921
PEU1	I notice it simple to learn ….banking services	1.00	0.891
PEU2	Interaction with Robots/Chatbots ….enjoyable	1.00	0.882
PEU3	It is appropriate to get AI technology …. to do	1.00	0.875
PR1	I hope my transaction information …. using AI technology	1.00	0.9
PR2	I believe AI technology…..my personal information	1.00	0.828
PR3	The risk associated ….. the transactions is low	1.00	0.871
PT1	I depend on AI technology…. of financial transactions	1.00	0.869
PT2	AI technology ….. secure all the time	1.00	0.858
PT3	There is… privacy…AI banking services	1.00	0.878
PB1	AI technology banking services … lot of comforts	1.00	0.935
PB2	The time spent is worthwhile…. digital banking	1.00	0.941
PB3	AI technology allows me …. banking services	1.00	0.918
CS1	I am generally convinced with AI technology….	1.00	0.927
CS2	I am happy with …. banking services	1.00	0.938
CS3	Overall I am fully satisfied …my bank	1.00	0.941

TABLE 13.13 *(Continued)*

Items		Initial	Extraction
CL1	I prefer AI technology.....other banking services	1.00	0.943
CL2	I will endorse AI technology-enabled digital banking	1.00	0.891
CL3	I mean to continue .....offered by my bank	1.00	0.951
	Extraction Method: Principal Component Analysis.		

All factors that could be extracted from the analysis are shown in Table 13.14, along with their eigenvalues. It is the percentage of variance attributable to each element, the cumulative variance of the factor, and the factors that came before it. The first factor corresponds to 30.82% of the variance, while the second factor constitutes 1.67% of the variance. All the other factors are negligible except for the first two components.

TABLE 13.14 Total Variance Explained

Component	Initial Eigenvalues			Extraction Sums of Squared Loadings			Rotation Sums of Squared Loadings
	Total	% of Variance	Cumulative %	Total	% of Variance	Cumulative %	Total
1	30.82	85.61	85.61	30.82	85.61	85.61	20.379
2	1.671	4.642	90.252	1.671	4.642	90.252	12.112
3	0.572	1.589	91.842				
4	0.396	1.1	92.942				
5	0.386	1.072	94.014				
6	0.247	0.686	94.7				
7	0.204	0.566	95.267				
8	0.184	0.51	95.777				
9	0.142	0.396	96.173				
10	0.129	0.359	96.531				
11	0.127	0.352	96.883				
12	0.113	0.315	97.198				
13	0.103	0.285	97.484				
14	0.097	0.27	97.753				
15	0.083	0.229	97.982				
16	0.08	0.222	98.204				
17	0.069	0.19	98.395				
18	0.063	0.174	98.569				

TABLE 13.14 *(Continued)*

Component	Initial Eigenvalues			Extraction Sums of Squared Loadings			Rotation Sums of Squared Loadings
	Total	% of Variance	Cumulative %	Total	% of Variance	Cumulative %	Total
19	0.061	0.169	98.738				
20	0.057	0.158	98.896				
21	0.046	0.129	99.025				
22	0.045	0.124	99.149				
23	0.041	0.113	99.262				
24	0.036	0.099	99.361				
25	0.034	0.095	99.456				
26	0.032	0.089	99.545				
27	0.031	0.085	99.63				
28	0.027	0.076	99.706				
29	0.024	0.066	99.772				
30	0.02	0.055	99.828				
31	0.016	0.046	99.873				
32	0.014	0.04	99.913				
33	0.012	0.032	99.945				
34	0.01	0.027	99.973				
35	0.008	0.021	99.994				
36	0.002	0.006	100				

13.6.3.3 DISCRIMINANT VALIDITY

Table 13.15 shows the results of all constructs using the Fornell and Larcker criterion that meet the discriminant validity (Figure 13.2). This technique evaluates factors' discriminant validity through the extracted square root of average variance extracted (Fornell and Larcker, 1981).

TABLE 13.15 Discriminant Validity: Fornell–Larcker Criterion

	ASS	CL	CS	CUS	EM	PB	PEU	PR	PT	PU	RE	RL
ASS	0.964											
CL	0.942	0.985										
CS	0.944	0.981	0.988									
CUS	0.953	0.939	0.939	0.971								
EM	0.921	0.894	0.899	0.876	0.975							
PB	0.937	0.956	0.949	0.952	0.846	0.988						

TABLE 13.15 *(Continued)*

	ASS	CL	CS	CUS	EM	PB	PEU	PR	PT	PU	RE	RL
PEU	0.928	0.929	0.932	0.902	0.902	0.886	0.975					
PR	0.742	0.723	0.72	0.704	0.801	0.687	0.795	0.967				
PT	0.906	0.895	0.894	0.876	0.911	0.847	0.925	0.831	0.962			
PU	0.934	0.925	0.922	0.951	0.834	0.965	0.872	0.658	0.835	0.981		
RE	0.96	0.943	0.951	0.945	0.937	0.921	0.923	0.75	0.903	0.922	0.969	
RL	0.957	0.941	0.946	0.97	0.891	0.933	0.914	0.715	0.901	0.929	0.947	0.977

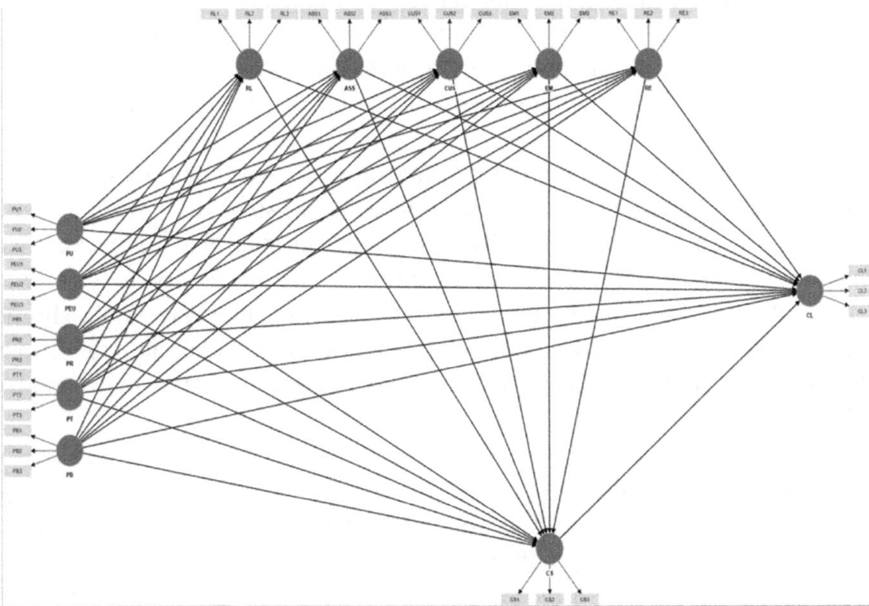

FIGURE 13.2 SmartPLS4 structural equation model.

13.7 DISCUSSION AND IMPLICATIONS

The modern world has made AI indispensable in all spheres of existence. Both industrialized and developing nations attempt to use AI in various fields. One of the most critical industries for any country is banking. Keskinbora (2018) argued that AI systems possess intelligence and problem-solving abilities that surpass those of humans, which may pose ethical and societal risks. Public values such as autonomy, privacy, security, human dignity, and justice have all been pressed (Yu et al., 2018).

Ethical governance becomes essential with these technological systems becoming increasingly commonplace.

Financial risks associated with climate change may immediately affect the counterparties and financial assets that central banks use to manage collateral and implement monetary policy. The 2017 UN AI Summit, held in Geneva, recognized that AI has the potential to accelerate humanity's transition toward a life of dignity, peace, and prosperity. It also recommended that AI—which powers self-driving cars and voice and facial recognition/acknowledgement in smartphones—serves more strategically to assist global efforts to end hunger and poverty, preserve the environment, and conserve natural resources. In particular, improved regulatory expectations and audit process facilitation between government agencies and organizations are critical to the broad adoption of sustainable AI.

As demonstrated by the study's findings, all AI technological acceptance factors and service quality dimensions strongly correlate with satisfaction and customer loyalty. This suggests that providing excellent banking services and advancing technology are necessary to win over satisfied and committed customers. It indicates that the bank should compel the establishment of personalized, attentive, and caring conditions for its users. It is also imperative for banks to devise tactics that increase client loyalty.

The aforementioned research findings are statistically significant and meet all threshold limits, including AVE, CR, discriminant validity, R^2 value, factor loadings, Cronbach's α, and variance inflation factor. As a result, the findings thoroughly understand the implications of extended TAM dimensions and present a comprehensive image of how changed TAM dimensions might influence customer satisfaction and loyalty.

This chapter gives academia and area experts' valuable insights to become acquainted with sustainable AI applications. For beginners, greater use of AI technology in all its forms (robotic process automation and chatbots) eases the achievement of sophisticated tasks, thus reducing human manual capability. For area experts related to various organizations, AI would be beneficial in avoiding technical glitches and simultaneously stimulating favorable opportunities for their future growth.

13.8 LIMITATIONS AND FUTURE SCOPE

As this chapter was restricted to two cities, additional research on other Indian cities may be undertaken, and by examining consumers' perspectives on AI, the results may be compared. Only the banking industry was the focus of this investigation. Future considerations will also extend to other financial sectors. This chapter uses a survey to collect self-reported data, which is a methodological necessity. These kinds of data could be ambiguous. It is advised that future researchers use different techniques for gathering data, like field experiments, and enlarge the sample size to improve accuracy. The survey sample size was limited to 300 banking respondents; in future research, even larger sample collections could be performed. The impact of AI technology stimuli is expected to be more widespread in this era of rapid technological advancement, urging more study. Future studies must investigate new aspects of AI technology stimuli and consider more extensive and varied case samples in addition to larger data sets. In this chapter, the EFA is studied, and a full chartered financial analyst would be suggested for model fit to continue.

13.9 CONCLUSION

The banking sector plays an essential role in the modern economic world, enhancing conventional banking services by making them more customer-centric, as they are the need of the hour. Thus, banks are exploring advanced technologies such as AI to provide convenient and user-friendly services.

It required time and effort to transform the banking industry completely. The banking sector underwent a dramatic transition from traditional banking (established in 1472) to AI-based banking (which began to emerge in 2017); this transition will become more noticeable in key banking domains such as core banking, efficiency in operations, and customer support. This research attempts to understand sustainable finance by combining AI with banking customers' intentions to remain loyal. As a result, the exploratory approach was used in the present inquiry, which used a design based on quantitative methods to examine the relationship among the independent variables—RL, ASS, CUS, EM, RE, PU, PEU, PT, PR, and PB—and the dependent variables—readiness for AI in the banking industry. The outcome demonstrates a positive correlation between CS and CL to accept AI in the banking sector for all predictors. Furthermore, the analysis showed that the acceptance of AI in the banking industry is not affected by the age or gender of the clientele. The results provide valuable insights for bank management as they develop future strategies, like algorithms powered by AI that collect and assess customer data, make relevant product recommendations that have already been approved, and offer customized financial advice. The study's findings can assist banking management in improving and modifying their marketing strategies to win over customers' trust and help them reduce the risk of transacting through digital technology. Consequently, more investigation is needed to evaluate the differences in customer loyalty intentions between the banking industry's early and late adopters of AI.

In this way, banks, practitioners, and environmentalists may find it easier to formulate their plans and future directions for development and sustainability with the present chapter's suggested conceptual framework.

KEYWORDS

- **artificial intelligence**
- **sustainability**
- **technology acceptance model**
- **service quality dimensions**
- **customer loyalty**

REFERENCES

Arndt, J. (1967). Role of product-related conversations in the diffusion of a new product. *Journal of Marketing Research, 4*, 291–295.

Bagozzi, R. P. (1990). The level of effort required for behavior as a moderator of the attitude-behaviour relation. *European Journal of Social Psychology, 20*(1), 45–59.

Bansal, H. S., et al. (2003). A three-component model of customer to service providers. *Journal of the Academy of Marketing Science, 32*(3), 234–250.

Bloemer, J., et al. (1998). Investigating drivers of bank loyalty: The complex relationship between image, service quality and satisfaction. *Journal of Bank Marketing, 16*(7), 276–286.

Boulding, W., et al. (1993). A dynamic process model of service quality: From expectations to behavioral intentions. *Journal of Marketing Research, 30*(1), 7–27.

Davis, F. D. (1989). Perceived usefulness, perceived ease of use, and user acceptance of information technology. *MIS Quarterly*, *13*(3), 319–339.

Douglas, K. M., et al. (2017). The psychology of conspiracy theories. *Current Directions in Psychological Science*, *26*(6), 538–542.

Fornell, C. G., Larcker, D. F. (1981). Evaluating structural equation models with unobservable variables and measurement error. *Journal of Marketing Research, 18*(1), 39–50.

Gillham, J. et al. (2020). *How AI can Enable a Sustainable Future*. PWC: UK. https://www.researchgate.net/publication/340386931_How_AI_can_enable_a_Sustainable_Future

Gremler, D. D.., et al. (2001). Generating positive word-of-mouth communication through customer-employee relationships. *International Journal of Service Industry Management*, *12*(1), 44–59.

Hair, J. F., Black, W. C., Babin, B. J., Anderson, R. E. (2010). *Multivariate Data Analysis*, 7th edn., Prentice Hall: Upper Saddle River, NJ, USA.

Hair, J. F., Ringle, C. M., Sarstedt, M. (2013). Partial least squares structural equation modeling: Rigorous applications, better results and higher acceptance. *Long Range Planning*, *46*(1–2), 1–12.

Hoehle, H. et al. (2012). The role of continuous trust in information systems continuance. *Journal of Computer Information Systems*, *52*(4).

Kaplan, A. M., Haenlein, M. (2007). A model to determine customer lifetime value in a retail banking context. *European Management Journal*, *25*(3), 221–234.

Katyal, S. (2019). Private accountability in the age of artificial intelligence. *66 UCLA Law Rev*, *54*, 125.

Keskinbora, K. H. et al. (2018). Ethical considerations on novel neuronal interfaces. *Neuroscience*, *39*, 607–613.

Khalid, J. et al. (2015). Factors affecting customer loyalty in banking sector: A study on banks in Bahawalpur. *International Journal of Accounting and Financial Reporting*, *5*(1), 239.

Kotler, P., et al. (2002). *Service Quality Influences on Customer Satisfaction* (p. 831). Institute of Research and Journals.

Kotler, P. Keller, K. L. (2009). *Marketing Management*, 14th edn., Pearson International Edition.

Lovelock, C. H., Wirtz, J. (2004). *Services Marketing* (p. 652). Pearson/Prentice Hall, 2004-Business & Economics.

Margherio, L. (1998). *The Emerging Digital Economy*. U.S. Department of Commerce.

McCarthy, J. (2007). From here to human-level AI. *Artificial Intelligence*, *171*(18), 1174–1182

Moorman, C., et al. (1992). Relationships between providers and users of market research: The dynamics of trust within and between organizations. *Journal of Marketing Research*, *29*(3), 314–328.

Newell, A., et al. (1958). Elements of a theory of human problem solving. *Psychological Review*, *65*, 157–166.

Oliver, R. L. (1997). *Satisfaction: A Behavioral Perspective on the Consumer*. The McGraw-Hill Companies, Inc.: New York.

Oliver, R. L. (1999). Whence consumer loyalty. *Journal of Marketing*, *63*, 33–34.

Parasuraman, A. et al. (1988). A multiple-Item Scale for measuring consumer perceptions of service quality. *Journal of Retailing*.

Parasuraman, A. et al. (1991). Refinement and reassessment of the SERVQUAL instrument. *Journal of Retailing, 67*(4), 420–450.

Parasuraman, A. et al. (1996). The behavioral consequences of service quality. *Journal of Marketing, 60*(2), 31–46.

Peha, J. M.,Khamitov, I. M. (2004). PayCash: A secure efficient internet payment system. *Electronic Commerce Research and Applications, 3*, 381–388.

Rana, D., Clement, J., Williams. (2017). Re-examining the unified theory of acceptance and use of technology. *Information Systems Frontiers, 21*, 719–734.

Ranaweera, C., et al. (2003). The influence of satisfaction, trust and switching barriers on customer retention in a continuous purchasing setting. *International Journal of Service Industry Management, 14*(4), 374–395.

Russell, S. J., Norvig, P. (1995). *Artificial Intelligence: A Modern Approach.* Prentice Hall.

Sarma, A. (2017). Digital banking in India: A review of trends, opportunities and challenges. *International Research Journal of Management Science and Tech., 8*(1).

Sheng, T., Liu, C. (2010). An empirical study on the effect of e-service quality on online customer satisfaction and loyalty. *Nankai Business Review International, 1*(3), 273–283.

Suleman et al. (2012). Internet banking service quality, e-customer satisfaction and loyalty: The modified e-SERVQUAL model. *TQM Journal.*

Turing, A. M. (1950). Computing machinery and intelligence. *Mind, New Series, 59*, 433–460.

Upton, J. (2013). *Encyclopedia of Behavioral Medicine* (pp.1580–1581).

Van Looy, B., et al. (1998). Dealing with productivity and quality indictors in a service environment: Some field experiences. *Journal of Service Industry Management, 9*(4), 359–376.

Weijters, B., et al. (2005). Customers' usage of self-service technology in retail setting. *Vlerick Leuven Gent working Paper Series 2005/19.*

Williams, A. et al. (May 2018). Sustainable AI in finance: Understanding the promises & perils. *Parker Fitzgerald,* https://parker-fitzgerald.com/?news=sustainable-ai-in-finance-understanding-the-promises-and-perils*

Wilson, H. J., Daugherty, P. R. (2018). *Collaborative Intelligence: Humans and AI are Joining Forces* (pp. 114–123). Harvard Business Review.

Yang, L., Zhao, Y., Wang, Y., Liu, L., Zhang, X., Li, B., Cui, R. (2015). *The Effects of Psychological Stress on Depression.* PMC, https://doi.org/10.2174/1570159x1304150831150507

Yu, H., Shen, Z., Miao, C., Leung, C., Lesser, V. R., Yang, Q. (2018). Building ethics into artificial intelligence. In: *Proceedings of the 27th International Joint Conference on Artificial Intelligence* (pp. 5527–5533).

Zeithaml, V. A., Parasuraman, A., Malhotra, A. (2002). Service quality delivery through web sites: A critical review of extant knowledge. *Journal of the Academy of Marketing Science, 30*, 362–375. https://doi.org/10.1177/009207002236911

Zeithmal, V. A., et al. (2009). *Service Marketing: Integrating Customer Focus across the Firm (5th ed.). Singapore*. McGraw-Hill and Irwin.

CHAPTER 14

Reshaping Global Financial Services Ecosystem: An Insight Into Drivers of Open Banking

VIJAY KUMAR JAIN,[1*] ADITI SINGHAL,[1] PREETI SHARMA,[1] BIPIN CHANDRA KANDPAL,[2] and SHIPRA AGARWAL[3]

[1]*Department of Management Studies, DIT University, Dehradun, Uttarakhand, India*

[2]*Uttaranchal Institute of Management, Uttaranchal University, Dehradun, Uttarakhand, India*

[3]*Department of Commerce, Graphic Era Deemed to Be University, Dehradun, Uttarakhand, India*

*Corresponding author

ABSTRACT

Banks are required to implement various innovative measures to cope with the budding needs of the customers and enhanced market competition. Open banking is the transformation of a traditional closed banking system to an open banking system and offers an excellent opportunity to meet the customer's requirement. Therefore, the objective of the chapter is to understand the drivers of open banking and prioritize them using an interpretive structural modeling approach. A total of 10 drivers of open banking were finalized using literature review and expert consultation. The result of the study reveals that regulation, customer push toward digital banking, and FinTech are the significant and key drivers of open banking. The findings of the study will help the banking industry to develop a new

Industry 5.0: Innovation and Adoption for Sustainability.
Tilottama Singh, Pradeep Suri, and Zulkarin Jahangir (Eds.)

DOI: 10.1201/9781779641878-14

financial system, improve its services, and transform customer experience completely.

14.1 INTRODUCTION

In the era of big data, banks are required to implement comprehensive digital transformations to cope with the budding needs of customers and to provide an enhanced experience of digital services to customers (Ramdani et al., 2023). One such comprehensive digital transformation is Open Banking. Open banking, a subcategory of open innovation (Stefanelli and Manta, 2022) is an emerging trend in the banking industry. From financial perspective, open banking denotes (1) customers have greater control over the banking data (Leong, 2020), (2) banks are required to share customer-centric data with third-party providers after obtaining consent of customers (Leong, 2020), and (3) banking data is shared through application programming interface (API) to develop various innovative applications (Preziuso et al., 2023).

From a market perspective, key open banking implications are (1) relatively low cost of entry for startups, (2) rethinking of business strategies to remain competitive, (3) promoting access to green finance, and (4) promoting financial inclusivity (Ibe-enwo et al., 2019; Mutambik, 2023). Open banking involves sharing of data through APIs with third-party providers. APIs are software that determine how one application interacts with another to facilitate an information exchange.

Open banking is an innovative framework that enables its participants (banking and nonbanking) to deliver their customers a wider array of financial products and services (Daiy et al., 2021). The services of open banking are tailored according to the expectations and needs of the customers (O'Leary et al., 2021; Mutambik et al., 2023) and consolidate the services under a single head, thereby providing infrastructure and collaborating with entities both within and outside the industry. Open banking stimulates innovation among the financial sector (Butola et al., 2022; Rawat et al., 2019).

Open banking has become a disruptive force in the financial sector, completely changing the way that both individuals and companies handle their money (Gonçalves and de Araujo, 2022; Vasić et al., 2023).

Its ability to enhance the client experience is one of its main advantages (Cosma et al., 2023). Those who avail the opportunity prior to others are rewarded in the era of open banking. This is one of the preliminary studies to explore the drivers of open banking. The key objective of the study is:

- to find and understand the drivers of open banking;
- to prioritize these drivers using an interpretive structural modeling (ISM) approach

The chapter is designed as follows. This chapter was planned in seven sections. Section 14.1 detailed the introduction of open banking. Section 14.2 talks about the literature review related to drivers of open banking. Sections 14.3–14.5 enlightened about the solution methodology used in the study. The formation of MICMAC analysis and ISM model is given in Section 14.6. Finally, discussion, conclusion, limitations, and implications are presented in Section 14.7.

14.2 LITERATURE REVIEW

14.2.1 MARKET COMPETITION

The financial industry has seen a surge in market competitiveness due to open banking, which has also encouraged innovation and given individuals more control over their financial data. In order to stay competitive, traditional banks are being forced to improve their services and client experiences (Zachariadis and Ozcan, 2022). This will ultimately benefit customers by resulting in better services, reduced costs, and more convenience. The market competition is one. Market competitiveness is one of the driving elements behind open banking (Chandrakumar et al., 2022). The ongoing change in regulation, the integration of the financial and technical sectors, and the emergence of new entities such as FinTech, payment apps, and neo-banks have increased competition for banks (Benešová and Husek, 2019). These nontraditional businesses target the same clientele as traditional banks and financial institutions but provide a broader range of services, increasing rivalry with banks. Banks must constantly innovate, develop new client channels, and cope with a shifting regulatory landscape (Odorovićsuni, 2023).

14.2.2 CUSTOMER EXPECTATIONS

The expansion of social and mobile technologies makes the customers more informed and knowledgeable, leading to the rapid change in customer expectations (Kasturi, 2023). Millennials across generations deserve a simple, faultless, and spontaneous experience with digital financial services (Mensah and Murorua, 2017; Iman et al., 2023). Banks are constantly under great pressure to understand, adjust, and adapt to this attitude of the customers to deliver experienced banking products, create loyalty among customers (Frow et al., 2011), building customer trust, and maintain market share (Rana and Sharma, 2019). To achieve the same, banks are shifting from closed models to open models to provide customers with better banking experiences. The introduction of open banking has created an ecosystem that has widened the boundaries putting the customers firmly in the driving seat and facilitating banks to offer their customers better and easier tailored services.

14.2.3 REGULATION

Regulations act as a facilitator for the rapid adoption of open banking revolution (Agarwal et al., 2017). Different regulatory approaches have been adopted toward open banking. These approaches can be categorized as mandatory, supported, and neutral approaches (Bodley and Brice, 2022). In a mandatory approach, regulators pass the legislation compelling the banks to adopt the practices of open banking. In a supported approach, also known as organic approach, adoption of open banking is encouraged without mandating the practice of open banking. In a neutral approach, adoption of the practice of open banking is led by the industry and not by regulatory legislation. The concept of open banking evolved out of regulatory requirements to provide customers with more choices, transparency, and to encourage innovation in the banking sector (Tsai and Peng, 2022). The regulation encourages banks to open their architecture for open banking, thereby allowing third-party providers to access customer details encouraging competition in the financial sector (Leong, 2020).

14.2.4 SECURITY

Security is an essential constituent toward the formation of a trustworthy open banking framework. Security of APIs has emerged as a priority for banks when 75% of hackers are targeting bank APIs in 2019 as per report of Akamai (Salem-Jackson and Aziz, 2023). Open banking enables banks to effectively monitor banking systems making it safer, secure against any threat, and making it valuable to all parties (Langford et al., 2024). Security builds up consumer confidence and encourages them to move toward open banking (Mallick et al., 2021). Open banking offers customers control over their data, allowing them to understand how their data are being used. The banks can ensure the security of API through the authentication of technical service providers and customers (Tsanakas, 2023). Thus, security is a key selling point for open banking.

14.2.5 SHARING OF DATA

Sharing of data leads to healthy competition among the players of the financial market, that is, challenger banks, neo-banks (Jejeniwa et al., 2024), FinTech, and big techs (Preziuso et al., 2023). Sharing data improves efficiency, leads to transparency in data as well as transparency in pricing and better decision making among the customers (Hjelkrem et al., 2022). Banks are duty-bound to protect the secrecy of the customer's information except when the disclosure is mandated by law, customers have given their consent either express or implied and for accomplishment of a public duty (Babina et al., 2024). The exceptions to the Banking Act are not well suited with the model of open banking. In the open banking paradigm, banks can share customer data with third-party provider on the direction of the customer. Open banking allows financial customers to effectively, securely, and easily share financial data with financial participants, that is, banks, FinTech, or technological giants such as Facebook (Hjelkrem et al., 2022). The disclosure of customer information is permissive in nature, nor directive, and banks are not obliged to entertain such permission (Odorović, 2023).

14.2.6 DATA MONOPOLY

Data serves as the backbone of any business operation (Liu et al., 2023). Given its crucial role in the global economy, companies such as Amazon, Google, and Facebook aim to command data control. This dominance over data is often referred to as data monopoly (Vezzoso, 2021). The initial premise of open banking models was the inability to access consumer data because traditional banks had monopolized client transactions (Plaitakis and Staschen, 2020). Open banking has the potential to enact substantial change by dismantling barriers between customers and their financial data within the sector (Ismail et al., 2021). Thus, offering customers a wider array of products and services at lower cost that suits their needs, lowers the entry barriers for new competitors, and enhances competition. Therefore, open banking can be a major instrument to end data monopoly (Briones de Araluze and Cassinello Plaza, 2023).

14.2.7 FINTECH

Traditional banks and financial institutions are repeatedly facing competition from new entrants like FinTech, payment apps, and neo-banks (Surekha et al., 2022). Banks and financial institutions are aiming toward open banking to counter rising competition. Banks can offer their customers more choice, better products, better banking experience, competitive differentiation, and faster growth by teaming up with FinTech firms. This is only possible when banks adopt an open banking approach. Leveraging digital advancements and connections, FinTech, or financial technology, has dramatically accelerated the proliferation of financial services (Preziuso et al., 2023). FinTech has significantly improved customer accessibility, convenience, and efficiency by smoothly integrating technology into conventional banking procedures (Gogia and Chakraborty, 2022). FinTech plays a major role in promoting pen banking by providing user-friendly online and mobile banking systems that let customers handle their finances and perform a variety of transactions with a few clicks on their computers or smartphones (Chan et al., 2022).

14.2.8 CUSTOMER PUSH TOWARD DIGITAL BANKING

The push toward digital banking has been driven by a growing customer preference for convenience, accessibility, and efficiency in managing their finances (Fratini Passi, 2022). With the widespread adoption of smartphones and the internet, customers are increasingly turning to digital channels for their banking needs, seeking the flexibility to conduct transactions and access services at any time and from anywhere (Tsai and Peng, 2022). The cost-effectiveness of digital banking, often associated with lower fees and charges compared to traditional banking, has appealed to customers seeking to save money (Rivero and Vives, 2022). An increase in customer demand for digital banking acts as a driver of open banking. Customers expect the same from banks (Karkkainen, 2023). This need for customers can be fulfilled through open banking. Through open banking, customers can access all financial services at one platform and personalize them for the customer.

14.2.9 TRANSPARENCY OF DATA

Every disruptive technology comes with a unique set of challenges. The challenges associated with technology drive the regulatory and other competent authorities to build transparency (Preziuso et al., 2023). Transparency of data improves the customer relationship (Kaul, 2019). Open banking facilitates banks to create better financial transparency (Parate et al., 2023). The process of making data available, intelligible, and useful to people or organizations is known as data transparency. In the financial context, data transparency refers to giving customers clear information about financial services, goods, fees, and terms and conditions (Nag and Manohar, 2024). Customers in traditional banking frequently have little control over how financial firms use or disclose their data.

14.2.10 TECHNOLOGY

The industry is facing rapid change due to technological evolution like the Internet of Things, big data analysis, robotics, artificial intelligence, Blockchain, machine learning, and cloud computing (Iman et al., 2023). Acceptance of technologies and digitization by customers act as a catalyst for the banking sector to adopt new technologies within banks for the purpose of payment (Mensah and Murorua, 2017), integration, and consumer delight (Patki and Sople, 2020). It is unrealistic for a bank to invest in any new technology due to factors such as availability of poor infrastructure (Özyeşil, 2022), security issues related to technology (Hefny et al., 2023), consumer ignorance toward technology (Tandon et al., 2014), and advertisement cost. Therefore, it is necessary for banks to collaborate with specialized institutions such as FinTech to remain ahead of other competitors. The flexibility in the adoption of emerging technologies will allow banks to survive in the marketplace. Thus, indirectly, they are encouraging open banking.

TABLE 14.1 Drivers of Open Banking

Variables	Description	References
Market competition	Consumer data will aid in eliminating the entry barrier for new players in the market as well as provide opportunities for consumers to choose among the services offered by various firms which best suit their needs	Zachariadis and Ozcan (2022), Chandrakumar et al. (2022), Benešová and Hušek (2019), Odorovićsuni (2023)
Customer expectations	The concept of open banking evolved out of regulatory requirements to provide customers with more choices, better products, transparency and encourage innovation	Kasturi (2023), Mensah and Murorua (2017), Iman et al. (2023), Frow et al. (2011), Rana and Sharma (2019)
Regulation	Consumer expectations change rapidly. Consumers expect more choices, better products, enhanced digital banking service and high value banking brands. Thus, encouraging open banking adoption	Agarwal et al. (2017), Tsai and Peng (2022), Bodley and Brice (2022), Leong (2020)
Security	Security is a key selling point of Open Banking. Open Banking enables banks to effectively monitor banking systems against any threat	Salem-Jackson and Aziz (2023), Langford et al. (2024), Mallick et al. (2021), Tsanakas (2023)
Sharing of Data	Sharing of data leads to competition among players in the financial sector. Open banking allows secure share of consumer data with third parties like FinTech and technological companies	Jejeniwa et al. (2024), Preziuso et al. (2023), Hjelkrem et al. (2022), Babina et al. (2024), Hjelkrem et al. (2022), Odorović (2023)
Data Monopoly	Data serves as the backbone of any business operation. Given its crucial role in the global economy, companies such as Amazon, Google, and Facebook aim to command data control	Liu et al. (2023), Vezzoso (2021), Plaitakis and Staschen (2020), Ismail et al. (2021), Briones de Araluze and Cassinello Plaza (2023)
FinTech	Banks can offer the customers more choice, better products, better banking experience, competitive differentiation, and faster growth by teaming up with FinTech firms. This is only possible when banks adopt an open banking approach	Surekha et al. (2022), Preziuso et al. (2023), Gogia and Chakraborty (2022), Chan et al. (2022)
Customer Push Towards using Digital Banking	Increase in customer demand for a digital banking act as a driver of open banking	Fratini Passi (2022), Tsai and Peng (2022), Rivero and Vives (2022), Karkkainen (2023)
Transparency of data	The need for building transparency for protecting customers' interests and rights	Preziuso et al. (2023), Parate et al. (2023), Kaul (2019), Nag and Manohar (2024)
Technology	Continuous evolving technologies drive customer expectations. The flexibility in adoption of emerging technologies will allow banks to survive in the marketplace. Thus, indirectly encouraging open banking	Iman et al. (2023), Mensah and Murorua (2017), Patki and Sople (2020), Özyeşil (2022), Hefny et al. (2023)

14.3 METHODOLOGY

ISM is a framework designed for synergistic learning. ISM is a powerful technique for transforming erratic behavior into an optimistic and well-designed configuration. It is commonly utilized for analyzing complex social and economic frameworks. Therefore, ISM provides a systematic and comprehensive framework to its customers to consolidate group decisions in the creation of systematic "first-cut" models. ISM provides an elementary understanding of complex situations and structures—a strategy to resolve the matter. The model is far more flexible than most of the other conventional quantitative modeling approaches, which require measurement of factors on a ratio scale.

In the study, ISM methodology is employed to analyze and demonstrate the correlation among different drivers of open banking. The drivers of open banking are recognized on the ground of the "literature review" and "opinion of experts." The works review and opinion by experts are to establish contextual relationships among drivers in the structural self-similarity matrix (SSIM), reachability, and partition matrix, and the same matrix is further used to construct the ISM model (Gupta and Jain, 2020) (Figure 14.1).

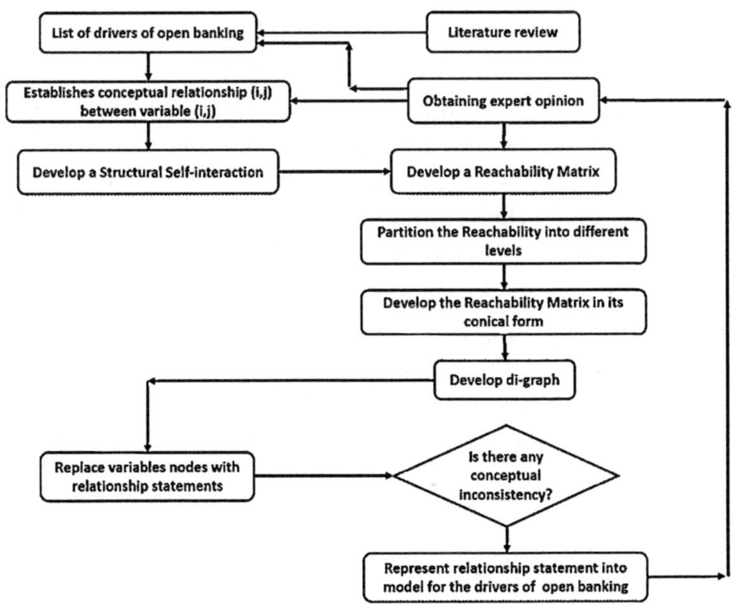

FIGURE 14.1 Flowchart of ISM methodology.

Source: Authors' own.

14.4 QUESTIONNAIRE DEVELOPMENT

Ten factors were identified as the drivers of open banking. Out of 10 drivers, seven drivers were recognized through literature, and three drivers were recognized after consulting the experts. These experts were professors, academicians, and managers of banks having in-depth knowledge of the banking industry.

14.5.1 DATA COLLECTION

The ISM methodology uses expert opinion by means of numerous management techniques such as brainstorming nominal technique decision tree, Delphi technique, didactic interaction, and so on to establish the contextual relationship between factors. Ten experts were consulted, out of which five are from the banking industry and five are esteemed professors and academicians to pinpoint the contextual relationship between the drivers of open banking. A "leads to" relationship was selected to develop contextual relationships among drivers of open banking.

14.5.2 SSIM

To establish the contextual relationship among drivers of open banking, expert's opinions were taken from the industry. On the basis of expert's opinion, SSIM establishes the contextual relationship between the factors. Based on contextual relationship, the four signifiers, that is, V, A, X, and O indicate the sort of association between the two variables. Assumption of structural self-interaction matrix are as follows:

V: Symbolize variable i leads to variable j
A: Symbolize variable j leads to variable i
X: Symbolize both variables i and j lead to each other
O: Symbolize variables i and j are unconnected to each other.

The subsequent statements will elucidate the use of signifiers in the SSIM matrix:

- Driver market competition (i) will lead to the driver technology (j) in the cell (1,10), so the relation between the two drivers will be symbolized by symbol "V" in the matrix.

TABLE 14.2 SSIM Matrix

	10	9	8	7	6	5	4	3	2	1
1	V	A	A	A	O	A	V	X	X	1
2	X	X	X	A	O	A	A	A	1	
3	V	V	A	X	V	V	V	1		
4	A	O	A	X	O	X	1			
5	X	X	O	A	O	1				
6	V	O	O	A	1					
7	X	X	X	1						
8	X	O	1							
9	V	1								
10	1									

- Driver technology (*j*) will lead to driver security (*i*) in the cell (4,10), so the relation between the two drivers will be symbolized by the symbol "A" in the matrix.
- Driver sharing of data (*i*) and transparency of data (*j*) in the cell (5,9) will lead to each other, so the relation between the two drivers will be symbolized by symbol "X" in the matrix.
- No relationship subsists among the driver's data monopoly (*i*) and customer push toward using digital banking (*j*) in the cell (5,8), so the relation between the two drivers will be symbolized by the signifier "O" in the matrix.

14.5.3 REACHABILITY MATRIX

The SSIM (Table 14.1) is transformed into a binary matrix. After such conversion, the matrix so formed is acknowledged as a reachability matrix (Table 14.2). The reachability matrix shows the direct correlation between the two factors. In Table 14.2, the four signifiers V, A, X, O are swapped with 0 and 1. The reachability matrix for drivers of open banking is displayed in Table 14.2. The conventions adopted regarding the formation of the reachability matrix are as follows:

- In case of V: (*i*, *j*) was switched to 1 and (*j*, *i*) was switched to 0
- In case of A: (*i*, *j*) was switched to 0 and (*j*, *i*) was switched to 1
- In case of X: (*i*, *j*) was switched to 1 and (*j*, *i*) was switched to 1

- In case of O: (*i*, *j*) was switched to 0 and (*j*, *i*) was switched to 0

Table 14.3 shows reachability matrix developed for the drivers of open banking to show the direct relationship between the two variables.

TABLE 14.3 Reachability Matrix

	1	2	3	4	5	6	7	8	9	10
1	1	1	1	1	0	0	0	0	0	1
2	1	1	0	0	0	0	0	1	1	1
3	1	1	1	1	1	1	1	0	1	1
4	0	1	0	1	1	0	1	0	0	0
5	1	1	0	1	1	0	0	0	1	1
6	0	0	0	0	0	1	0	0	0	1
7	1	1	1	1	1	1	1	1	1	1
8	1	1	1	1	0	0	1	1	0	1
9	1	1	0	0	1	0	1	0	1	1
10	0	1	0	1	1	0	1	1	0	1

The following statements will explain the transformation of SSIM matrix into reachability matrix:

- Driver market competition (*i*) will lead to driver technology (*j*) in the cell (1,10). The relationship between the two drivers will be represented by the signifier "V" in the SSIM matrix. The cell (1, 10) depicted by "1," whereas cell (10, 1) depicts by "0" in its reachability matrix.
- Driver technology (*j*) will lead to driver security (*i*) in the cell (4,10). The relationship between the two drivers will be represented by signifier "A" in the SSIM matrix. The cell (4, 10) depicted by "0," whereas cell (10, 4) depicts by "1" in its reachability matrix.
- Driver sharing of data (*i*) and transparency of data (*j*) in the cell (5,9) will lead to each other. The relationship between the two drivers will be represented by the signifier "X" in the SSIM matrix. The cell (5,9) depicted by "1," whereas cell (9,5) depicts by "1" in its reachability matrix.
- No relationship exists between the driver's data monopoly (*i*) and customer push toward using digital banking (*j*) in the cell (5,8).

The relationship between the two drivers will be represented by the signifier "O" in the matrix. The cell (5,8) depicted by "0," whereas cell (8,5) depicts by "1" in its reachability matrix.

14.5.4 LEVEL PARTITIONS

The reachability set of every single variable comprises itself and other variables; likewise, an antecedent set of every single variable comprises itself and other variables. The intersection set is derived from reachability and antecedent set. Those with identical reachability and intersection sets were given level I and acquired the highest level in the interpretive structure modeling ladder. The variables conquering the highest level were cast off from the residual variables. This process is known as iteration. The iteration will last until all variables have obtained the levels in the ISM hierarchy.

TABLE 14.4 Final Iteration (Level Partition)

Variables	Reachability Set	Antecedent Set	Intersection Set	Levels
1	1,2,3,4,10	1,2,3,5,7,8,9	1,2,3	IV
2	1,2,8,9,10	1,2,3,4,5,7,8,9,10	1,2,8,9,10	I
3	1,2,3,4,5,6,7,9,10	1,3,7,8	1,3,7	VI
4	2,4,5,7	1,3,4,5,7,8, 10	4,5,7	II
5	1,2,4,5,9,10	3,4,5,7,9,10	4,5,9,10	V
6	6,10	3,6,7	6	IV
7	1,2,3,4,5,6,7,8,9,10	3,4,7,8,9,10	3,4,7,8,9,10	VII
8	1,2,3,4,7,8,10	2,7,8,10	2,7,8,10	VII
9	1,2,5,7,9,10	2,3,5,7,9	2,5,7,9	V
10	2,4,5,7,8,10	1,2,3,5,6,7,8,9,10	2,5,7,8,10	III

In the study (Table 14.4), 10 drivers of open banking, reachability set, antecedent set, intersection set, and levels were displayed. The level partition of 10 drivers is accomplished in seven iterations. Table 14.4 indicates that customer expectations are obtained the first level. Hence, it is located at the topmost level in the ISM model (Figure 14.2). This iteration will continue until all 10 variables obtain the levels. The obtained level contributes to the formation of the ISM model.

14.5.5 FORMATION OF ISM MODEL

The ISM model was generated from the reachability matrix. If the correlation subsists among the variables "*j*" and "*i*," it will be indicated by directing an arrow from "*i*" to "*j*." The model so formed is known as a digraph. In the study, FinTech, customer push toward digital banking, and regulation are the most significant variables that encourage banks to adopt open banking. The three variables FinTech, customer push toward digital banking, and regulation obtained the bottom level in the ISM model, whereas customer expectation has obtained the highest level in the ISM model.

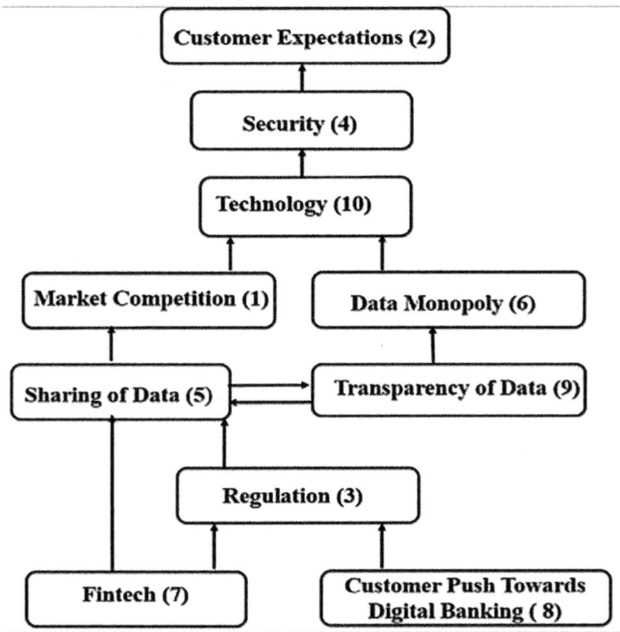

FIGURE 14.2 ISM-based model of drivers.

14.6 DISCUSSION AND CONCLUSION

The conclusion obtained from the ISM model shows a correlation among the drivers of open banking. In the ISM model, the level partition of drivers of open banking is carried out so that policymakers can be aware of various drivers of open banking. Knowledge of drivers of open banking will be helpful for policymakers to formulate policies for open banking. In

TABLE 14.5 MICMAC Analysis

10						7				
9				3						
8		IV—Independent (driver)								
7				8				III—Linkage		
6					9	5			10	
5							1		2	
4		I—Autonomous					4			
3								II—Dependent		
2			6							
1										
	1	2	3	4	5	6	7	8	9	10

the study, 10 drivers of open banking are identified. With the aid of identified drivers, the ISM model was formulated, and the interaction among the drivers is identified through MICMAC analysis (Table 14.5).

The ISM model revealed (Figure 14.2) that regulation (3), customer push toward digital banking (8), and FinTech (7) are the significant and the key drivers of open banking. These drivers have strong driving power. So should be treated as key motivators to encourage open banking. These three drivers influence the sharing of data (5) and transparency of data (9), which in turn influence the driver market competition (1). The drivers' market competition (1) and data monopoly (6) affect the technology (10), security (4), and customer expectation (2), which are positioned at the highest level in the ISM model (Figure 14.2). The drivers market competition (1), security (4), data monopoly (6), and customer expectation (2) have weak driving power but strong dependence power.

Banks and financial institutions are aiming toward open banking to counter rising competition. Banks can offer their customers more choice, better products, and better banking experience, competitive differentiation, and faster growth by teaming up with FinTech companies. Collaboration between FinTech and banks is possible when banks adopt an open banking approach.

The model of open banking evolved out of the regulatory requirement to provide customers with more choices, transparency, and to encourage innovation in the banking sector. The regulation encourages banks to open for open banking, boosting banks to allow third-party providers to access customer details, and encourages competition in the financial sector. An increase in customer demand for digital banking acts as a driver of open banking. Customers expect the same from banks. This need for customers can be fulfilled through open banking. Through open banking, customers can access all financial services at one platform and personalize them for the customer.

The changes in regulation, customers push toward digital banking, and FinTech will generate the ripples in the ocean that will be soon exaggerated and reach the coast of mainstream banking. It cannot be predicted when the wave will exactly break through, but it can be sooner than banks have imagined. Thus, the banks not progressing toward an open banking framework may miss this new wave. So, in a nutshell, it can be said that banks need to open to open banking to remain competitive, significant, and a winner in today's digital economy.

14.7 MANAGERIAL IMPLICATIONS

Open banking is an innovation that is transforming the banking industry and is expected to dramatically reshape the banking landscape over the years to come. Banks embracing open banking prior to others will gain competitive advantages by driving new revenue openings, adding value for their customers, and detecting new techniques of doing business. This study has a significant implication and is of great value to the bank managers, decision, policy makers, and researchers. By analyzing the interrelationship among the drivers of open banking, management of banks can chalk out the strategies to implement open banking in the banks. Management will become aware of the driver which have the maximum influence and then can act accordingly. Thus, assisting the management in the implementation of open banking in a fast and organized manner.

Open banking will help banks to develop new applications, products, and services to provide customers with a better banking experience, formulating unique strategies for approaching customers and maintaining trust and being first to provide physical as well as digital services. Further, an open banking system will integrate banks with the digital ecosystem by innovating and exploring a wide range of digital solutions. Open banking makes it possible for parties to securely share financial information, which promotes innovation, competitiveness, and better customer experiences. Customers will benefit from more convenience and financial management, as well as increased access to a greater range of financial goods and services that are catered to their individual needs. But the increased exchange of private and sensitive financial data also gives rise to privacy and data security concerns, calling for strong legal frameworks and technological protections.

14.8 LIMITATIONS AND DIRECTIONS FOR FUTURE RESEARCH

Regardless of the contribution made, the research is not free from shortcomings. As an open banking regime has been introduced recently, the lack of available literature in the domain has been the main limitation of the study. Second, the model recommended is conceptually based on the available literature and professional's opinion. The professional's opinion changes with time and situation. Moreover, further studies need to test and corroborate these outcomes using other relevant models. Future research may examine the social and ethical ramifications of open banking, including questions of permission, data ownership, algorithmic bias, and financial inclusivity and create frameworks for responsible innovation and governance.

KEYWORDS

- **digital transformation**
- **open banking**
- **FinTech**
- **application programming interface**
- **data privacy**
- **interpretive structural modeling approach**

REFERENCES

Agarwal, S., Alok, S., Ghosh, P., Ghosh, S., Piskorski, T., Seru, A. (2017). Banking the unbanked: What do 255 million new bank accounts reveal about financial access? *Columbia Business School Research Paper*, (17–12).

Babina, T., Fedyk, A., He, A., Hodson, J. (2024). Artificial intelligence, firm growth, and product innovation. *Journal of Financial Economics*, *151*, 103745.

Benešová, D., Hušek, M. (2019). Factors for efficient use of information and communication technologies influencing sustainable position of service enterprises in Slovakia. *Entrepreneurship and Sustainability Issues*, *6*(3), 1182.

Bodley, A., Brice, S. (2022). Fintech and payments. In: *Payment Services* (pp. 5–36). Edward Elgar Publishing.

Briones de Araluze, G., Cassinello Plaza, N. (2023). The relevance of initial trust and social influence in the intention to use open banking-based services: An empirical study. *SAGE Open*, *13*(3), 21582440231187607.

Butola, P., Dube, P., Jain, V. K. (2022). A study on impact of credit risk management on the profitability of Indian banks. *International Journal of Management and Sustainability*, *11*(3), 103–114.

Chan, R., Troshani, I., Rao Hill, S., Hoffmann, A. (2022). Towards an understanding of consumers' FinTech adoption: The case of open banking. *International Journal of Bank Marketing*, *40*(4), 886–917.

Chandrakumar, T., Parthasarathy, S., Sridevi, S., Ramya, T. (2022). Consumer's perception of digital transactions in India: Sentiment analysis using R. *International Journal of Business Information Systems*, *41*(2), 193–206.

Cosma, S., Rimo, G., Torluccio, G. (2023). Knowledge mapping of model risk in banking. *International Review of Financial Analysis*, *89*, 102800.

Daiy, A. K., Shen, K. Y., Huang, J. Y., Lin, T. M. Y. (2021). A hybrid MCDM model for evaluating open banking business partners. *Mathematics*, *9*(6), 587.

Fratini Passi, L. (2022). Open banking and digital transformation in Italy: The current situation and the challenges ahead. *Journal of Payments Strategy and Systems*, *16*(4), 358–368.

Frow, P., Payne, A., Wilkinson, I. F., Young, L. (2011). Customer management and CRM: Addressing the dark side. *Journal of Services Marketing*, *25*(2), 79–89.

Gogia, J., Chakraborty, D. (2022). Open banking: A revolution in the Tech-Fin industry. *International Journal of Electronic Banking*, *3*(2), 100–120.

Gonçalves, A. V., de Araujo, F. O. (2022). The implementation of Open Banking in Brazil: An overview from the perspectives of professionals from large retail banks. *Brazilian Journal of Operations and Production Management*, *19*(4), 1661–1661.

Gupta, P., Jain, V. K. (2020). Interpretive structural modeling of GIoT enablers. *Journal of Information Technology Research (JITR)*, *13*(2), 129–140.

Hefny, M. H. M., Helmy, Y., Abdelsalam, M. (2023). Open banking API framework to improve the online transaction between local banks in Egypt using blockchain technology. *Journal of Advances in Information Technology*, *14*(4), 729–740.

Hjelkrem, L. O., De Lange, P. E., Nesset, E. (2022). The value of open banking data for application credit scoring: Case study of a Norwegian bank. *Journal of Risk and Financial Management*, *15*(12), 597.

Ibe-enwo, G., Igbudu, N., Garanti, Z., Popoola, T. (2019). Assessing the relevance of green banking practice on bank loyalty: The mediating effect of green image and bank trust. *Sustainability*, *11*(17), 4651.

Iman, N., Nugroho, S. S., Junarsin, E., Pelawi, R. Y. (2023). Is technology truly improving the customer experience? Analysing the intention to use open banking in Indonesia. *International Journal of Bank Marketing*, *41*(7), 1521–1549.

Ismail, F. B., Ling, L. S., Kadir, A. A., Al Hosaini, A. A. H. (2021). Employee satisfaction and performance of E-HRM system in Malaysia banking sector. *Annals of the Romanian Society for Cell Biology*, 6721–6729.

Jejeniwa, T. O., Mhlongo, N. Z., Jejeniwa, T. O. (2024). AI solutions for developmental economics: Opportunities and challenges in financial inclusion and poverty alleviation. *International Journal of Advanced Economics*, *6*(4), 108–123.

Karkkainen, T. (2023). *The FinTech Evolution: Technology Push, Market Pull Dynamics, and Inertia*. Tatja Karkkainen.

Kasturi, S. (2023). Evolving consumer expectations and the future of digital banking. *Journal of Digital Banking*, *8*(1), 37–48.

Kaul, A. (2019). Structural capital: Concept and its application in service sector with special reference to information technology and banking sector. *Amity Business Review*, *15*(1) .

Langford, W. S., Thomas, H. W., Feldman, M. P. (2024). Banking for the other half: The factors that explain banking desert formation. *Economic Development Quarterly*, *38*(2), 71–81.

Leong, E. (2020). Open banking: The changing nature of regulating banking data-A case study of Australia and Singapore. *Banking and Finance Law Review*, (35.3), 443–469.

Liu, H., Bai, Y., Huang, Z., Qiao, H., Wang, S. (2023). Private banking development in China under two organizational structures: Economic analysis from an organizational innovation perspective. *Financial Innovation*, *9*(1), 116.

Mallick, A. K., Mishra, A. K., Vyas, I. (2021). Forecasting stock prices of five major commercial banks in India and stress testing: A multivariate approach. *Journal of Public Affairs*, *21*(3), e2239.

Mensah, B., Murorua, M. (2017). Sustaining networks in the new age: Turning Open Banking into a sustainable competitive advantage. *Africa Expansion Project 2017*.

Mutambik, I. (2023). Customer experience in open banking and how it affects loyalty intention: A study from Saudi Arabia. *Sustainability*, *15*(14), 10867.

Mutambik, I., Lee, J., Almuqrin, A., Zhang, J. Z., Homadi, A. (2023). The growth of social commerce: How it is affected by users' privacy concerns. *Journal of Theoretical and Applied Electronic Commerce Research*, *18*(1), 725–743.

Nag, I., Manohar, S. (2024). Blockchain technology in peer-to-peer transactions emphasizing data transparency and security in banking services. In: *Driving Decentralization and Disruption with Digital Technologies* (pp. 21–35). IGI Global.

Odorović, A. (2023). Open banking: Between cooperation and competition. *Анали Правног факултета у Београду*, *71*(1), 65–91.

Odorovićsuni, A. (2023). Open banking: Between cooperation and competition. *Annals of the Faculty of Law in Belgrade*, *71*(1), 65–91.

O'Leary, K., O'Reilly, P., Nagle, T., Filelis-Papadopoulos, C., Dehghani, M. (2021, January). The sustainable value of open banking: Insights from an open data lens. In: *54th Hawaii International Conference on System Sciences,* Kauai, Hawaii, USA, 4–8 January 2021 (pp. 5891–5901). University of Hawai'i at Manoa.

Özyeşil, M. (2022). Open banking concept. *Current Financial Studies*, 19.

Parate, S., Josyula, H. P., Reddi, L. T. (2023). Digital identity verification: Transforming KYC processes in banking through advanced technology and enhanced security measures. *International Research Journal of Modernization in Engineering Technology and Science*, *5*(9), 128–137.

Patki, A., Sople, V. (2020). Indian banking sector: Blockchain implementation, challenges and way forward. *Journal of Banking and Financial Technology*, *4*(1), 65–73.

Plaitakis, A., Staschen, S. (2020). Open banking: How to design for financial inclusion. *Consultative Group to Assist the Poor (CGAP) Working Paper*.

Preziuso, M., Koefer, F., Ehrenhard, M. (2023). Open banking and inclusive finance in the European Union: Perspectives from the Dutch stakeholder ecosystem. *Financial Innovation*, *9*(1), 111.

Ramdani, A., Jufri, A. W., Jamaluddin, J., Yustiqvar, M. (2023, April). Increasing student science literacy: Learning studies using Android-based media during the Covid-19 pandemic. In: *AIP Conference Proceedings* (Vol. 2619, No. 1). AIP Publishing.

Rana, G., Sharma, R. (2019). Assessing impact of employer branding on job engagement: A study of banking sector. *Emerging Economy Studies*, *5*(1), 7–21.

Rawat, P., Jain, V. K., Chamola, P. (2019). Identifying enablers of social entrepreneurship: Interpretive structural modelling. *SCMS Journal of Indian Management*, *16*(1), 117–131.

Rivero, D., Vives, X. (2022). Open banking: Promise and trade-offs 16. *European Economy*, (1), 45–56.

Salem-Jackson, K., Aziz, A. (2023). Why it is important for B2B brands to find their why: A case study of Akamai Technologies. *Journal of Brand Strategy*, *12*(3), 226–232.

Stefanelli, V., Manta, F. (2022). The rise of digital finance: Empirical evidence on fintech firms, banks and customers.

Surekha, N., Sangeetha, R., Aarthy, C., Kavitha, R., Anuradha, R. (2022). Leveraging blockchain technology for internet of things powered banking sector. In: *Blockchain based Internet of Things* (pp. 181–207). Springer Singapore: Singapore.

Tandon, D., Tandon, K., Malhotra, N. (2014). An evaluation of the technical, pure technical and scale efficiencies in the Indian banking industry using data envelope analysis. *Global Business Review*, *15*(3), 545–563.

Tsai, C. H., Peng, K. J. (2022). *Regulating Open Banking: Comparative Analysis of the EU, the UK and Taiwan*. Routledge.

Tsanakas, E. (2023). Open banking: Application difficulties and API security, under PSD2.

Vasić, D., Barać, D., Radenković, M. (2023). Leveraging open banking: Challenges and opportunities. In: *Digital Transformation of the Financial Industry: Approaches and Applications* (pp. 217–232).

Vezzoso, S. (2021). Competition policy in transition: Exploring data portability's roles. *Journal of European Competition Law and Practice*, *12*(5), 357–369.

Zachariadis, M., Ozcan, P. (2022). Open banking: How platforms and the API economy change competition in financial services. In: *Global Fintech* (pp.57–72).

CHAPTER 15

Culinary Intelligence Unleashed: Navigating the Impact and Moderating Role of Artificial Intelligence in the Restaurant Industry

EKTA SODHA[1] and ABHILASHA CHAUHAN[2*]

[1]*Uttaranchal Institute of Management, Uttaranchal University, Dehradun, Uttarakhand, India*

[2]*Jaypee Business School, Jaypee University of Information Technology, Waknaghat – Solan, Himachal Pradesh, India*

*Corresponding author

ABSTRACT

The restaurant industry is a prominent area of technological development, acutely impacted by artificial intelligence (AI). This chapter elaborates on the multifaceted effects of AI, emphasizing its tremendous influence on operations, customers, and the competition structure. Virtual assistants and chatbots streamline customer service and supply chain management in restaurants by reducing queues, limiting food waste, encouraging the use of eco-friendly procurement systems, and improving demand predictions through AI-powered predictive analytics and machine learning. More comprehensive research and discussion are requested in the conclusion of the study in question. Because of the importance of having a full understanding of AI's moderating effect on the restaurant industry, until AI and the food sector may work in harmony, technological factors and prudent use of AI's great potential are presumably essential.

Industry 5.0: Innovation and Adoption for Sustainability.
Tilottama Singh, Pradeep Suri, and Zulkarin Jahangir (Eds.)

DOI: 10.1201/9781779641878-15

15.1 INTRODUCTION

The restaurant industry is at the forefront of that trend. Artificial intelligence (AI) is transforming age-old practices and the customer experience. AI is quickly becoming an "important driving force of the contemporary service system" that has an impact on various sectors, such as the restaurant business. Technological improvements have not only revolutionized the way much of the work inside the kitchen is done but also the method by which people prepare and consume food. The discussion aims to discuss the impact that AI and the corresponding technology have had on the restaurant industry.

The impact of AI has the potential to drastically alter restaurant operations and consumer behavior in the 21st century (Berezina et al., 2019). AI's impact extends beyond fleeting fads and is evident in everything from personalized recommendations to more streamlined operations. Robots with AI capabilities are commonplace. The goal of these robots is to be human-centered innovators with social awareness and emotional intelligence. Restaurants use smart technologies for a range of purposes due to the complexity of their operations (Addanki et al., 2022). Customers and service providers alike gain from new, readily available platforms for service delivery. There is a real possibility of disruptions in the race to capitalize on technology in this competitive environment, especially in the restaurant industry, which has a history of either resisting or embracing change more slowly (Alt and Information Systems Institute, Leipzig University, 2021). The Internet of Everything, Internet connectivity, and big data are driving a revolution in production and consumption. More sophisticated carryout and delivery packaging; more restaurant spaces devoted to these operations; limited-service restaurants that frequently use kiosks; technology used more sparingly to reduce expenses and enhance management efficacy; and increasingly stringent federal regulations protecting the privacy of personal information. Most orders for delivery and takeout will be placed online. When disruptions are severe enough to cause an agent to reevaluate how it will adapt to its changing environment, that is when disruptions occur (Singh et al., 2022). To help understand technological innovations (such as AI technologies), employee engagement, retention, and productivity trends, and their impact on service quality and customer satisfaction under well, this chapter is a theoretical framework. It helps them to develop service quality and customer satisfaction from these levers and to see how these factors relate to growth and profitability (Yoon and Yu, 2022). It seeks to encourage evidence-based investment decisions to facilitate the adoption of new technologies in the restaurant industry, as well as improvements in talent and talent management strategies (Lee et al., 2021). The chapter concludes with a discussion of the contribution and its implications for practice and research. Thanks to AI, people are becoming more productive, increasing business productivity. But it also requires additional skills such as technical skills, social–emotional intelligence, and creative thinking. The most challenging area of technology development is managing and nurturing digital talent such as AI technology. However, when digital analytics and AI tools are used to monitor and manage human talent performance, the result is increased Human Resource Management (HRM) talent attraction, development, engagement, and productivity. This is referred to as AI-enhanced e-HRM. Given the potential impact on customer satisfaction and service quality, research in the hospitality and tourism industry is interested in understanding more about translating this understanding into the restaurant industry. This framework provides managers with the analytical tools needed to examine and understand the relationship between productivity, employee retention, technology development, service expectations, and stakeholders. Using these levers role, they can identify the relationship between the development and profitability of these factors to better improve customer satisfaction and service. It aims to advance the field of talent and talent management techniques as well as evidence-based investment and technology adoption decisions in the restaurant industry. The contribution and its implications for practice and research are covered in the chapter's conclusion.

15.2 REVIEW OF LITERATURE

AI is increasingly being integrated into many industries, driving curiosity and innovation across a wide range of sectors. Restaurant marketing has emerged in recent years as a key area for AI applications, transforming customs, and operations.

15.2.1 AUTOMATION IN OPERATIONS

AI has significantly superior the automation of repetitive duties in restaurant operations. AI-pushed answers, according to research, beautify workflows, reduce operating costs, and grow normal productivity. Robotic kitchen assistants and automated stock control systems are examples of those solutions. Technological advancements are converting the manner kitchens paintings, which has consequences for both fast-meal chains and upscale eating places.

15.2.2 PERSONALIZED CONSUMER EXPERIENCES

The literature often discusses how AI can improve customer experience. AI-powered recommendation systems can increase customer satisfaction and loyalty by designing menus and promotions based on analysis of customer preferences, actions, and behaviors. Restaurants can adjust their supply chains because AI can process large amounts of data in real time.

15.2.3 CHATBOTS AND VIRTUAL ASSISTANTS

Several studies show the prevalence of virtual assistants and chatbots in the restaurant industry. These AI-powered solutions accelerate customer interactions by managing reservations, taking orders, and providing menu item information. According to research, the integration of chatbots reduces orders.

15.2.4 ONLINE ORDERING AND DELIVERY SERVICES

Customers are moving in the direction of online ordering and delivery offerings made possible with the aid of AI-pushed systems, according to research. Customers are locating online ordering to be increasingly handy, and AI is helping to streamline the process with its user-friendly interfaces and personalized tips.

15.2.5 PERSONALIZED RECOMMENDATIONS AND MENUS

AI algorithms that provide personalized menus and recommendations can be used to identify digital shopper behavior. Research shows that when making decisions, consumers engage with and rely on AI-based recommendations, which can increase customer satisfaction and loyalty.

15.2.6 SOCIAL MEDIA INFLUENCE AND ENGAGEMENT

Restaurants are using AI to influence customer behavior on social media through targeted interactions and advertising. AI-driven social media data analytics and consumer attitudes, sentiments, and preferences have received much emphasis in the literature. According to Choi et al. (2021), this understanding allows restaurants to adapt their digital marketing strategy as needed.

15.2.7 PREDICTIVE ANALYTICS FOR DEMAND FORECASTING

The restaurant industry has used AI to forecast demand due to its ability to analyze historical data and predict future trends. Research has shown that predictive analytics can reduce food waste for products that are delivered to optimize the inventory. Restaurants can use AI-powered predictive models to more accurately forecast customer needs, allowing them to use resources more efficiently.

15.2.8 DATA PRIVACY CONCERN AND TRUST

Customers' movements also are motivated by the aid of consideration in AI structures and issues about information privacy. Studies show that if customers agree that their records are dealt with securely and transparently, they are much more likely to interact with AI-pushed platforms. Establishing trust in AI packages is crucial for fostering long-term customer relationships and promoting digital engagement.

15.2.9 AUGMENTED REALITY AND VIRTUAL EXPERIMENTS

Augmented reality is an AI-powered technology that enhances the digital customer experience. Based on the research, customers are more likely to interact with restaurants that offer sophisticated, enhanced experiences—such as virtual menu previews or interactive dining simulations—that can influence their choices and satisfaction levels.

15.2.10 DYNAMIC PRICING AND INCENTIVES

Augmented reality powered by AI improves digital customer experience. Research shows that patrons are more likely to associate with restaurants that offer sophisticated, enhanced experiences—such as interactive dining simulations or virtual menu previews—which can influence their choices and satisfaction levels.

15.3 CONTRIBUTION OF STUDY

- This study will show how much AI technology has changed the restaurant industry.
- AI tools have dramatically increased restaurant efficiency. AI-driven inventory management, automated order processing systems, and predictive maintenance algorithms have reduced processing time, reduced errors, and provided more time for routine tasks.
- One of AI's many advantages for the restaurant industry is its ability to offer individualized customer experiences.
- AI tools have made big data analytics more accessible to restaurants. Through the analysis of operational data, market trends, and customer behavior, restaurants can make well-informed decisions regarding pricing strategies, resource allocation, and menu optimization.
- The use of AI tools in the kitchen has revolutionized culinary practices. The accuracy and consistency of food production are increased by robotic chefs powered by AI, intelligent cooking appliances, and automated food preparation systems. This guarantees that the dishes will always be of a high enough caliber to satisfy customers while also lowering labor costs.

15.4 AIMS AND OBJECTIVES

The fundamental question that motivates this research keeps coming up. Still, it is hard to forecast where we are headed and how quickly because business, politics, technology, economies, and hospitality all change so quickly. That is the location. There is a lot of information available about the restaurant industry, including discussions about front-end, back-end, and changes that will support the industry's future. Depending on your source and their position on the rate of technological advancement, it may be difficult to determine a clear and succinct representation of the industry's technology integration.

Our capacity for future forecasting will rely primarily on the user's needs and convenience of use. The questions that professionals in the field have worked so hard to address about what lies ahead are so complex that there is no true quantitative or qualitative answer, and that is precisely what draws us to this work and makes it so enticing. Regarding our personal plans and career objectives, we felt that the next 10 years would be the time that we give up the hard and exhausting daily work schedules that come with building brands and restaurants.

Right now, we want to focus on helping people architect through brand identification, establishing their own four businesses, and working incredibly hard to achieve their goals with the support of our professional experience. What are the objectives and top concerns for restaurants? What are the operating models for these establishments, and how forward-thinking will these issues be overall? The next set of questions focuses on how we can increase sustainability on a global scale. Will there be no consequences at all for killing animals, destroying rainforests, and releasing methane into the atmosphere?

15.5 ANALYZING INTELLIGENCE FOR THE HOSPITALITY INDUSTRY

Businesses operating in the digital economy should be able to transition to more digital modes of working, managing, organizing, and facilitating change in various organizational processes as a result of technological advancements. The combination of robotics and AI is predicted to have a profound effect on the structure, nature, and working conditions of both the human and global economies. Therefore, to continue drawing in and keeping talent, restaurants will need to change. AI about intelligent agents is the application of intelligent machines designed to act, react, learn, and train like humans. Its goal is to perform tasks that would normally require human labor and to augment human intelligence agents to have greater real-time data storage and analysis capabilities than humans. By enhancing hotel and restaurant operations and management systems, AI adds intelligence to the restaurant industry, redefining it as an intelligent restaurant sector. This technological innovation has fundamentally altered and upended the status quo in the study of tourism and hospitality. Through technological platforms, customers can now customize their requests, make reservations, and purchase hotel and tourism-related goods and services directly, eliminating the need for them to deal with hotels, booking agents, or official travel agencies. It is becoming more difficult to have a competitive edge in the hospitality sector as reservations and travel become easier and more affordable, and as costs and services become more uniform. The focus these days is on the volume of data and information about travel and lodging that is made available through digital channels

15.6 TECHNOLOGICAL INTERVENTION

Chatbots instantly answer customer questions, help with ordering, and offer menu recommendations. This has resulted in improved reservation administration, better customer support, and faster responses to customer inquiries (Prentice et al., 2020a, 2020b). Intelligent product designs use historical data, local events, and regional weather data to predict consumer demand. This helps restaurants to better prepare food and staffing levels (Ai et al., 2021). AI-driven systems provide real-time inventory tracking, automation of reordering processes, and detection of user patterns. AI searches for user preferences, preorders, and popular items through apps or interactive menus with personalized menu recommendations (by Bhushan, 2021). Robots powered by AI help with food preparation, ensuring accuracy, precision, and faster cooking times. AI systems search social media sites for mentions, reviews, and trends about restaurants for guests (Gonzalez et al., 2022). AI-enabled cameras can identify and recognize customers, allowing for preference management, personalized greetings, and integration into loyalty programs (Koo et al., 2021). To reduce their reliance on physical cards or cash, AI enables contactless payments, such as mobile wallets and facial recognition. Through integrated systems or voice-enabled devices, AI-powered voice assistants fulfil customer orders, answer questions, and provide feedback (Rawal et al., 2022). These technological advances show that AI devices are transforming the restaurant industry in many ways, from customer interaction to operational efficiency, ultimately leading to innovative and seamless dining experiences (Khan, 2020).

15.7 ARTIFICIAL INTELLIGENCE IN RESTAURANT

Globally, acceptance of the use of AI robotics is increasing, with applications shifting from industrial manufacturing to service sectors. They have been replaced by many robots. When it comes to directing, receiving, and providing services, robots can easily step in for humans. Many jobs, including dishwashing, chopping vegetables, sorting dishes, and even serving and cleaning food, can be performed by robots in restaurants. The roles and features of the robots used in restaurants vary. Industrial robots make up the majority of the robots in the back of the house. These are the robots that serve as the restaurant's front-door greeters. A robot that consists of an automated dumb waiter and a base is called a waiter robot. There are now three stages rather than just one for waiters, and they are human. In addition, they assist by ensuring that the food is served promptly, clean, and warm. A drawbridge door opens as soon as the robot gets to the table, lifting the tray that was meant to be delivered to the opening. The food is then brought out of the door and served to the customer. When tables are positioned at a particular height in restaurants, a particular type of motor is connected to the dumbwaiter's base and raised to the necessary height so that the robots can access it. A preestablished mathematical model and programming code are developed and applied to both the old and new robot models in ROS (robot operating system). Visualized maps that display the whole restaurant table configuration are used to operate the robots. This visual map can be made from a ground plan or with the aid of SLAM (simultaneous localization and mapping) algorithms. Under different circumstances, the server's role is to direct the customer to the reserved table; to do this, the server only needs to enter the table number and select "lead" from the main menu. The table number then keeps showing on the robot's screen as it moves. All the customer has to do to get the robot back to standby at the front is select OK on the screen when they arrive. A self-delivering robot can also recycle the waste that is gathered from the plate. To summon the robot after their meal is finished, the customer simply needs to press the bell next to the dining table.

To proceed, the customer must select "return to origin." Massive amounts of data are generated by restaurant software systems for managing everything from placing orders for food delivery and staffing shifts to taking reservations and maintaining inventory for paying bills. Restaurants gain from AI because it streamlines the supply chain, lowers waste, and allows for home delivery. Chatbots can assist restaurants by handling order taking, payment processing, and automated, customized customer service. With minimal setup costs and no need for app transfers. Regular price and item updates can be done manually, but digital menu boards are far simpler to use.

A chatbot is an online application that simulates human chat conversational exchanges. Through machine learning and natural language processing, chatbots can comprehend the conversational data that is provided to them and respond to it in a way that is similar to that of a human. Chatbots can handle payment processing, order placement and tracking, and answering commonly asked questions. A chatbot learns new things and is exposed to new data each time it engages with a customer. The key advantages of using chatbots in restaurants are their ability to work around the clock without taking sick or vacation days, and their ability to assist customers promptly when needed.

AI-based biometric technologies are used to generate detailed user profiles, enhance security and user recognition, grant or deny access, personalize more, and approve or deny payment transactions—all to reduce misleading circumstances. There are four main parts to the design of biometric systems. Devices that read and scan biometric identifiers, like faces and fingerprints, are the first. The software, which uses a secure algorithm to transform the scanned biometric data into a digital code, is the second part.

Encoded biometric data stored in a secure database for future comparison will be the fourth and final step. Restaurants use biometric technology such as facial recognition and fingerprint recognition. Fingerprinting is a useful and highly reliable method. Fingerprinting technology not only avoids duplication of employees but also assures accurate maintenance of customer and employee records. Since faces are often used to identify people, a facial recognition system can be very helpful for restaurants.

15.8 RESTAURANT PROCESSES AND SYSTEMS

The restaurant industry has been heavily influenced by AI, which is transforming many facets of it. AI technology is transforming the restaurant industry by improving customer experience, increasing productivity, and supporting customer service, back-end processing and support AI-powered chatbots and virtual assistants. Booking through various channels, which messaging apps, websites, manage orders, and customer inquiries, can be answered quickly to increase customer engagement. Algorithms powered by AI analyze customer preferences and behavior to provide personalized menu recommendations that enhance the dining experience and even increase revenue. Customers can order through voice commands within AI-powered voice recognition systems, streamlining the ordering process, human—reducing investment requirements. Automated kiosks with AI capabilities can speed up ordering processes, reduce wait times, and reduce errors in the ordering process AI can assist eating places lessen waste and boom profitability via analyzing ancient facts to forecast call for styles and improve stock management. When inventory levels are low, AI systems can mechanically reorder supplies, ensuring that restaurants have the necessary components without requiring human intervention. Food practice, cooking, and plating may be performed using robots ready with AI, to improve kitchen output and ensure regular food first-class. AI lowers downtime and the possibility of unplanned disruptions to kitchen operations. AI algorithms can make sure that the right range of workers is available when needed by optimizing staff schedules in reaction to ancient information, top hours, and special occasions. AI gives interactive modules, actual-time comments, and skill enhancement to aid in worker education. By analyzing considerable datasets and deriving treasured insights approximately client choices, AI analytics gear allows restaurants to make educated decisions concerning menu changes, promotions, and advertising and marketing strategies. By tracking key overall performance signs such as sales traits, patron delight, and operational performance, AI enables records-driven selection-making. Customers can have quicker, more handy right of entry to stable, efficient contactless price alternatives with the help of artificial intelligence. AI-driven algorithms assess patron facts to create focused advertising campaigns and raise the effectiveness of loyalty and merchandising programs. AI solutions can display social media for consumer reviews, enabling restaurants to engage with their consumers and directly respond to damaging remarks.

15.9 INNOVATION WITH DIGITAL SERVICES

Innovations in digital services have radically changed the way businesses and organizations deliver and communicate services. Digital services use technology to enhance user experience, increase productivity, and streamline processes. Data analytics, AI, and cloud computing are just a few examples of these technologies. Many industries use this innovation, including entertainment, healthcare, education, and finance. One of the key features of the new product is the digitization of traditional services, which allows for remote reach and personalized experiences. These services are available anywhere, anytime, through web platforms, mobile apps, and smart devices. Automation and AI-powered solutions increase productivity and free up organizational resources for critical and complex tasks by automating repetitive tasks. Additionally, digital businesses that implement strong security measures to protect user data tend to prioritize data security and privacy. With technology constantly evolving and capable of adapting to new trends and changing needs, digital businesses are dynamic. From an overall perspective, this innovation not only increases operational efficiency but also encourages a more integrated and user-focused approach to service delivery.

15.10 ARTIFICIAL INTELLIGENCE IN CUSTOMER EXPERIENCE

Apart from the modern outcomes on clients enjoy, the incorporation of AI gear within the restaurant sector additionally highlights the importance of data security and privacy. Data safety is important due to the boom of virtual offerings and the huge volumes of customer statistics being accrued. AI-powered restaurants are aware of how essential it is to have strong safety protocols in the region to defend sensitive purchaser statistics. These groups appoint present-day access controls, authentication protocols, and encryption to shield the confidentiality and integrity of consumer records. Since technology is a constantly evolving sector, normal safety audits and updates are carried out to prevent the rise of hazards. Furthermore, the flexibility of the era allows the restaurant commercial enterprise to rapidly adapt to new trends and convert consumer needs. AI-pushed solutions grow operational performance and allow a greater incorporated and person-centric method to carrier transport. By analyzing the large volumes of information generated with the aid of AI systems, restaurants may be able to discover greater accurately the tastes, behaviors, and patterns of their clients. When used correctly, these records allow organizations to customize their items and offerings, imparting clients with a greater one-of-a-kind and satisfying experience. In conclusion, the usage of AI technologies highlights the dedication to fact security even as well as revolutionizing the purchaser revel. It is important to strike a balance between innovation and defending the security and privacy of client information because the restaurant industry adopts new technologies, ensuring an unbroken and secure eating experience for all.

15.11 INNOVATIVE MENU DESIGN AND CULINARY CREATIVITY

Innovation has emerged as a key factor influencing menu layout and culinary inventiveness within the dynamic restaurant enterprise. This chapter investigates how innovative menu design, driven by advances in culinary technology and the use of AI, may have a modern effect. The fusion of creativity and era is changing the way restaurants create menu designs. The incorporation of information-pushed insights is a primary catalyst for innovation in menu design. Chefs and restaurateurs may gain important insights into the changing global culinary via using AI algorithms to observe huge volumes of records, including patron alternatives, nutritional traits, and cultural effects. Restaurants can adapt their menus to shape the converting desires of their clients by way of reading patron behavior and options. Creative menu design goes above and beyond widespread selections to give each patron a custom-designed ingesting revel. In this, AI-powered advice engines are essential. They have a look at patron statistics and make meal pointers based totally on personal tastes and previous eating reports. This improves purchaser happiness at the same time as also promotes the invention of new tastes and gastronomic delights. Due to AI's capacity to interpret statistics in real time, eating places can dynamically regulate their menus. Weather styles, seasonality, and emerging food developments all smoothly consist of menu offerings. Restaurants' agility lets them live ahead of the curve by presenting buyers with a dynamic and continuously evolving eating revel. AI handles the fact-based factors of menu layout, allowing chefs to specify their creativity inside the kitchen.

15.12 OVERVIEW OF TECHNOLOGY

Industry spurred technological change in how to stay connected and interact with the method AI technology. AI technology reduced time orientation, list control, etc. The production increases AI personalizes customer experiences to tailor interactions based on individual preferences and behaviors through chatbots, virtual assistants, and recommendation systems. AI algorithms analyze large data sets to gain predictive insights into demand forecasting, menu planning, and marketing strategy, helps informed decision-making in the domain.

Robotic assistants powered by AI simplify food production, ensure consistency, reduce cooking time, and reduce labor costs. AI enables contactless order payment systems, including smartphone apps, QR codes, and voice-activated interfaces, to respond to growing safety and security requirements. AI-powered training simulations support employee onboarding and skill development, while they organize optimization tools based on requirements planning. AI assists management to improve security measures, ensure safe online transactions, fraud prevention, and data protection. AI technology enables personalized services, customer identification, and loyalty program integration.

15.13 TECHNOLOGICAL BENEFITS

AI will increase productivity and reduce human mistakes by automating repetitive tasks like stock manipulation, order processing, and scheduling (Kler et al., 2022). Chatbots pushed through AI enhance customer service effectiveness through right away responding to inquiries right away, scheduling appointments, and updating orders—especially during peak hours. AI algorithms assist restaurants in optimizing staffing, inventory, and menu offerings using forecasting customer calls based totally on historical trends (Pillai and Sivathanu, 2020). AI gear provides real-time insights into stock levels, permitting correct stock control and decreasing the risk of overstock or stockouts. By examining client choices and behavior, restaurants can improve the overall dining experience by offering tailor-made menu pointers, promotions, and targeted advertising (Prentice et al., 2020a, 2020b). Robots powered by AI simplify food preparation, ensure consistency, reduce cooking time, and improve kitchen efficiency. Restaurants can use AI technology to effectively manage their online reputation to take immediate action on customer reviews and comments on social media, and resolve any issues that arise (Doborjeh et al., 2022). Contactless communication is safer and more efficient thanks to AI, improving overall security as well as reducing the need for physical contact with employees and customers, market development, and productivity (Ozdemir et al., 2023). AI-enabled voice-activated systems accelerate order-taking processes, increase peak-time production, and reduce employee workload

(Cain et al., 2019). These technological developments show how AI will play a key role in shaping dining experiences in the future. Together, they create a convenient, efficient, and customer-focused restaurant business. (Law et al., 2023).

15.14 SUGGESTIONS

For your website and mobile applications, use AI-powered chatbots to manage reservations, provide faster customer service, and provide customized recommendations, thereby improving customer satisfaction and service greater than. Use AI algorithms to analyze historical data, weather, and local events to create accurate demand forecasts. As a result, there is less waste and much better operational efficiency and inventory. Use AI to automate inventory management to track inventory in real time and place orders automatically. This ensures better stock management, reduces the chance of stockouts, and uses less feed. Use AI-driven recommender systems to learn about customer preferences and provide personalized menu suggestions.

This enhances the overall dining experience and increases customer satisfaction and loyalty. Consider using AI-powered robot kitchen assistants to ensure food consistency, speed up food preparation, reduce cooking time, and increase kitchen efficiency. Consider using robotic, intelligent kitchen assistants to help with food preparation, reduce cooking time, maintain consistency, and increase kitchen productivity. Welcome to AI-enhanced contactless ordering and payment systems for a safer and more convenient dining experience. It includes QR code menus, smartphone ordering applications, and secure contactless payment methods. AI-powered training simulations can enhance employees' knowledge of order processing, crisis management, and customer service. This ensures a uniform training program and overall increases employee productivity.

15.15 CONCLUSION

AI equipment has revolutionized the restaurant industry. AI is rapidly becoming an integral part of the modern dining experience, transforming customer service and increasing operational efficiency. AI-powered technology has made restaurant operations easier and logistics more efficient. Examples of this technology include predictive analytics for inventory management, automated ordering systems, and personalized recommendation machines. AI has greatly improved customer engagement and satisfaction. Using chatbots, virtual assistants, and facial recognition technology, customers can now place orders with ease and convenience. AI's ability to analyze customer preferences and behaviors has made it easier to create more personalized dining experiences, increase customer loyalty, and repeat business. While these advances have many benefits, they are important to remember that there are some shortcomings, particularly with respect to ongoing needs and long-term usage. Restaurants in the hospitality industry must strike a balance between using AI technology and preserving human hands. The AI has a dynamic and ever-evolving impact on the restaurant industry. Facing technological advances, restaurants must embrace innovation as well as consider societal and ethical implications. Chatbots are designed for immediate customer service through AI-enabled solutions, demand forecasting through predictive analytics, and efficiency through robotic kitchen assistance fascinating potential uses.

KEYWORDS

- **artificial intelligence**
- **culinary innovation**
- **restaurant industry**
- **dining technology**
- **chatbots**
- **robotics**
- **operational efficiency**
- **sustainable practices**

REFERENCES

Addanki, M., Patra, P., Kandra, P. (2022). Recent advances and applications of artificial intelligence and related technologies in the food industry. *Applied Food Research*, *2*(2), 100126. https://doi.org/10.1016/j.afres.2022.100126

Ai, N., Zheng, J., Chen, X., Kawamura, K. (2021). Neighborhood-specific traffic impact analysis of restaurant meal delivery trips: Planning implications and case studies in

Chicago. *Journal of Urban Planning and Development, 147*(2), 05021013. https://doi.org/10.1061/(asce)up.1943-5444.0000682

Alt, R., Information Systems Institute, Leipzig University. (2021). Digital transformation in the restaurant industry: Current developments and implications. *Journal of Smart Tourism, 1*(1), 69–74. https://doi.org/10.52255/smarttourism.2021.1.1.9

Berezina, K., Ciftci, O., Cobanoglu, C. (2019). Robots, artificial intelligence, and service automation in restaurants. In: *Robots, Artificial Intelligence, and Service Automation in Travel, Tourism and Hospitality* (pp. 185–219). Emerald Publishing Limited.

Bhushan, S. (2021). The impact of artificial intelligence and machine learning on the global economy and its implications for the hospitality sector in India. *Worldwide Hospitality and Tourism Themes, 13*(2), 252–259. https://doi.org/10.1108/what-09-2020-0116

Cain, L. N., Thomas, J. H., Alonso, M., Jr. (2019). From sci-fi to sci-fact: The state of robotics and AI in the hospitality industry. *Journal of Hospitality and Tourism Technology, 10*(4), 624–650. https://doi.org/10.1108/just-07-2018-0066

Doborjeh, Z., Hemmington, N., Doborjeh, M., Kasabov, N. (2022). Artificial intelligence: A systematic review of methods and applications in hospitality and tourism. *International Journal of Contemporary Hospitality Management, 34*(3), 1154–1176. https://doi.org/10.1108/ijchm-06-2021-0767

Gonzalez, R., Gasco, J., Llopis, J. (2022). Information and communication technologies in food services and restaurants: A systematic review. *International Journal of Contemporary Hospitality Management, 34*(4), 1423–1447. https://doi.org/10.1108/ijchm-05-2021-0624

Khan, M. A. (2020). Technological disruptions in restaurant services: Impact of innovations and delivery services. *Journal of Hospitality & Tourism Research, 44*(5), 715–732. https://doi.org/10.1177/1096348020908636

Kler, R., Elkady, G., Rane, K., Singh, A., Hossain, M. S., Malhotra, D., Ray, S., Bhatia, K. K. (2022). Machine learning and artificial intelligence in the food industry: A sustainable approach. *Journal of Food Quality, 2022*, 1–9. https://doi.org/10.1155/2022/8521236

Koo, C., Xiang, Z., Gretzel, U., Sigala, M. (2021). Artificial Intelligence (AI) and robotics in travel, hospitality and leisure. *Electronic Markets, 31*(3), 473–476. https://doi.org/10.1007/s12525-021-00494-z

Law, R., Lin, K. J., Ye, H., Fong, D. K. C. (2023). Artificial intelligence research in hospitality: A state-of-the-art review and future directions. *International Journal of Contemporary Hospitality Management, 36*(6), 2049–2068.

Lee, M., Kwon, W., Back, K.-J. (2021). Artificial intelligence for hospitality big data analytics: Developing a prediction model of restaurant review helpfulness for customer decision-making. *International Journal of Contemporary Hospitality Management, 33*(6), 2117–2136. https://doi.org/10.1108/ijchm-06-2020-0587

Ozdemir, O., Dogru, T., Kizildag, M., Erkmen, E. (2023). A critical reflection on digitalization for the hospitality and tourism industry: Value implications for stakeholders. *International Journal of Contemporary Hospitality Management, 35*(9), 3305–3321. https://doi.org/10.1108/ijchm-04-2022-0535

Pillai, R., Sivathanu, B. (2020). Adoption of AI-based chatbots for hospitality and tourism. *International Journal of Contemporary Hospitality Management, 32*(10), 3199–3226. https://doi.org/10.1108/ijchm-04-2020-0259

Prentice, C., Lopes, D., Wang, S. (2020a). Emotional intelligence or artificial intelligence-an employee perspective. *Journal of Hospitality Marketing & Management*, *29*(4), 377–403.

Prentice, Catherine, Weaven, S., Wong, I. A. (2020b). Linking AI quality performance and customer engagement: The moderating effect of AI preference. *International Journal of Hospitality Management*, *90*(102629), 102629. https://doi.org/10.1016/j.ijhm.2020.102629

Rawal, Y. S., Soni, H., Dani, R., Bagchi, P. (2023). A review on service delivery in tourism and hospitality industry through artificial intelligence. In: *Proceedings of Third International Conference on Computing, Communications, and Cyber-Security* (pp. 427–436). Springer Nature Singapore: Singapore.

Singh, S., Singh, G., Dhir, S. (2022). Impact of digital marketing on the competitiveness of the restaurant industry. *Journal of Foodservice Business Research*, *27*(2), 109–137. https://doi.org/10.1080/15378020.2022.2077088

Yoon, J., Yu, H. (2022). Impact of customer experience on attitude and utilization intention of a restaurant-menu curation chatbot service. *Journal of Hospitality and Tourism Technology*, *13*(3), 527–541. https://doi.org/10.1108/jhtt-03-2021-0089

CHAPTER 16

The Transformation of the Industry 4.0 Paradigm to Industry 5.0

PRITI SHARMA,[1*] RUPA KHANNA MALHOTRA,[1] and RAKESH KUMAR[2]

[1]*Department of Commerce, Graphic Era University, Dehradun, Uttarakhand, India*

[2]*Uttaranchal University, Premnagar, Dehradun, Uttarakhand, India*

*Corresponding author

ABSTRACT

The revolution in the industrial sector, which is referred to as Industry 4.0, leads to the integration of advanced digital technology into manufacturing and digital operations. Further, it is distinguished by ubiquity connection, rapid connectivity data transfer, smart manufacturing, better usage of online available resources, artificial intelligence, and printing with 3D printers. Manufacturers and other businesses have adopted these technologies in order to improve the experience of customers while increasing profits. The goal of merging technical and human capabilities into the Industry 5.0 is to enhance the prosperity of all social components, rather than simply profiting. The forthcoming transformation in commerce and product development is going to be unlike any other, with innovative human–robot interactions and an emphasis on societal well-being. Continuous the shift to an industry that is more flexible, human-centered, and sustainable is being prompted by technological advances and research. The concept of Industry 5.0 is discussed in this chapter, along with a comparison of the same with the Industry 4.0.

Industry 5.0: Innovation and Adoption for Sustainability.
Tilottama Singh, Pradeep Suri, and Zulkarin Jahangir (Eds.)

DOI: 10.1201/9781779641878-16

16.1 INTRODUCTION

Value generation and delivery have undergone a dramatic paradigm change with the shift from Industry 4.0 to a new era of advancement and collaboration of man and machines, along with sustainable development. The Fourth Industrial Revolution (2023), or Industry 4.0, brought connection, automation, and digitization to production operations. The industrial sector saw a boost in productivity, efficiency, and innovation as a result of these developments (Carvalho and Cazarini, 2020). However, Industry 4.0 placed less of an emphasis on human contact and cooperation and more on streamlining manufacturing processes and maximizing the capabilities of robots. However, Industry 5.0 goes a step further by emphasizing the value of customization and human interaction in the production process (Ghobakhloo et al., 2023). It acknowledges that although technology may increase production and efficiency, people are ultimately better at problem-solving and working together because they are more creative, adaptive, and empathetic.

A harmonic balance between technology and human talents is the goal of the shift to Industry 5.0 (Carvalho and Cazarini, 2020). The sources claim that Industry 4.0 provided a framework for the incorporation of technologies like artificial intelligence (AI) and big data, including the things on the internet into industrial processes. Large-scale data gathering and analysis to inform insights and decision-making were made possible by this integration. Future industries hope to create welcoming workspaces and flexible yet sturdy production lines. They aim to adopt environmentally friendly consumer and manufacturing practices, along with innovative methods. The goal is to foster strong collaboration between people, tools, robots, and our planet in the Industry 4.0 and 5.0 frameworks. Looking at the 4th industrial generation as an economic growth strategy, it stands on two essential pillars. This innovation of techniques from Industry 4.0 to Industry 5.0 entails the green growth of the smart industries to maintain sustainable development, with the aim of creating an industry where there will be a balanced interaction of man with machines.

16.2 FROM INDUSTRY 1.0 TO INDUSTRY 5.0

The progression of industries from 1.0 to 5.0 involves remarkable technological adjustments and thinking shifts in methods of making things. Each industry overhaul spurred big changes in creating products, boosting speed, mechanization, and broad economic progress. Industry 1.0 started from the late 1700s and lasted to the early 1800s, and underwent big changes, such as steam power and machines were introduced. These powered mass production in industries like textiles and mining. Textiles and mining saw the first use of steam and water power for mass production. There was a big move from farming-based economies to the beginnings of industries. The factory system started. Production on a large scale increased. The economy saw good growth. And there was a gradual shifting from farming to starting industries. Factories were lined properly, and mass production was in full swing. In this phase, the economy grew as productivity increased.

Industry 2.0 took place between the late 1800s and early 1900s. In this phase, power switched to electricity supported by assembly lines. Telegraph and telephone was brought in to ease communication. Power was mainly supplied by electricity, and manufacturing was aided by assembly lines. The telegraph and telephone were new communication tools at that time. It triggered more automation and electrification within manufacturing processes. There was an uptick in uniformity and efficiency. Automation became even more prevalent with electricity powering production. This new age paved the way for more uniformity and productivity.

Around the mid-20th century, Industry 3.0 took place. Several important innovations helped advance manufacturing technology. Electronics and automation played a key role. The emergence of computers and programmable logic controllers (PLCs) allowed manufacturing processes to become more automated. Computers and PLCs allowed electronics and automation. In recent times, the emergence of computers and PLCs has streamlined many industrial processes. PLCs have led to the widespread adoption of computers automated many production processes and transformed manufacturing. The introduction of numerical control allowed machines to be precisely programmed and controlled by software rather than manually operated. This led to improvements in precision and consistency compared to human labor. Robots were also increasingly used for tasks that are of a routine nature or tasks that do not require any human intelligence. The automation of manufacturing processes through computer control has streamlined production in many industries. With enhancements in modern manufacturing techniques, there is potential for increased precision and control during production processes.

Industry 4.0 in the 21st century is an era that involves key innovations in technology, which enabled entirely new applications and possibilities. Cyber-physical systems are those in which computing and physical processes are combined, allowing the physical world to be monitored and controlled digitally over the internet. Smart factories with interconnected systems for real-time data exchange. Increased automation, efficiency, and customization lead to faster industrial processes and create a boom in the industrial sector.

Nowadays, world is moving toward Industry 5.0 for better teamwork between people and machines. Unlikely Industry 4.0, where everything is completely dependent on machines and technology, it is indeed a rise in sharing, individuality, and adaptability. The outcome of using this approach results in a strategy that focuses on the person. The partnership between people and robots is key. This gives industries a new way of thinking where they think about custom-made and flexible ways of making things. It is not just about top-notch tech but is focused toward creativity, solving problems, and skills for dealing with people.

16.2.1 UNDERSTANDING INDUSTRY 5.0: THE NEXT PARADIGM SHIFT

Widespread connection, quick information transfer, intelligent manufacturing, the Internet of Things, AI, and 3D printing are the hallmarks of the Fourth Industrial Revolution (4IR). Retailers and service providers have adopted these technologies in an effort to improve consumer experiences and boost revenue. The Fifth Industrial Revolution (5IR), on the other hand, is centered on human–machine cooperation. Instead of only concentrating on profit, the Fifth Industrial Revolution aims to promote society well-being for all members by merging technical and human assets.

There is going to be a revolution in retail and services research like never before, with new human–robot collaborations and a focus on the welfare of society. The transition of the industry to one that is resilient, human-centered, and sustainable is being driven by research and innovation. Industry 5.0 includes a number of supporting technologies, such as mass customization, hyper-personalization, cobots, and digital twins, that facilitate industrial change.

Every sector has been influenced by the digital revolution, including marketing and commercial organizations of all sizes. Thanks to technical advancements brought forth by societal and industrial revolutions, marketers have created creative methods. Industry 5.0 presents a mass-customization-focused new marketing paradigm that gives rise to a brand-new class of "super-empowered customers," or "consumer 5.0."

16.3 LITERATURE REVIEW

Handan (2023) discusses the growth of Industry 5.0, a centered around people, resilient, and environmentally friendly sector that includes supporting technologies. It focuses on the influence of digital transformation in marketing and innovative strategies supported by technological assistance from industrial and societal revolutions. He shows that consumer 5.0 is a new type of empowered customer shaped by mass customization and digital transformation, as presented in the paper. This chapter has practical implications in the sense that businesses should modify their marketing strategies to meet consumer 5.0's preference for personalized experiences and products. It also implies that enterprises should capitalize on the enablers of Industry 5.0-technologies like digital twins, cobots, mass customization, and hyper-personalization as drivers in delivering better customer experiences to foster innovation.

Industry 5.0 prioritizes adaptability, environmental responsibility, and human involvement in the manufacturing process. There has been a notable surge in the popularity of research on Industry 5.0, as seen by the large number of papers and citations. The study looks at a variety of angles and subjects, including big data, AI, and human aspects. It also emphasizes how crucial using digital platforms and personalization is to becoming competitive (Madsen et al., 2023). A rising interest in subjects like human-centricity, smart manufacturing, sustainable development, and the integration of technologies like 6G is shown by the research trends in Industry 5.0. Researchers are especially curious to learn more about how Industry 5.0 will facilitate coexistence and connection between humans and machines (Larissa, 2023).

Ghobakhloo et al. (2023) provided the key bibliometric trends with respect to the research in the concept of Industry 5.0 and identified three significant advancements. The quantity of articles and citations has increased dramatically as related research and new concepts have been introduced. The latest version of Industry is an innovative framework that promotes environmental responsibility, adaptability, and human-centeredness. It acknowledges that industry must accomplish social objectives beyond growth and job creation by protecting the environment and putting industrial workers' welfare first (Zofia et al., 2023).

Industry 5.0, a more coordinated strategy that prioritizes human–machine cooperation, has emerged as a result of the COVID-19 pandemic's emphasis on the value of flexibility in production systems (Bader, 2023). In order to solve moral dilemmas and advance social justice and sustainability in industrial settings, the European Commission is pushing Industry 5.0 (Gómez et al., 2023). Technology 5.0 combines modern technologies, such as machine learning and the World Wide Web of Things, with analytic algorithms. Its acceptance is critical for future growth opportunities. Industry 5.0 seeks to accelerate and advance the advancement of Industry 4.0 and address the difficulties that the business community is now facing by including people in the manufacturing process and taking into consideration their cognitive and intellectual talents (Monika and Zieminski, 2022).

16.3.1 INDUSTRY 4.0 AND 5.0—POINT OF DISTINCTION

Industry 4.0 is the term used to address the digitization, automatic techniques, and data-driven methods used to increase the productivity and quality of industrial operations. It also encompasses other fields, such as education, where Education 4.0 seeks to modernize and adapt learning to the digital era, equipping students with problem-solving abilities and readying them for the challenges of Industry 4.0 (Udvaros et al., 2023).

On the other hand, Industry 5.0 aims to take a more human approach in focusing on humans and machines working together, where technology is used as an enhancement of its respective user's capabilities. Industry 4.0 generally involves automation and incorporates advanced technologies; however, the relationships established between humans and machines are limited to simplistic transactions only. Industry 5.0 depicts a cooperative relationship in which humans and machines work closely together as partners. This may include robots working alongside humans, where each has different strengths.

Despite the focus of Industry 4.0 on efficiency and optimization through automation in mass production, it is predicted that Industry 5.0 will emphasize individualization together with flexibility because they are engaged usually at relatively small quantities to satisfy consumer's needs.

The transition includes changes in manufacturing processes to improve individual customers' needs and enables more flexible, adaptable production. Industry 5.0 might be focused on decentralization of decision-making processes, which can go beyond centralized cloud-based systems. The bottom-up approach, of course, entails the real-time data processing tools at edge places to take more leeway in developing even faster decision-making, reducing latency in critical applications. Industry 4.0 centers on the digitalization and interconnectivity of individual industrial sectors alone. Alternatively, Industry 5.0 may further expand the scope of integration and collaboration among various industries. This may result in wider cross-industry collaborations and the emergence of complex ecosystems.

Soft skills like problem-solving, creativity, and interpersonal communication are probably going to be increasingly important in Industry 5.0. Human workers can be urged to work on activities that are the essence of their emotional intelligence and creativity, while repetitive and routine tasks may be automated. Although Industry 4.0 for the first time suggested cyber–physical systems, in industry attention here is in designing systems where cyber, physical, as well as humans act simultaneously.

Industry 5.0 is also expected to focus on value creation in a more integrated way, taking into account social and sustainable development issues alongside economic factors alone. This should include a wider perspective on the role of industrial processes in affecting communities, nature, and general welfare. The interaction of humans with machines serves as a technological bridge while moving toward Industry 5.0 (2023).

First off, industry-to-industry transfer is still in its early stages, and the differences mentioned below are based on reviews of similar field-related research publications. A clearer picture of what defines Industry 5.0 will become apparent as the industry environment continues to evolve. Industry 5.0, built on the foundation of Industry 4.0, envisions a resilient, human-centered, sustainable European industrial sector that respects environmental boundaries and places a high priority on the welfare of industrial workers (Zofia et al., 2023).

16.3.2 TRANSITION OF INDUSTRY FROM 4.0 TO 5.0

In order to develop smart factories that increase production efficiency, lower costs, and improve product quality, Industry 4.0 integrates technologies (Da Silva, 2023). Besides the idea of paradigm shift from the concept was in the early stage of formation. But if we generalize from the development of ideas about industry, by allowing us to project a possible period between 4.0 and 5.0, industries transformation phase. Industries need to assess themselves and be ready for the shift. Organizations will have to evaluate their existing technological competencies and suitability for the transition. This entails evaluating the infrastructure, Industry 4.0 technology, and the organization's level of digitalization.

Industries also need to find out where they are lacking behind. The holes and deficiencies in contemporary applications of Industry 4.0 should be highlighted. These can be areas where human–machine partnership does not work effectively; customization and flexibility are missing, or the level of integration of sophisticated technologies is poor. Industries also have to create a proper plan that reflects the goals and objectives of Industry 5.0. This could include linking technology investments to larger business strategies, identifying critical performance indicators, and establishing markers of the migration.

One of the most important aspects in this transformation will be to upgrade existing Industry 4.0 technologies and introduce new ones as well. This may include achievements in such areas as AI, robotics, human–machine interfaces, and other technologies that make it possible to cooperate with humans. Since the focus of Industry 5.0 is on a much closer cooperation between humans and machines, there will be demand for retraining workers for new skills. Workers will have to learn new abilities, including teamwork and solving issues, as well as managing human–machine engagements.

Organizations can implement pilot programs and demonstrations to try new technologies or workflows in a safe manner. Parkinson's law enables challenges to be identified, processes fine-tuned, and best practices formed before complete implementation. The engagement with the industry partners, suppliers, and other stakeholders shall contribute to that endeavor. Facilitating best practices, harmonizing standards, and nurturing a collaborative environment can hasten the shift and allow for seamless implementation of Industry 5.0 values. In the course of this changeover, data governance and security will likely come back into focus. With the emergence of Industry 5.0, there is a need for organizations to implement effective data governance frameworks and ensure that cybersecurity measures are adopted in order to protect information from unwanted uses by other parties involved in online business transactions.

The Industry 5.0 transition of companies will also require careful attention to changing regulatory frameworks. This could mean collaborating with regulators to address any emerging issues regarding human–machine collaboration, data privacy, and ethical deployment of technology. The transformation from the Fourth to Fifth Industry is likely to be cyclical. Organizations will continuously evaluate the situation, learn from pilot programs, and fine-tune on running of technologies operating with humans in collaboration.

It should also be noted that the actual transition process will vary depending on the specific industry, nature of technologies, and rate at which technological innovations are developing. Companies that manage to survive this shift will be those who carefully map out and invest in technology, prioritize workforce training, and upgrade their facilities according to Industry 5.0 practices.

16.4 CHALLENGES OF INDUSTRY 5.0

There was no widespread acceptance of Industry 5.0, and the focus centered on conversations surrounding Industry. However, if we speculate about the challenges that Industry 5.0 might face based on the trends and issues in Industry 4.0, here are some potential challenges that industries may face in the future.

Integration and compatibility issues: Industry 5.0 is likely to create new levels of integration between physical and digital systems that may permeate other sectors aside from the manufacturing sector. Smooth integration of various technologies and systems could pose a great challenge.

Cybersecurity concerns: However, as the level of connectivity rises and complex technologies are implemented, cyber threats may escalate. Safeguarding sensitive information, intellectual property, and securing the interdependent systems will be paramount.

Workforce adaptation: With more automation and digitization prevalent in Industry 5.0, there will most likely be an adaptable workforce that is well-skilled. It may also be hard to prepare the current and future workforce for these changes from an educational, training as well as skill development viewpoint.

Data privacy and ethics: Data privacy and ethical issues become a greater concern in Industry 5.0 due to growing reliance on data. For companies to attract, as well as retain trust from its customers and stakeholders will have to put in place effective data protection systems that are also ethical.

Regulatory compliance: Regulating entities could find it difficult to keep up with technological advancements. Challenges ranging from the formulation and implementation of regulations that govern appropriate and honest use cases for emerging technologies in Industry 5.0 may become prevalent.

Infrastructure development: Industry 5.0 will likely necessitate large-scale renovations of the existing infrastructure, such as communication networks, sensors, and other digital technologies. This construction can pose a significant challenge, especially in areas where this infrastructure is poorly developed.

Cost of implementation: For enterprises, using Industry 5.0 technology may result in significant upfront expenses. However, justifying and managing these investments may pose a challenge for businesses, especially startups with little autonomy.

Environmental impact: With the increased automation and interconnection of industries, there can be anxiety about industrial energy use levels leads to degrading environmental aspects due accumulation of electronic waste, as well as other sustainable issues. Striking a balance between technology and preservation of the environment will not be an easy task.

Global collaboration and standards: It may be difficult to attain global coordination and create universal standards pertaining to Industry 5.0 technologies. All these differences in regulations, technology adoption rates, and cultural factors might prevent the rise of a homogeneous global methodology.

Public perception and acceptance: Such technologies will have to be accepted by the public. It will be necessary to address issues of job displacement, privacy, and social consequences in order for the massive acceptability of these technologies.

16.5 EFFECT OF INDUSTRY 5.0 ON THE WORLD'S ECONOMY

Industry 5.0 revolves around the collaboration of machinery and manpower, flexible industrial methods, and modern technology. This integration helps promote productivity and efficiency in manufacturing as well as other industries. The capability to tailor and changes production in real settings may improve overall functional performance. This might strengthen the international supply chains due to being collaborative and interconnected. More transparency, real-time data sharing, and dynamic production processes can enable organizations to be more effective in handling disruptions, thus minimizing the effects of events such as natural disasters or geopolitical issues.

Industry 5.0 could potentially change job-nature, as it moves in a more humanistic direction. Although some routine and repetitive tasks will be automated, the need for skills such as collaboration, problem-solving, creativity, and human–machine interactions management will grow. It is believed that Industry 5.0 will promote intersectoriality and the development of collaborating ecosystems across industries. Through its collaborative nature, economic growth and competitiveness are increased because there will also be new business models, innovative solutions, as well as shared resources.

The emergence of Industry 5.0 technologies could lead to boosted investment in research and development. Companies and governments might spend some funds to create new technologies that will lead them toward technological advances, which in turn increases the economic growth of a country. The earlier companies and countries adopt Industry 5.0, the better their chances of gaining a competitive advantage through the successful implementation of its principles. First movers can capitalize on efficiency, agility, and collaboration gains, which makes early adoption an advantage in the global market.

Industry 5.0 may speed up the digitalization of emerging markets, which would open opportunities for these economies to skip the traditional industrialization phases. Economic development, job creation, and competitiveness worldwide may be achieved through the integration of modern technologies. Industry 5.0 transformation may impact the world trade architecture. Trade dynamics can be altered due to the changing point of demand for certain products and services that are affected by increased efficiency in production, customization capabilities, as well as cross-industry collaboration.

The importance placed by Industry 5.0 on sustainability can produce some beneficial consequences in the world economy, as it encourages eco-friendly activities. Waste reduction, energy efficiency, and responsible resource use can improve the sustainability and resilience of the world economy. Though Industry 5.0 creates many opportunities, it also presents challenges such as worries about job replacement of lost jobs, computer network vulnerability, and its genuine unethical implications. This is critical for ensuring that the advantages of Industry 5.0 are distributed universally and do not increase economic disparities.

This comes without saying that the true effect of Industry 5.0 will depend on how successfully legislators and organizations handle the difficulties posed by cross-systems collaboration and the application of technologies. Moreover, regional differences, political regulations, as well as geopolitical issues will influence the global industry impact of Industry 5.0, and these issues are required to be addressed to use the concept widely.

16.5 CONCLUSION

An industrial revolution is usually sparked by revolutionary technological advancements that have fundamentally altered the way the industry operates. There are cultural and economic repercussions to these shifts. A few are desired and purposeful, while others are undesirable and unintentional. Industry 4.0, like its predecessors, is centered around technology. But Industry 5.0 is one that has been driven by ethics.

The former depends on the latter to remind them of the importance of values, responsibility, and basic society requirements as the ultimate goals, and the latter depends on the former to drive technology forward and provide answers. The goal of the study was to highlight the keystones of Industry 5.0, which include resilience, sustainability, are to enhance working conditions by improving production efficiency, resource management, and placing an emphasis on ecological and human issues. However, given the technical breakthroughs involved while framing the basics in the earlier industrial revolution, the emerging concept in the field is pushing for shifts in society and human behavior as a result of 4.0-era technological achievements.

Industry 5.0 is all about people first in making things. It builds on Industry 4.0's style. It moves from a tech and productivity focus to caring about people and their specific needs. Now, people boss technology around rather than being told what to do by it. This makes technology helpful to us. The brains behind Industry 5.0 are researchers. Their goal is to fix problems from Industry 4.0 and the digital age. But not all companies and industries are using it yet. Still, it has a wider scope as people are becoming more aware toward the social environment. Now they know that they need human and social interference for bringing up a better world and a brighter future.

KEYWORDS

- **Industry 4.0**
- **Industry 5.0**
- **industrial revolution**
- **sustainable development**
- **sustainable development goals**

REFERENCES

Bader, A. (2023). Technological modernizations in the industry 5.0 era: A descriptive analysis and future research directions. *Processes*, *11*(5), 1318. https://doi.org/10.3390/pr11051318

Carvalho, N. G. P., Cazarini, E. W. (2020). Industry 4.0-what is it. In: *Industry 4.0-Current Status and Future Trends*. IntechOpen.

Da Silva, J. G. (2023). Indústria 4.0: Componentes e princípios de design. *Revista de Gestão e Secretariado (Management and Administrative Professional Review)*, *14*(4), 5772–5784, https://doi.org/10.7769/gesec.v14i4.2021

Fourth Industrial Revolution. (2023). https://doi.org/10.1002/9781119865391.ch2.\

Ghobakhloo, M., Iranmanesh, M., Tseng, M. L., Grybauskas, A., Stefanini, A., Amran, A. (2023). Behind the definition of Industry 5.0: A systematic review of technologies, principles, components, and values. *Journal of Industrial and Production Engineering*, *40*(6), 432–447. https://doi.org/10.1080/21681015.2023.2216701

Gómez, A. M. M., Lama, J. R., Ruiz. (2023). Industry 5.0 values. A bibliometric analysis of the new industrial paradigm from the social approach. *Dyna*, *5*, https://doi.org/10.6036/10834

Handan, M. (2023). Industry 5.0, digital society, and consumer 5.0. In: *Advances in Human and Social Aspects of Technology Book Series* (pp. 11–33). https://doi.org/10.4018/978-1-6684-8397-8.ch002

Larissa, T. (2023). Industry 5.0 and digital ecosystems: Scientometric research of development trends. https://doi.org/10.1007/978-3-031-32719-3_42

Madsen, D. Ø., Berg, T., Di Nardo, M. (2023). Bibliometric trends in industry 5.0 research: An updated overview. *Applied System Innovation*, *6*(4), 63–63. https://doi.org/10.3390/asi6040063

Monika, R., Zieminski, M. (2022). Industry 5.0 in industrial and academic applications. *International Journal of Innovative Technology and Exploring Engineering*, *11*(12), 22–25. https://doi.org/10.35940/ijitee.l9325.11111222

Udvaros, J., Miklós, G., Gubán, Á., Sándor, Á. (2023). Industry 4.0 from the perspective of education 4.0. *International Journal of Advanced Natural Sciences and Engineering Researches*, 7(4), 230–234. https://doi.org/10.59287/ijanser.705

Zofia, G.-S., Luis Ochoa, S., Agata, N., Anna, S-S. (2023). From industry 4.0 paradigm towards industry 5.0. vide. *Tehnoloģija. Resursi*, https://doi.org/10.17770/etr2023vol2.7192

CHAPTER 17

Comparative Analysis of Influence of Nonperforming Assets on the Market Price per Share and Return on Net Worth of State Bank of India Before and After the Implementation of Legislation, "Insolvency and Bankruptcy Code, 2016"

SANGEETA BAGGA,[1*] RUCHITRA KAPARWAN,[1] DEEPAK SAHNI,[2] PRITI SHARMA,[1] and VASIM AHMAD[3]

[1]*School of Commerce, Graphic Era Hill University, Dehradun, Uttarakhand, India*

[2]*School of Management and Commerce Studies, Shri Guru Ram Rai University, Dehradun, Uttarakhand, India*

[3]*Uttaranchal Institute of Management, Uttaranchal University, Dehradun, Uttarakhand, India*

*Corresponding author

ABSTRACT

The banking system is the foundation upon which every economy is built and thrives, making a resilient banking sector crucial for sustainable financial ecosystem development. A primary challenge facing this sector is the elimination of nonperforming assets (NPAs), which directly impair a financial institution's profitability and solvency. This chapter conducts a

Industry 5.0: Innovation and Adoption for Sustainability.
Tilottama Singh, Pradeep Suri, and Zulkarin Jahangir (Eds.)

DOI: 10.1201/9781779641878-17

comparative analysis of the influence of NPAs on the market price per share (MPS) and return on net worth (RNW) of the State Bank of India (SBOI) before and after the implementation of the Insolvency and Bankruptcy Code (IBC), 2016. Utilizing secondary data from SBOI's annual reports from 2010–2011 to 2021–2022, the study employs regression analysis to assess this impact across two six-year periods. The findings reveal a nuanced picture: NPAs demonstrated an insignificant negative influence on MPS both before and after the IBC's enactment. However, a stark contrast was observed concerning RNW. While NPAs had an insignificant negative impact on RNW in the pre-IBC period, this influence became substantial and significantly negative in the post-IBC era. This indicates that the IBC, while designed as a recovery mechanism, has heightened the direct negative impact of NPAs on the bank's profitability metrics, even as their effect on share prices remained muted. The study fills a critical research gap by providing a focused, temporal analysis of the IBC's impact on key financial indicators of India's largest public sector bank.

17.1 INTRODUCTION

17.1.1 STATE BANK OF INDIA

The State Bank of India (SBOI) represents India's multinational financial and banking services statutory entity, boasting a legacy exceeding two centuries. Established in 1806, the SBOI is regarded as the oldest commercial bank in the nation. Initially, it operated under the designation of Bank of Calcutta. Subsequently, a royal charter was bestowed upon the institution 3 years later, thereby rebranding it as the Bank of Bengal. This institution was one of the three prominent presidency banks, alongside the Bank of Bombay (founded in 1840) and the Bank of Madras (established in 1843), which were, at that time, co-owned by private stakeholders and the provincial government. The aforementioned presidency banks consolidated in 1921, transitioning into the Imperial Bank of India (IBI), which subsequently emerged as the preeminent commercial entity in the country.

In 1955, the Government of India, in conjunction with the Reserve Bank of India (RBI) (established in 1935), the nation's central banking authority, assumed joint ownership of IBI, which was subsequently rebranded as the SBOI. Following the enactment of the SBOI (Subsidiary Banks) Act 4 years later, banks previously managed by distinct princely states were integrated as subsidiaries of SBOI. In 2007, the stake held by the RBI in

SBOI was transferred to the government. Since its nationalization, SBOI has effectively addressed the requirements of India's developing economy by implementing rural development initiatives, microcredit programs, financing substantial agricultural and industrial projects, and securing government loans.

The SBOI Act, 1955, Section 33, gave the bank the authority to grant term financing to both small and large companies. In addition, the State Bank group was given broad authority to carry out regular banking operations. Since then, the State Banks' Group's operational model has undergone various changes. The bank has made investments in merchant banking, innovative banking, capital venture banking, and so on, in recent years. Its operations involve conducting business internationally with the assistance of its overseas branches. As a result, the SBOI operates in multiple capacities, including standard banking operations and creative assistance to financial intermediaries in the nation's capital and money markets.

With an outlined motive to build a resilient and sustainable financial ecosystem, it is essential that the banking sector of the economy must be strengthened. Banking sector is essentially considered as the backbone of any economy, and this stands true for the Indian Economy as well. One of the main purposes for enactment of Insolvency and Bankruptcy Code, 2016 has been to enable the banking sector to become resilient and make significant contribution for the sustainable development of the entire financial ecosystem.

17.1.2 NONPERFORMING ASSETS AND THEIR CLASSIFICATION

According to the RBI's regulations, loan assets are classified as nonperforming assets (NPAs) in the following situations:

1. Interest and/or principal installments on term loans are not paid for at least 90 days after the date of becoming due.
2. In the case of overdraft or cash credit loans and advances, the account is deemed "out of order" if the unpaid balance exceeds the authorized limit or drawing power for a period of 90 days; if no repayments or credits are made for a period of 90 days as of the balance sheet date or if the repayments or credits are insufficient to pay the interest that is due during that same period.
3. A bill that has been discounted or purchased is not paid for a period of time longer than 90 days after the due date.

4. If agricultural advances are made for short-term crops, the principal or interest is not paid for two consecutive crop seasons after the due date.
5. If agricultural credit is extended for long-term crops, the principal or interest is not paid for a single crop season after the due date.

The RBI's guidelines classify NPAs into the following three categories:

1. Substandard assets: A loan asset that has not performed for a maximum of 12 months is considered substandard.
2. Doubtful assets: A loan that has at least a 12-month history of substandard performance.
3. Loss assets: A loan that has not been fully written off but is still in the loss category.

17.1.2.1 KINDS OF NONPERFORMING ASSETS

1. Gross NPAs
2. Net NPAs

17.1.2.1.1 Gross Nonperforming Assets

The total amount of loan and advances that the RBI has classified as NPAs as of the balance sheet date is known as gross nonperforming assets (GNPAs). A bank's total GNPA number is a measure of the quality of its loans. This category includes all nonstandard assets, including substandard, questionable, and lost assets. The following formula is to be used to compute it (RBI, n.d.):

$$\text{GNPAs ratio} = \frac{\text{Gross NPAs}}{\text{Gross advances}}$$

17.1.2.1.2 Net Nonperforming Assets

The assets for which the bank has deducted the provision for NPAs are known as net nonperforming assets (NNPAs). Because there are a lot of NPAs on bank balance sheets in India, the NNPAs figure accurately reflects the banks' underlying financial burden. Furthermore, it takes time to reduce and eliminate debt. Under RBI guidelines, banks have to make a large amount of provisions for NPAs. This causes a sizable difference to exist between gross and NNPAs.

The RBI provides for the following method to be adopted for computation of NNPAs:

$$\begin{aligned}\text{Net NPA} = \text{Gross NPA} - (&\text{Balance in interest suspense account}\\ &+\text{DICGC / ECGC claims received and held pending adjustment}\\ &+\text{Part payment received and kept in suspense account}\\ &+\text{Total provisions held})\end{aligned}$$

The NNPAs ratio can be calculated as follows (RBI, n.d.):

$$\text{NNPAs ratio} = \frac{\text{Gross NPAs} - \text{Provisions}}{\text{Gross advances} - \text{Provisions}}$$

17.1.3 MARKET PRICE PER SHARE

The cost of a single share listed and traded on a stock exchange is simply referred to as the "market price per share (MPS)." This price is not set in stone as it changes as a result of many market forces that act on it throughout the trading day.

17.1.4 NET WORTH

The assets value *less* the amount due, that is, the liabilities is equal to the net worth. It can also be calculated by adding reserves and surplus to the bank's capital.

17.1.5 RETURN ON NET WORTH

Return on net worth (RNW) or return on shareholders' funds measures the company's ability to generate profits proportionate to the amount invested by shareholders.

$$\text{Return on net worth or shareholders' fund} = \frac{\text{Profit after tax}}{\text{Shareholders' funds or net worth}} \times 100$$

17.1.6 INSOLVENCY AND BANKRUPTCY CODE, 2016

In order to address NPAs brought on by corporate debtors, the Insolvency and Bankruptcy Code (IBC) was passed in 2016.

As per Section 6 of the code, in the event that a corporate debtor is unable to make payments, the corporate debtor or any operational or financial creditor may initiate the process of corporate insolvency resolution. A request is made to the National Company Law Tribunal (NCLT) in order to start a corporate insolvency resolution process (IRP) against a corporate debtor. After receiving the application, the NCLT has 14 days to evaluate whether there is a default; if so, the NCLT allows the application. Within 180 days after the date the application was submitted, the NCLT is required to offer a resolution plan. This deadline may be extended up to 90 days in order to ensure that the matter is settled on time but no more. The corporate debtor's property is defended and kept valuable by the NCLT through the efforts of its resolution professionals. Additionally, the code was updated in 2019 and Section 12 now has a proviso stating that the settlement process must be finished in 330 days. This modification ensures that all parties involved in the insolvency resolution process are in agreement in order to prevent unwarranted delays. If within 330 days of the application's admission, a settlement plan is not agreed, and the delinquent firm will be liquidated.

17.1.7 GUIDELINES PROVIDED BY RESERVE BANK OF INDIA FOR ADMINISTERING THE INSOLVENCY AND BANKRUPTCY CODE TO NONPERFORMING ASSETS

In 2018, the RBI issued a circular in response to much debate from the past. Every bank and financial institution having loan exposure more than Rs 2000 Cr is mandated by the RBI to start a corporate IRP against the defaulting business if the bank is unable to impose a resolution plan within 180 days of the default. Even if an institution missed a single payment, they still had to notify RBI and start working on the resolution plan. The circular mandated that banks identify and categorize NPAs as special mention accounts right once. In the wake of this circular, the existing mechanisms—such as strategic debt reduction and corporate debt restructuring—have been discontinued.

17.1.8 PETITION SUBMITTED IN OPPOSITION TO THE 2018 CIRCULAR FROM THE RESERVE BANK OF INDIA

According to the circular, a significant issue was identified with the 1-day default norm. Because it did not meet certain requirements, such as framing a schedule of 118 days, without taking into account the challenges encountered in other industrial zones, complaints have been filed by the textile industry, the power industry, and others, claiming that it is unfair and unreasonable. Moreover, as required under Section 35AA of the Banking Regulation Act, no prior authorization was requested from the Central Government. Section 35AA of the Banking Regulation Act allows the RBI to mandate the establishment of an IRP for any financial company in the event of a failure. It had to get approval from the central government, nevertheless, prior to issuing a circular under section 35AA. The Supreme Court, therefore, deduced this circular violated section 35AA of the Banking Regulation Act.

17.1.9 REVISED CIRCULAR ISSUED BY RESERVE BANK OF INDIA IN 2019

In 2019, the RBI revised its NPA resolution circular, giving financial institutions 30 days to classify an account as a NPA. This came about 2 months after the Supreme Court struck down the RBI's 2018 Circular, which required the bank to initiate the IRP even in cases of a single day default. Under the new circular, lenders would be fully empowered to develop and implement resolution strategies. The IBC process may start with a 30-day default. The intercreditor agreement, which specifies a majority of decision-making standard, must be signed by the lenders. Additionally, the RBI increased the creditor approval rate to 100%. All other lenders must be bound by any decision made by a lender, representing 60% of all lenders and 75% of the total outstanding credit facility's value.

17.2 REVIEW OF LITERATURE

Singh (n.d.) concluded that banking is crucial to the expansion and growth of the economy. Public sector banks in India have taken the lead in providing financial services in the most remote regions of the nation and in coordinating resources from far rural areas. Public sector banks have borne the brunt of the social agenda without receiving any remuneration. Therefore, the government is justified in routinely recapitalizing public sector banks in order to preserve the credibility of these institutions, which represent about 70% of all banking activity in the nation. Research must be done, meanwhile, on the development of fine-grained standards that are suitable for assessing the performance of various Indian banks without restricting loans to profitable industries.

Satpal (2014) observed that India's banks have long faced significant challenges as a result of NPAs. It is not simply a problem for the banks; the economy is also affected. Since interest income from funds lent is a major source of revenue for Indian banks, money held in NPAs directly affects the profitability of the bank. According to this report, NPAs in public sector banks is significantly higher than in private sector banks. Even though the government has made a number of actions to lower the NPAs, more work needs to be done to resolve this matter. When compared to foreign banks, the NPAs of Indian banks are elevated. Having zero NPAs is not feasible in the slightest. The bank's management ought to expedite the process of recovery. Large borrowers are the obstacles in recovery, not small borrowers, and a stringent policy needs to be implemented to address this issue. Since the priority sector is the one that generates the majority of problems, the government should also increase the provisions for the quicker resolution of pending cases and decrease the requirement for forced loans to this sector. Therefore, the issue of NPAs requires significant action; else, NPAs would result in killing the profits of Indian banks that are completely detrimental to the expanding Indian economy.

Chaudhary and Sharma (2011) concluded that the moment is opportune to implement appropriate and strict steps to resolve the NPA issue. It is necessary to create an effective management information system. The bank employees who approve the advances should be incentivized to take action to stop the advances from becoming NPAs by providing them with appropriate training on documentation and handling of securities. For public banks to compete with private banks, they need to focus on how they operate. Banks need to know how to properly choose borrowers and projects and analyze financial statements.

Bhardwaj and Chaudhary (2018) concluded that NPAs are a major problem for Indian banks as well as the nation's economy as a whole. Since interest income from lent funds is a major source of income for Indian banks, money held in NPAs directly affects the profitability of the bank. They came to the conclusion that, in comparison to international banks, the amount of NPAs in Indian banks is quite large. Even though the government has made a number of actions to lower the NPAs, effective provisions are required to address this issue. Since the occurrence of NPAs is inevitable, the managerial bodies of banks may seek for the methods for speeding up the recovery process. Large borrowers are the obstacle in recovery, not small borrowers, and a stringent policy needs to be implemented to address this issue. Since the priority sector is the one that generates the majority of problems, the government should also increase the provisions for the quicker resolution of pending cases and decrease the requirement for forced loans to this sector.

Deshmukh and Dongre (2018) observed that opaque accounting policies and insufficient procedures and practices of auditing, lack of monitoring on the part of banks, lack of vision by bank in sanctioning, reviewing and enhancing credit limits, the legal environment, political interference and target advances under various government schemes, and also the loan waiver scheme of the government are the causes identified for mounting NPAs and further deterioration of the bank's condition in recovering loan.

Mohanty (2017) stated that over the course of the study's 10 years, the CBE's profits have increased, with certain years seeing particularly strong profitability. According to an International Monetary Fund (IMF) working paper, on the determinants of commercial bank profitability in sub-Saharan Africa (SSA), bank profits are high in SSA in comparison to other regions. This profitability is compared to the average profitability of banks on the sub-Saharan African continent.

Tripathi and Mewada (2016) clarified that only a sound system for managing risks and credit evaluation will be able to resolve the NPAs problem. Prudential criteria must be incorporated into the framework for managing accounts so as to lessen, if not completely eliminate, the issue of NPAs. It is appropriate to eliminate the NPAs at the maturing phase of credit by setting up of exact and legitimate credit examination systems. In this sense, it is especially crucial for bank to keep the amount of NPA as low as may be expected under the conditions. As a reason to the fact that NPAs is one type of barrier to bank success and negatively affects bank performance in this manner, managing NPAs in a bank is crucial.

Pandey et al. (2020) concluded that the Indian economy as a whole and especially the Indian banks both have been aided greatly from this code. There were high expectations for its success ever since it launched. After a few years, its successful implementation has indicated the advancements in both NPAs recovery and turnaround time. They further stated that the Insolvency Code has benefited the Indian economy and will ultimately support the government's "Invest in India" and "Make in India" efforts, resulting in a large number of job possibilities.

Karmakar and Padhi (2017) stated following a review of the historical data that was accessible, it was discovered that there has been a significant rise in NPAs across several institutions. The ratio of NPAs in the banks has not decreased despite the efforts of the RBI and other parties. The bank needs to learn how to lessen the growing issue of NPAs. Since the priority sector is the one that generates the majority of problems, the government should also increase the provisions for the quicker resolution of pending cases and decrease the requirement for forced loans to this sector. Because of this, the problem of NPAs needs to be given a lot of attention; if it is not, NPAs will keep killing bank profitability, which is disastrous for the developing Indian economy. If the NPAs dropped, the bank would see a rise in revenue and be able to help the nation's capital recycling process run more smoothly. By offering enhanced financial services, this would contribute to the expansion of the nation's banking industry and economy.

17.3 RESEARCH GAP

While undergoing the literature review, no research studies or papers were found that analyzed the influence of NPAs on the MPS and the RNW of the public sector banks of India, especially before and after the IBC, 2016 was implemented. The objective of this chapter is to analyze the influence of NPAs on the MPS and the RNW of the largest public sector bank of India, that is, SBOI before and after the IBC, 2016 was enacted. During the review of the literature, no such study that looked at the impact of NPAs from this angle was discovered.

17.4 OBJECTIVES OF THE STUDY

- To analyze the influence of NPAs on the MPS of SBOI during pre- and postenforcement era of IBC, 2016.
- To examine the influence of NPAs on the RNW of SBOI during pre- and postenforcement era of IBC, 2016.

17.5 LIMITATIONS OF THE STUDY

1. This study is limited to SBOI.
2. The annual reports of the SBOI were used to identify the NPAs of aforementioned bank.
3. The MPS of the SBOI has been taken as on 31st March of the respective financial years under study.
4. Based on data retrieved from the SBOI's annual reports, the bank's return on net worth has been calculated.
5. Seven years have passed since the IBC came into force in 2016. Nevertheless, the specifics of the annual report for 2022–2023 were not yet finalized at the time of the study; therefore, six years before and after the enforcement were taken into account.

17.6 SOURCES OF DATA

Most of the data collected and examined for this study is secondary in nature. This paper's data came from a number of sources, including the SBOI's annual reports, RBI publications, a variety of journals, books, and websites that discussed contemporary banking issues, as well as past research studies.

17.7 METHODOLOGY OF STUDY

We considered into account SBOI's NPAs for this study based on information from its annual reports for the years we were examining. Secondary data forms the basis of this research article. The significance of SBOI in the Indian economy is examined in this research. It provides more clarification on the IBC, 2016 rules that apply to NPAs as well as the conceptual framework of NPAs. This study uses regression analysis through SPSS to examine the impact of NPAs on the MPS and RNW of SBOI for the

6 years prior to and 6 years following the implementation of the IBC or 2010–2011 to 2015–2016 and 2016–2017 to 2021–2022, respectively. For the intent of this analysis, numerous reputable research journals with research papers and articles have been consulted.

17.8 TIME PERIOD OF THE PAPER

For this study article, 12 years' data of SBOI, that is, from 2010–2011 to 2021–2022, have been gathered and examined.

17.9 GROSS NONPERFORMING ASSET RATIO, MARKET PRICE PER SHARE, AND RETURN ON NET WORTH OF STATE BANK OF INDIA FROM 2010–2011 TO 2021–2022 (TABLES 17.1 AND 17.2)

TABLE 17.1 GNPA Ratio, MPS, and RNW of SBOI Before IBC, 2016 (Amount in Rupees Crores)

Year	GNPA Ratio (in %)	Market Price Per Share (in Rs.)	Return on Net Worth (in %)
2010–11	3.28	2765.30	12.72
2011–12	4.44	2096.35	13.95
2012–13	4.75	2072.75	14.26
2013–14	4.95	1917.70	11.08
2014–15	4.25	267.05	8.48
2015–16	6.50	194.25	6.90

Source: Data compiled from Annual Reports of SBOI and www.investing.com

TABLE 17.2 GNPA Ratio, MPS, and RNW of SBOI After IBC, 2016 (Amount in Rupees Crores)

Year	GNPA Ratio (in %)	Market Price Per Share (in Rs.)	Return on Net Worth (in %)
2016–17	6.90	293.40	5.57
2017–18	10.91	249.90	–2.99
2018–19	7.53	320.75	0.39
2019–20	6.15	196.85	6.24
2020–21	4.98	364.30	8.04
2021–22	3.97	493.55	11.31

Source: Data compiled from Annual Reports of SBOI and www.investing.com

17.10 PRELIMINARY ANALYSIS FOR APPLICATION OF REGRESSION ANALYSIS

A Shapiro–Wilk test has been applied through SPSS to check the normality of data as it is considered to be appropriate in case of small sample sizes.

TABLE 17.3 Test of Normality

Variable	Before IBC 2016	Df.	Sig.	After IBC 2016	Df.	Sig.
	Shapiro–Wilk Statistic			Shapiro–Wilk Statistic		
Gross NPA	0.839	6	0.127	0.960	6	0.820
MPS	0.941	6	0.670	0.943	6	0.685
RNW	0.911	6	0.444	0.958	6	0.805

As depicted in Table 17.3, it is observed that the *P*-value estimated by the Shapiro–Wilk test for all the variables is more than 0.05; hence, the null hypothesis for this test that the data is normally distributed is accepted, and it is concluded that the data for all the three variables, that is, GNPA ratio, MPS, and RNW before and after IBC, 2016 is normally distributed.

I. **Research hypothesis for influence of NPAs on MPS during the two periods**

   H1 = There is significant difference in influence of NPAs on MPS of SBOI before and after the enactment of IBC, 2016.

   Regression analysis is applied through SPSS to test the hypothesis for difference in influence of NPAs on MPS before and after the implementation of IBC, 2016.

   The hypothesis tests the difference in influence of NPAs on the MPS of SBOI during the two periods. For this purpose, the dependent variable, that is, MPS is regressed on independent variable, that is, gross NPA, for both the periods under study.

17.10.1 APPLICATION OF REGRESSION ANALYSIS

17.10.1.1 BEFORE INSOLVENCY AND BANKRUPTCY CODE, 2016

To test the hypothesis that MPS is impacted significantly by GNPA ratio, the dependent or response variable, that is, MPS is regressed on the independent or predictor variable, that is, GNPA ratio. The results obtained through SPSS show that $F\ (1,4) = 2.801$, $P > 0.05$, which indicates that GNPA ratio does not have significant influence on MPS. These results clearly represent the negative effect of GNPA on MPS, which is not significant in nature ($b = -646.118$, $P > 0.05$). Further, $R^2 = 0.412$ depicts that only 41.2% variation in MPS is explained by GNPA ratio for pre-IBC era. Table 17.4 presents the summary of the findings.

TABLE 17.4 Summary of Regression Analysis for Influence of GNPA on MPS (Pre-IBC Period)

Hypothesis	Regression Weights	Beta Coefficient	*R*	R^2	*F*	P Value	Hypothesis Supported
H_1	G1 → MPS1	−646.118	−0.642	0.412	2.801	0.170*	No

Notes: $^*P > 0.05$, G: GNPA ratio, MPS: market price per share.

TABLE 17.5 Coefficients for Regression of MPS on GNPA Ratio (Pre-IBC Period)

	Unstandardized Coefficients		Standardized Coefficients	Sig.
	B	Std. Error	Beta	
Constant	4585.756	1850.555		0.068
Mean of GNPA Ratios	−646.118	386.078	−0.642	0.170

Notes: Regression equation: $X = 4585.756 - 646.118Y$, where X = MPS and Y = GNPA ratio.

As there is only one independent variable, therefore, the value of standardized beta coefficient is equivalent to coefficient of correlation, that is, $R = -0.642$.

The regression equation derived from Table 17.5 is

$$X = 4585.756 - 646.118Y$$

where X = MPS and Y = GNPA ratio

In Table 17.5, the value of constant = 4585.756 and coefficient of GNPA ratio= −646.118. This implies that the GNPA has a nonsignificant negative influence on the bank's market price per share. Based on the facts above, it is determined that NPAs do not have any significant influence on the marketability of shares of SBOI before the implementation of IBC, 2016.

17.10.1.2 AFTER INSOLVENCY AND BANKRUPTCY CODE, 2016

To test the hypothesis that MPS is impacted significantly by GNPA, the dependent or response variable, that is, MPS is regressed on the independent or predictor variable, that is, GNPA. The results of regression analysis through SPSS indicate that, as F (1,4) = 2.904, $P > 0.05$, which indicates that GNPA has insignificant negative influence on MPS (b = –0.002, $P > 0.05$). Further, R^2 = 0.397 depicts that only 39.7% variation in MPS is explained by GNPA ratio. Table 17.6 presents the summary of the findings.

TABLE 17.6 Summary of Regression Analysis for Influence of GNPA on MPS (Post-IBC Period)

Hypothesis	Regression Weights	Beta Coefficient	R	R^2	F	P Value	Hypothesis Supported
H_1	G2 → MPS2	−26.821	−0.630	0.397	2.630	0.180*	No

Notes: *$P > 0.05$, G: GNPA, MPS: market price per share.

TABLE 17.7 Coefficients for Regression of MPS on GNPA Ratio (Post-IBC Period)

	Unstandardized Coefficients	Standardized Coefficients	Sig.	
	B	**Std. Error**	**Beta**	
Constant	500.565	117.270		0.013
Mean of GNPA ratios	−26.821	16.537	−0.630	0.180

Notes: Regression equation: $X = 500.565 - 26.821Y$, where X = MPS and Y = GNPA ratio

As there is only one independent variable, therefore, the value of standardized beta coefficient is equal to correlation coefficient, that is, R = –0.630.

The regression equation derived from Table 17.7 is

$$X = 500.565 - 26.821Y$$

where X = MPS and Y = GNPA ratio

In Table 17.7, the value of constant = 500.565 and coefficient of GNPA ratio = –26.821. According to the analysis, the GNPA ratio has not so significant detrimental influence on the bank's market price per share. Based on the findings above, GNPAs do not have considerable influence on the marketability of shares of SBOI after implementation of IBC, 2016.

17.10.1.3 COMPARISON OF INFLUENCE OF NONPERFORMING ASSET ON MARKET PRICE PER SHARE BEFORE AND AFTER THE IMPLEMENTATION OF INSOLVENCY AND BANKRUPTCY CODE, 2016

As we intend to assess the difference in the influence of NPAs on the MPS of SBOI before and after the IBC, 2016 came into force, we must analyze the numerous observations obtained through regression analysis for the two periods under consideration, which are tabulated in Table 17.8.

TABLE 17.8 Comparative Analysis of GNPA's Influence on MPS Before and After IBC, 2016

Period	*R*	R^2	*F*	*P* Value
2010–16	−0.642	0.412	2.801	0.170*
2016–22	−0.630	0.397	2.630	0.180*

Note: $^{*}P > 0.05$.

The correlation coefficients between GNPA ratio and MPS determined through SPSS for the two periods indicate that the negative degree of correlation indicates that as the GNPA ratio increases, MPS decreases. The coefficient of correlation remains negative during both the periods, but it has reduced slightly after IBC, 2016 compared to during the period before IBC, 2016 came into force. The coefficient of determination, that is, R^2 depicts that the variation in market price of shares accounted due to NPAs is 41.2% before IBC, 2016, which falls to 39.7% after IBC, 2016. Hence, it can be concluded that NPAs do not have a significant influence on the market price of shares of the SBOI during pre-IBC and post-IBC eras, and it is further observed that this influence is reduced during post-IBC, 2016 era by 1.5% compared to pre-IBC, 2016 era.

II. Research hypothesis for influence of NPAs on RNW during the two periods

H1 = There is significant difference in influence of NPAs on RNW of SBOI before and after the enactment of IBC, 2016.

Regression analysis is applied through SPSS to test the hypothesis for difference in influence of NPAs on RNW before and after the implementation of IBC, 2016.

The hypothesis tests the difference in influence of NPAs on the RNW of SBOI during the two periods. For this purpose, the dependent variable, that is, RNW is regressed on independent variable, that is, GNPA ratio, for both the periods under study.

17.10.2 APPLICATION OF REGRESSION ANALYSIS

17.10.2.1 BEFORE INSOLVENCY AND BANKRUPTCY CODE, 2016

To test the hypothesis that RNW is impacted significantly by GNPA ratio, the dependent or response variable, that is, RNW is regressed on the independent or predictor variable, that is, GNPA ratio. The results obtained through SPSS show that, $F(1,4) = 2.062$, $P > 0.05$, which indicates that GNPA ratio has insignificant negative influence on RNW ($b = -1.657$, $P > 0.05$). Further, $R^2 = 0.340$ depicts that only 34% variation in RNW is explained by GNPA ratio. Table 17.9 presents the summary of the findings.

TABLE 17.9 Summary of Regression Analysis for Influence of GNPA on RNW (Pre-IBC Period)

Hypothesis	Regression Weights	Beta Coefficient	R	R^2	F	P Value	Hypothesis Supported
H_1	G1 → RNW1	−1.657	−0.583	0.340	2.062	0.224	No

Notes: $P < 0.05$, G: GNPA ratio, RNW: return on net worth.

As there is only one independent variable, therefore, the value of standardized beta coefficient is equivalent to coefficient of correlation, that is, $R = -0.583$.

The regression equation derived from Table 17.10 is

$$X = 19.012 - 1.657Y$$

where X = return on net worth and Y = GNPA ratio

TABLE 17.10 Coefficients for Regression of RNW on GNPA Ratio (Pre-IBC Period)

	Unstandardized Coefficients	Standardized Coefficients	Sig.	
	B	Std. Error	Beta	
Constant	19.012	5.531		0.026
Mean of GNPA Ratios	−1.657	1.154	−0.583	0.224

Notes: Regression equation: $X = 19.012 - 1.657Y$, where X = return on net worth and V = GNPA ratio.

In Table 17.10, the value of constant = 19.012 and coefficient of GNPA ratio = –1.657. This implies that the GNPA ratio has an insignificant negative influence on the bank's RNW. Based on the facts above, it is determined that NPAs have not so significant negative influence on the RNW of SBOI before the implementation of IBC, 2016.

17.10.2.2 AFTER INSOLVENCY AND BANKRUPTCY CODE, 2016

To test the hypothesis that RNW is impacted significantly by GNPA ratio, the dependent or response variable, that is, RNW is regressed on the independent or predictor variable, that is, GNPA ratio. The results of regression analysis through SPSS indicate that, as F (1,4) = 48.063, $P < 0.05$, which indicates that GNPA has significant negative influence on RNW ($b = -0.002$, $P > 0.05$). Further, $R^2 = 0.923$ depicts that 92.3% variation in RNW is explained by GNPA ratio. Table 17.11 presents the summary of the findings.

TABLE 17.11 Summary of Regression Analysis for Influence of GNPA on RNW (Post-IBC Period)

Hypothesis	Regression Weights	Beta Coefficient	*R*	R^2	*F*	*P* Value	Hypothesis Supported
H_1	G2 → RNW2	−2.073	−0.961	0.923	48.063	0.002	Yes

Notes: $P < 0.05$, G: GNPA ratio, RNW: return on net worth.

As there is only one independent variable, therefore, the value of standardized beta coefficient is equal to correlation coefficient, that is, $R = -0.961$.

The regression equation derived from Table 17.12 is

$$X = 18.729 - 2.073Y$$

where X = RNW and Y = GNPA ratio

In Table 17.12, the value of constant = 18.729 and coefficient of GNPA ratio = –2.073. According to the analysis, the GNPA has significant detrimental impact on the bank's RNW. Based on the findings above, NPAs have considerable impact on the RNW of SBOI after implementation of IBC, 2016.

TABLE 17.12 Coefficients for Regression of RNW on GNPA Ratio (Post-IBC Period)

	Unstandardized Coefficients	Standardized Coefficients	Sig.	
	B	**Std. Error**	**Beta**	
Constant	18.729	2.120		0.001
Mean of GNPA Ratios	−2.073	0.299	−0.961	0.002

Notes: Regression equation: $X = 18.729 - 2.073Y$, where X = RNW and V = GNPA ratio.

17.10.2.3 COMPARISON OF INFLUENCE OF NONPERFORMING ASSET ON RETURN ON NET WORTH BEFORE AND AFTER THE IMPLEMENTATION OF INSOLVENCY AND BANKRUPTCY CODE, 2016

As we intend to assess the difference in the influence of NPAs on the RNW of SBOI before and after the IBC, 2016 came into force, we must analyze the numerous observations obtained through regression analysis for the two periods under consideration, which are tabulated in Table 17.13.

TABLE 17.13 Comparative Analysis of GNPA's Influence on RNW Before and After IBC, 2016

Period	***R***	R^2	***F***	***P* Value**
2010–16	−0.583	0.340	2.062	0.224*
2016–22	−0.961	0.923	48.063	0.002**

Notes: $^*P < 0.01$, $^{**}P < 0.05$.

The correlation coefficients between GNPA ratio and RNW determined through SPSS for the two periods indicate that the negative degree of correlation has increased considerably during post-IBC era compared to pre-IBC era. The coefficient of determination, that is, R^2 depicts that the variation in RNW accounted due to NPAs is 34% before IBC, 2016, which has risen to 92.3% after IBC, 2016. It is further observed that this impact is increased during post-IBC, 2016 era by 58.3% compared to pre-IBC, 2016 era. Hence, it can be concluded that NPAs do not have a significant influence on the market price of shares of SBOI during pre-IBC era, but the said influence is significant during post-IBC era.

17.11 FINDINGS AND CONCLUSION

- As noted and concluded in this study, for SBOI, the correlation coefficients obtained through SPSS for the two periods show a negative degree of correlation between the GNPA ratio and MPS, meaning that as the GNPA ratio rises, MPS falls. While the coefficient of correlation is still negative for both periods, it has somewhat decreased with the implementation of IBC, 2016 in comparison to the period prior to its implementation. The market price variation of shares accounted for by NPAs varies from 41.2% before IBC, 2016 to 39.7% after IBC, 2016, according to the coefficient of determination or R^2. Therefore, it can be said that both the pre-IBC and post-IBC eras of NPAs' influence on the market price of the SBOI's shares are negligible, and that NPAs' influence is lessened by 1.5% in the post-IBC, 2016 era compared to the pre-IBC, 2016 era.
- Further, the post-IBC era exhibits a much stronger negative degree of relationship with the GNPA ratio than the pre-IBC era, according to the correlation coefficients between the two variables as calculated by SPSS. The coefficient of determination, or R^2, indicates that the variation in RNW accounted for by NPAs is 34% before IBC, 2016 and has risen to 92.3% after IBC, 2016. Furthermore, it is noted that this influence has increased by 58.3% in the post-IBC, 2016 era compared to the pre-IBC, 2016 era. Therefore, it can be claimed that during the pre-IBC era, NPAs had no effect on the RNW of SBOI. However, during the post-IBC era, this impact became evident.

KEYWORDS

- **nonperforming assets**
- **State Bank of India**
- **Insolvency and Bankruptcy Code**
- **market price per share**
- **return on net worth**

REFERENCES

Bhardwaj, P., Chaudhary, I. (2018). A study of non-performing assets of commercial banks and it's recovery in India. *International Journal of Creative Research Thoughts (IJCRT) www.Ijcrt.Org*, *5*(2), 1365–1373. www.ijcrt.orgwww.ijcrt.org

Chaudhary, K., Sharma, M. (2011). Performance of Indian public sector banks and private sector banks: A comparative study. *International Journal of Innovation, Management and Technology*, *2*(3), 249–256. NPA, Profitability, private Sector banks. Public Sector banks, I.

Deshmukh, R., Dongre, A. P. (2018). Comparative analysis of non-performing assets of selected commercial banks in India. *International Journal of Scientific Research and Review*, *7*(12), 451–481.

Karmakar, S., Padhi, B. (2017). Non-performing assets and Indian banking sector: A descriptive study. *International Journal of Computational Engineering Research (IJCER) Non-Performing*, *07*(12), 66–72.

Mohanty, A. K. (2017). Factors affecting bank profitability of commercial bank of Ethiopia: An econometric analysis. *International Journal of Management, IT & Engineering*, *7*(8), 118–136.

Pandey, M., Vigg, M., Agarwal, C., Malyadri, P. (2020). Impact of the insolvency and bankruptcy code on NPA resolution. *Journal of Xi'an University of Architecture & Technology*, *XII*(IV), 923–932.

Satpal. (2014). A Comparative study of non performing assets in public and private sector banks in the new age of technology. In: *Research Article International Journal of Current Engineering and Technology* (Vol. 4, Issue 4). http://inpressco.com/category/ijcet

Singh, C. (n.d.). *An Essay on Banking and Macroeconomics: Role of Public Sector Banks in India.*

Tripathi, D. J. P., Mewada, S. (2016). Indian banking sectors: Rising trends & preventive measures of NPA's. *IOSR Journal of Business and Management*, *18*(10), 38–48. https://doi.org/10.9790/487x-1810013848

CHAPTER 18

Merger and Acquisition in Banking Sector of India in 2020: A Critical Analysis

RUCHITRA JOSHI KAPARWAN,[1*] SANGEETA BAGGA,[1]
DIVYA NEGI GHAI,[2] PRITI SHARMA,[1] and VASIM AHMAD[3]

[1]*Department of Commerce, Graphic Era Hill University, Dehradun, Uttarakhand, India*

[2]*SMCS, Shri Guru Ram Rai University, Dehradun, Uttarakhand, India*

[3]*Uttaranchal Institute of Management, Uttaranchal University, Dehradun, Uttarakhand, India*

*Corresponding author

ABSTRACT

In today's global economy, building strong, resilient banks is more important than ever. Since India's economic reforms in 1991, its banking sector has seen massive changes. One of the biggest trends has been mergers and acquisitions, where banks join forces to become larger and more competitive. The idea is that by combining, they can operate more efficiently, save costs, and create a more stable financial system. This research takes a close look at the major wave of bank mergers that swept through India's public sector in 2020. In this consolidation, 10 government-owned banks were merged into four larger entities. We wanted to see if these newly formed banks actually achieved the benefits everyone was talking about. Did they become more profitable? Did their financial health improve? By comparing

Industry 5.0: Innovation and Adoption for Sustainability.
Tilottama Singh, Pradeep Suri, and Zulkarin Jahangir (Eds.)

DOI: 10.1201/9781779641878-18

their performance before and after the mergers, our analysis found encouraging results. The combined banks saw their total business and assets grow, their branch and ATM networks expand, and crucially their levels of bad loans (nonperforming assets [NPAs]) decrease. In short, the evidence suggests that these mergers have successfully helped create a more robust and sustainable banking system for India, better equipped to compete on the world stage.

18.1 INTRODUCTION

Indian banks have demonstrated resilience against worldwide economic downturns. In addition, the financial industry and technology are ever-evolving, and banks frequently face intense competition to win over a sizable clientele. This is where the banks' use of mergers and acquisitions, or "M&As," becomes crucial. A merger occurs when two distinct companies combine to form a single organization, with one of them ceasing to exist as a corporation. The act of assuming control of a smaller business's share capital from a larger entity is known as an acquisition. The Indian banking sector in the past gone through the various changes related to policies; when the macroeconomic environment is stable, the banking sector will be able to operate successfully and profitably. The banks have seen drastic transformation over the past several years due to shifting regulations, new technology for competing in the global market, and mergers operations that have significantly altered the banking industry's size, structure, and activity. The two main categories of Indian banks are: (1) scheduled and (2) nonscheduled banks. The banks covered by the Reserve Bank of India (RBI) Act, 1934's Second Schedule are known as scheduled banks. Nationalized banks are another classification for the scheduled banks.

The banks in India had become a significant tool for facilitating the growth of the Indian economy by the 1960s. The Indian Prime Minister then revealed his plans in a paper titled "Stray Thoughts on Bank Nationalization" to the All-Party Meeting Committee's annual conference. The Indian government then issued the Banking Companies (Acquisition and Transfer of Undertakings) Ordinance, 1969, which went into force at midnight on July 19, 1969, nationalizing the country's 14 largest commercial banks. These institutions held 85% of the country's bank deposits. This ordinance was subsequently ratified by the President and formally enacted as a law on August 9, 1969. In 1980, six additional commercial banks were nationalized in a second wave. Approximately 91% of India's banking industry was under the authority of the Indian government after the second series of nationalizations (Ray, 2018). The examination of the amalgamation of three public sector banks in India is the focus of the essay. Punjab National Bank is a prominent Indian multinational bank and provider of financial services. United Bank of India and Oriental Bank of India were the two banks that made up this merger's other two banks in addition to the owning bank. After the year 2000, there were several mergers in the Indian banking industry. Acquisitions and mergers are advantageous to the merging organizations in a variety of ways. Some of them include profitability, synergies, growth in market share, heightened competitiveness, risk diversification, and increased shareholder value. The government just began its massive merger push with the intention of producing world-size banks to compete in the global market and, second, a mechanism to protect weak banks with rising nonperforming assets (NPAs) from the impacts of liquidation (Abdulazeez et al, 2020).

18.2 LITERATURE REVIEW

Sushant Sant, Mousumi Bhattacharya (June 2020): The sample size for this study is 3 years prior to and following unification, using secondary data from the BRICS countries. The information obtained from the Bloomberg Terminal was subjected to statistical analysis. Resources: The accounting ratio and the Kolmogorov–Smirnov test were utilized to ascertain whether the data were regularly distributed. To find out if any items defied the normal distribution, a test was calculated. Except for Brazil, there is no appreciable effect on any of the many criteria of the combined institutions. Nonetheless, the majority of the ratios in Brazil have decreased when compared to the period prior to the merger. South African and Chinese banks have benefited from the capital adequacy ratio (Al-Hroot et al., 2020).

Panagiotis Liargovas, Spyridon Repousis (January 2020): In this research paper, the following points were considered:

1. Sample size: Nine cases of merger and acquisition.
2. Secondary data: Balance sheet, annual reports, and financial statement. Tools: For event study—market model method, Average Abnormal Returns (AARs), and Cumulative Average Abnormal

Returns (CAARs). Operating performance—accounting ratios such as profitability ratio, productivity ratio, liquidity ratios, and *T*-test. The analysis shows that operating performance does not improve following mergers and acquisitions, while there are controversial results when comparing merged banks with the group of nonmerging banks.

Sri Ayan Chakraborty (June 2020): This paper examines the merger of Allahabad Bank with an Indian Bank, which occurred on September 16, 2019. The hope is that the merged bank will be more profitable, have fewer NPAs, and have a strong national brand following the merger. Financial and statistical methods [such as mean, analysis of variance (ANOVA), and *t*-test] were utilized for data analysis (Sharma and Modi, 2014). The analysis shows that by combining the strengths of both banks, the combined entity performed better overall and grew to become India's seventh-largest public sector bank. Furthermore, data indicates that Indian Bank outperforms Allahabad Bank in terms of revenue, profitability, and NPA reduction.

Ishwarya J (September 25, 2019): The trend of merger and acquisition activity in the Indian banking sector is examined in this research report. Financial metrics are used in the pre- and postmerger examination of State Bank of India (SBI) and its affiliates. The research findings indicate that while the profitability ratio of merged banks is likely to rise in the near future, there has been no significant change (Yogashree, 2020). However, when cost efficiencies are taken into account, the results are impressive. According to the study, there is a lack of beneficial consequences from the tendency of merging weak banks with strong banks. Therefore, the government should exercise caution when formulating rules for mergers in the Indian banking industry, particularly when encouraging mergers as economies of scale (Anand and Singh, 2008).

Sonia Singh; Subhankar Das (March 2018): The contradictory outcomes of the merger suggest that the policies and plans developed for the process are essential to achieving the intended outcome of a successful merger and acquisition. Regarding how a merger announcement affects stock prices, the study finds that while it is possible for expectations to diverge from actual performance, securities values are predicted to rise in the years after the merger. The study takes into account nonfinancial factors as well, such as how the combined banks' workers perceive themselves. It has been observed that employee involvement and transparency in the process instill trust in them, and as a result, the staff members appear to be (Goyal and Joshi, 2012).

G.M. Reshembai (December 2016): The combined bank's expansion was facilitated by the acquisition and merger of Kotak Mahindra Bank and ING Vysya Bank, according to the study of the transaction. In the wake of the merger, the company's nationwide reach has expanded, as has the number of deposits and advances, staff members, ATM networks, clients, and products and services offered. By using financial statistics and simple random sampling, the study concluded that the bank's financial performance is improving as interest income, profitability, liquidity, and share price all rise. The paper also addresses the challenges that the bank encountered after the merger, including as managing NPAs, paying employees on different pay scales, and handling variations in deposit rates. In conclusion, we may say that in this study project, the economies of scale are a central issue of debate (Kumar et al., 2023).

18.3 RESEARCH GAP

After considering the various research papers, the researcher identify that no research study was found related to mergers, which took place in 2020 taking into consideration the aspect of sustainability of banking ecosystem in long run, its positive social impact due to broader asset base, and growing deposit and advances, further building its capacity to deal with global banking economic system.

18.4 OBJECTIVES

1. To peruse the mergers occurred in India in the year 2020.
2. To estimate the pre- and postmerger performance of the resulting financial entities.
3. To conduct a detailed analysis of the effects of mergers on the Indian economy and its sustainability in long run.

18.5 RESULT ANALYSIS METHODS

Journals, academic publications, news announcements, the Internet, and official bank websites are some of the secondary sources the study examines to gather data on banks' performance. The study's nature is descriptive. The pre- and postmerger means are analyzed for every bank using SPSS.

Mergers Occurred in 2020	
Name of the Bank	**Mered with**
Punjab National Bank	Oriental Bank of Commerce and United Bank of India
Indian Bank	Allahabad Bank
Union Bank of India	Andhra Bank and Corporation Bank
Canara Bank	Syndicate Bank

18.6 ANALYSIS OF MERGED ENTITIES

18.6.1 PUNJAB NATIONAL BANK

The Punjab National Bank (PNB) was brought about by the Swadeshi movement, which was founded in 1984 by L Harkishan Lal, S Dayal Singh, Majithia L Chand, and L Dholan Das. PNB was the initial bank established using Indian capital. The bank has continued to grow since its establishment and has endured during the most difficult periods in history. The Indian banking industry was given a new direction in 1964 when 14 banks, including PNB, were nationalized. The bank merged seven times up until 2020; the most recent being in April 2020 when it joined forces with United Bank of India and Oriental Bank of Commerce (Table 18.1).

TABLE 18.1. PRE–POST ANALYSIS OF PNB

Parameters (in Crores)	Premerger	Postmerger
Total business	1,134,279	1,175,674
Total assets	774,949	830,666
No. of branches	6989	7040
No. of ATM	9255	13,506
Operating profit	12,995	14,739
Net profit	9975	336
ROA	1.25	0.04
CAR (basel III)	9.73	14.14
CASA ratio	43.51	44.05

Source: Website of PNB bank.

18.6.2 INDIAN BANK

The Swadeshi movement during the British Raj led to the founding of Indian Bank on August 15, 1907. One of the key figures involved in its establishment was V Krishnaswamy Iyer. The Indian government further nationalized the bank in 1969. Indian banks are essential to the growth of rural banking in India and help the country's farmers (Table 18.2).

TABLE 18.2 Analysis of Indian Bank Before and After the Merger

Parameters (in Crores)	Premerger	Postmerger
Total business	458,113	902,081
Total assets	310,052	626,005
No. of branches	2825	5744
No. of ATM	3892	5428
Operating profit	6498	11,396
Net profit	753	3005
ROA	0.26	0.50
CAR (basel III)	13.21	15.71
CASA ratio	35.4	42.89

Source: Official website of Indian Bank.

18.6.3 UNION BANK OF INDIA

Seth Sita Ram Poddar established Union Bank of India on November 11, 1919, with its headquarters located in Mumbai. The government owns roughly 89% of the total share capital of the bank (Table 18.3).

TABLE 18.3 Pre- and Post-Analysis of Union Bank of India

Parameters (in Crores)	Premerger	Postmerger
Total business	458,113	902,081
Total assets	310,052	626,005
No. of branches	4288	6007
No. of ATM	11,800	13,300
Operating profit	6498	11396
Net profit	753	3005
ROA	0.26	0.50
CAR (basel III)	11.78	12.81
CASA ratio	35.4	42.89

Source: Website of Union Bank of India.

18.6.4 CANARA BANK

Canara Bank was established in July 1906 in Mangalore, a small port city in Karnataka, by the brilliant thinker and humanitarian Shri Ammembal Subba Rao Pai. Canara Bank is widely known for its emphasis on client satisfaction. The bank has seen several phases of its growth trajectory in the course of its 100-year existence. After being nationalized in 1969, Canara Bank saw spectacular growth, becoming a major participant on the national stage in terms of geographic reach and clientele groups. The bank's commercial diversification throughout the 1980s was notable (Table 18.4).

TABLE 18.4 Pre- and Post-Analysis of Canara Bank

Parameters (in Crores)	Premerger	Postmerger
Total business	458,113	902,081
Total assets	310,052	626,005
No. of branches	6311	10,403
No. of ATM	8851	13,406
Operating profit	6498	11,396
Net profit	753	3005
ROA	0.26	0.50
CAR (basel III)	13.65	15.71
CASA ratio	35.4	42.89

Source: Official website of Canara Bank.

Following taking into account all of the relevant variables, the descriptive analysis reveals that after the merger, the organizations were able to realize anticipated advantages, including higher overall revenue, total assets, branch and ATM counts, and so forth. The study includes a ratio analysis as well as an examination of the pre- and postmerger means worth of understudy banks to ascertain whether the merger actually produces the economic progress that is claimed.

18.7 COMPARISON OF PRE- AND POSTMERGER RATIO: ACROSS THE BANKS

Ratio	Codes
Net profit ratio	ACR 1
Return on assets	ACR 2
Return on equity	ACR 3
Advances-to-deposit ratio	ACR 4
Net NPA's to net advances ratio	ACR 5

1. ACR 1: Net profit ratio

ACR 1	Pre	Post
Canara Bank	0.1	0.04
Punjab National Bank	–0.19	0.03
Indian Bank	0.02	0.08
Union Bank of India	–0.09	0.04

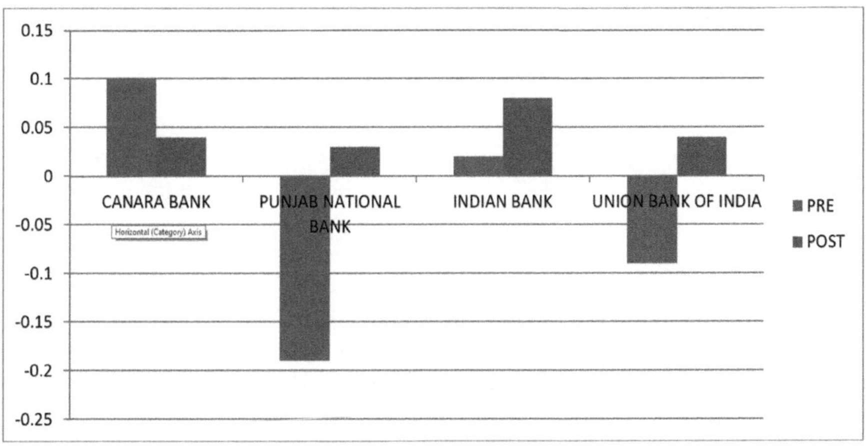

FIGURE 18.1 ACR 1: net profit ratio.

Figure 18.1 depicts the evolution of all banks' pre- and postmerger ratios in 2020. With the exception of Canara Bank, all banks' net profit ratio has been growing after the merger.

Report:

ACR 1	Pre	Post
Mean	**–0.0400**	**0.500**

Interpretation: ACR 1 is the code for the net profit ratio. A mean value is calculated to assess the merger's effects both before and after. The results showed that the NPR improved from before (M = –0.04) to after (M = 0.05).

2. ACR 2: Return on equity

ACR 2	Pre	Post
Canara Bank	0.6	0.23
Punjab National Bank	–1.25	0.15
Indian Bank	0.12	0.5
Union Bank of India	–0.59	0.27

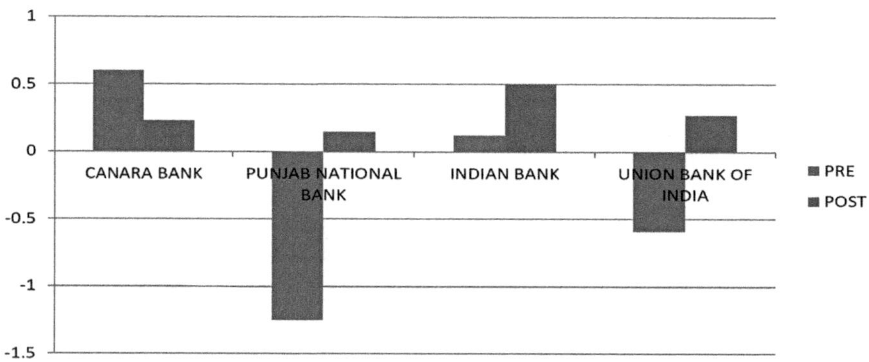

FIGURE 18.2 ACR 2: return on assets.

Figure 18.2 depicts the evolution of all banks' pre- and postmerger ratios in 2020. With the exception of Canara Bank, all banks' net profit ratios have been growing after the merger.

Report:

ACR 2	Pre	Post
Mean	–0.2800	0.2875

Interpretation: ROA (ACR 2) is the code for this. A mean value is employed in order to compute the impact of the merger both before and after. As per the findings, the net profit ratio had raised from –0.2800 before to –0.2875 after.

3. ACR 3: Return on equity

ACR 3	Pre	Post
Canara Bank	1.4	6.71
Punjab National Bank	–24.2	2.41
Indian Bank	2	2.5
Union Bank of India	–15.57	6.68

In 2020, Figure 18.3 shows the progression of all banks' before and after merger ratios. Following the merger, all banks' return on equity has increased.

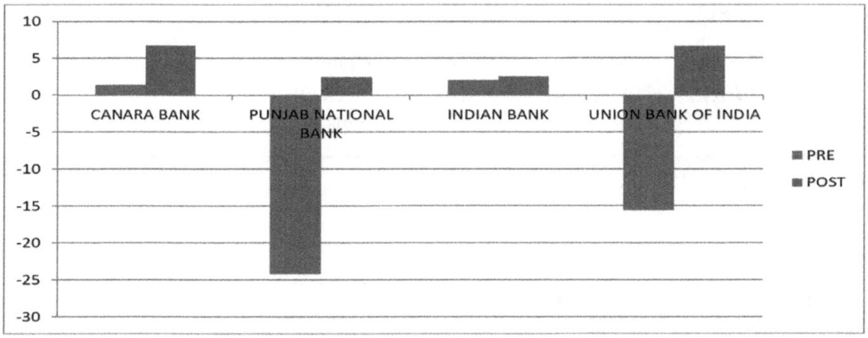

FIGURE 18.3 ACR 3: return on equity.

Report:

ACR 3	Pre	Post
Mean	–0.2800	0.2875

Interpretation: ROE is coded as ACR 3. To analyze the effects of the merger both before and after, mean values are calculated. The return on equity grew significantly between the pre- and postcrisis periods (M = –0.2800 and M = 0.2875).

4. ACR 4: Advance to deposit ratio

ACR 4	Pre	Post
Canara Bank	71.4	63.22
Punjab National Bank	67.79	61
Indian Bank	74.88	67.65
Union Bank of India	71.39	64.1

In the postmerger period, all banks' advance to deposit ratios decreased, as seen in Figure 18.4.

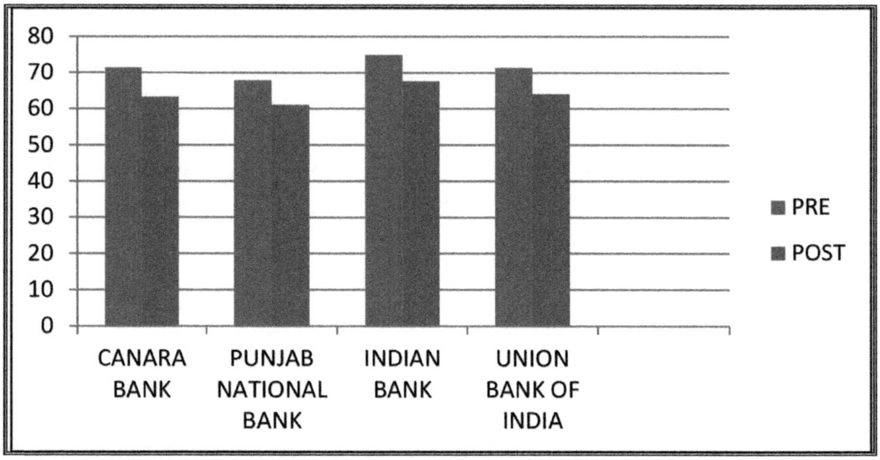

FIGURE 18.4 ACR 4: advance to deposit ratio.

Report:

ACR 4	Pre	Post
Mean	71.365	63.992

Interpretation: The advance to deposit ratio is represented by the code ACR 4. The consequences of the merger are assessed both prior to and following by computing mean values. From before (M = 71.36) to after (M = 63.99), it was evident that the advance to deposit ratio had dropped. This implies that the bank made advances to its clients solely using deposits from itself. The merger has reduced the reliance on funding from other banks or the stock market.

5. ACR 5: Net NPA's to net advances

ACR 5	Pre	Post
Canara Bank	5.37	3.82
Punjab National Bank	6.56	5.73
Indian Bank	3.75	3.37
Union Bank of India	6.84	4.59

Figure 18.5 shows the decreasing movement of NPA's after merger in all the banks. It is a positive indicator because the lower NPAs indicate that the bank's loans are of good standard.

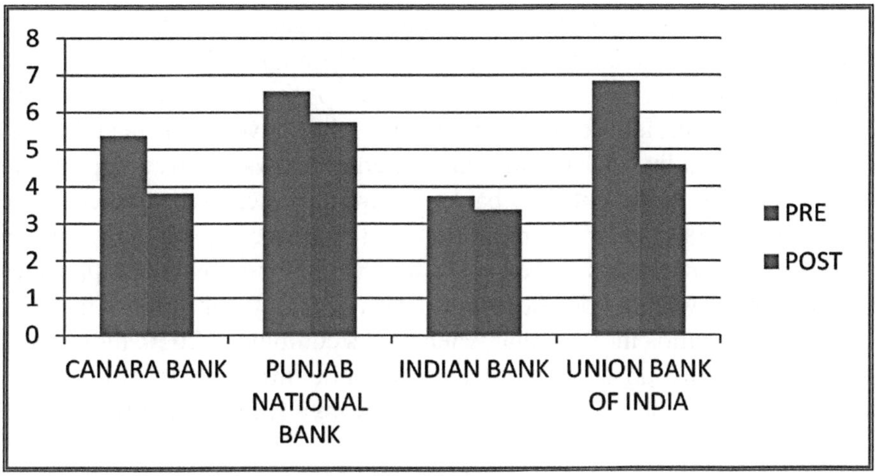

FIGURE 18.5 ACR 5: net NPA's to net advances.

Report:

ACR 5	Pre	Post
Mean	5.630	4.377

Interpretation: Net NPAs to net advance is coded as ACR 5. Mean values are computed in order to evaluate the impact of the merger both before and after. There was a noticeable drop in the ratio of net NPAs to net advance from before (M = 5.630) to after (M = 4.37). In this instance, a lower mean value is preferred since a higher percentage of NPAs adversely affects the liquidity and profitability of the business. The companies' reduced valuation after the merger makes it more likely that they will produce a profit in the future (Mondal et al., 2017).

Analysis: Prior to and during the merger, the analysis across the banks indicates that after merger, the results show the increasing impact on the ratios under the study, that is, net profit ratio, which reveals the profitability after all costs, return on asset, which indicates the capital intensity of entity, and return on equity, which measures the financial performance of the banks, whereas advance to deposit ratio shows the decreasing values after merger;

this ratio indicates the liquidity of banks by comparing total deposit to total loans and provides the appearance that the bank made advances to its customers using just its own deposits; therefore, it has a lower reliance on outside finance. This study's final ratio, net NPAs to net advances, indicates a decrease in the study following the merger. It acts as an indicator of the overall quality of the bank's loan portfolio. High NPAs lead to low funds, which in turn result in lower interest revenue. Another problem with high NPAs is that banks aim for fewer NPAs since they have a detrimental effect on their profitability. As the ratio after merger shows positive impact and win–win situation for both the banks in terms of increased market share, strengthen to face global competition and improved profit will help in providing better services and efficient survival in long run, which provides sustainable growth and development in banking sector in India. The findings of all the ratios indicate that when banks combine in 2020, those businesses will have a positive social and economic influence, from the union as anticipated by the study (Nagaria and Lavanya Kumari, 2020).

18.8 CONCLUSION

The chance to build a more resilient and diversified financial organization is one of the advantages of bank mergers. The combined company will be competitive in the global market since it will have combined the best features of the two banking systems.

The merger is a very effective instrument for bank expansion and growth. The banking sector can see tremendous expansion in operations, reduced considerably costs, and reduced competition through mergers and acquisitions since these transactions eliminate rivals from the banking sector. Therefore, it is reasonable to conclude that mergers are one of the most important instrumental tools for strategic and sustainable approach toward social and economic development in banking sector in India.

KEYWORDS

- **merger 2020**
- **banking sector**
- **Indian banks**

REFERENCES:

Abdulazeez, D. A., Suleiman, O., Yahaya, A. Impact of merger and acquisitions on the financial performance of deposit money banks in Nigeria. *Arabian Journal of Business and Management Review*.

Al-Hroot, Y. A., Al-Qudah, L. A., Alkharabsha, F. I. (2020). The Impact of horizontal mergers on the performance of the Jordanian banking sector. *Journal of Asian Finance, Economics and Business*, *7*(7), 49–58.

Anand, M., Singh, J. (2008). *Impact of Merger Announcements on Shareholders' Wealth: Evidence from Indian Private Sector Banks*. Sage Publication.

Goyal, K.A., Joshi, V. (2012). Merger and acquisition in banking industry: A case study of ICICI Bank Ltd. *International Journal of Research in Management*, *2*(2).

Kumar, R., Goel, R., Singh, T., Mohanty, S. N., Gupta, D., Alkhiyyat, A., Khanna, R. (2023). Sustainable finance factors in Indian economy: Analysis on policy of climate change and energy sector. *Fluctuation and Noise Letters*, *23*(2), 2440004–2440339 https://doi.org/10.1142/S0219477524400042

Mondal, G. C., Pal, M. K., Ray, S. (2017). Influence of merger on performance of indian banks: A case. *An International Peer-reviewed Journal*, *32*, ISSN 2422-846X.

Nagaria, S., Lavanya Kumari, R. (2020). Bank crisis: The current scenario of merger and acquisition in India. *International Journal of Recent Technology and Engineering (IJRTE)*, *8*(5), ISSN: 2277-3878.

Ray, S. (2018). Impact of merger on efficiency of acquiring Bank. *An Analytical: (Paper)*, *9*(3). ISSN 2222-2847 (Online).

Sharma, A., Modi, G. N. (2014). A comparative study of financial performance. ISSN - 2250-1991.

Yogashree, C. (2020). Effect of amalgamation on financial performance: A case study. *International Journal of Science and Research (IJSR)* ISSN: 2319-7064.

CHAPTER 19

Workplace Spirituality and Mediating Effect of Trust for Organizational Performance

SATISH CHANDRA MISHRA* and PANKAJ KUMAR

Department of Management Studies, DIT University, Dehradun, Uttarakhand, India

*Corresponding author

ABSTRACT

Organizations, in their endeavor to improve their performance, are shifting to a hybrid workforce model by creating digital workplaces with the provision to work from remote, where employees with solitary working conditions feel isolated as there is no face-to-face communication in the present technologically driven environment. Industry 5.0 is more concerned about human beings, due to smart technologies, and human–robotic coworking, where human interaction with each other has decreased and increased human-related problems, which is detrimental to employees, organizations, and society in the long run. This study, based on literature, considered models of WPS that impact organizational performance (OP) and where trust mediates OP, and concludes that there is a positive and significant impact of WPS on OP and on employee's work–life balance and well-being. As the outcome of the study, a model is developed based on theoretical and empirical evidence that will be beneficial for the performance of the present organization and take care of employee well-being and their work-life balance, the major concern of Industry 5.0. The academic

Industry 5.0: Innovation and Adoption for Sustainability.
Tilottama Singh, Pradeep Suri, and Zulkarin Jahangir (Eds.)

DOI: 10.1201/9781779641878-19

and managerial implications and recommendations for future studies, and the limitations have been highlighted.

19.1 INTRODUCTION

There is a growing awareness among managers and academics of the importance of spirituality in the workplace, for having the potential to improve organization performance, and that organizational resilience for survival is strengthened in times of crisis, and that it has solutions to many of the human resource problems being faced by the present organizations. Industry 5.0 is more concerned about human being and, in the present time, the use of smart manufacturing, AI, robotics, cloud, Internet, big data, cyber security, digital twins, printing, and augmented reality have increased but have raised concerned about employees work life balance and well-being, as employees lack face-to-face interaction with each other. Industry 5.0's use of AI, robots, and the Internet has greatly benefited businesses by increasing productivity and efficiency; it has also created new difficulties for worker well-being (Hakan, 2023). The impact of human–robot coworking industrial systems on human psychological implications is not known, since new technologies have psychological effects on human being is a matter of concern and needs to be investigated (Ikumapayi et al., 2023).

In the third millennium, spiritually based organizations that are classified as organizational fourth waves are crucial as spirituality has the power to enhance people's lives, inspire environmental consciousness, and encourage them to look beyond materialism (Garg, 2020; Vasconcelos, 2015). Research on workplace spirituality (WPS) in businesses has revealed that it improves business success (Garcia-Zamor, 2003; Jurkiewicz and Giacalone, 2004). Garg (2020) reports that a number of businesses, including Apple Computers, Ford Motor Company, Yahoo, Hewlett-Packard, AT&T, DuPont, Google, IBM, and Yahoo, have seen increases in productivity as a direct result of implementing a spiritual workplace culture. During the worldwide pandemic, WPS showed how organizational resilience for survival is bolstered during a crisis. A company must be spiritually led because it offers solutions to workers who, because of remote work environments and hybrid work models, feel demotivated and alone in the absence of face-to-face interaction (Pew Research Center, 2020; Yin and Marhous, 2022).

The study of spirituality in the workplace is still in its early stages as an area of academic research in the 21st century. WPS plays a crucial role in both public and private organizations by fostering trust, commitment, engagement, workforce agility, and empowerment. This is particularly important in the current era of globalization and digitalization, where organizations strive for survival and excellence. Retaining talented and committed employees has become a significant concern for the human resource department. The study aims to explore the action of WPS on organizational performance (OP) and the mediating role of trust. Industry 5.0 emphasizes the importance of social goals in smart manufacturing systems, prioritizing the well-being of individuals and promoting sustainable development for all (Leng et al., 2022). Implementing policies and fostering innovation that prioritize the well-being of individuals can have a profound impact on the sustainability of the industrial revolution and human society. One effective approach is to provide spirituality training to employees, which can enhance their skills, enable them to handle heavy workloads, and effectively manage both their work and relationships in the workplace (Saharan and Sharma, 2022).

19.2 SIGNIFICANCE OF THE STUDY

The result indicates that WPS, OP, and trust are related, and trust plays a crucial role as a mediator for OP. Various factors such as stress, job satisfaction, work attitude, organizational commitment, job engagement, innovative work behavior, job performance, work agility, knowledge sharing behavior, employee engagement (EE), employee productivity, organizational citizenship behavior, employee sustainable behavior and commitment, trust, and empowerment all have an impact on OP. The recognition will help the organizations to understand the importance of integrating spirituality into their operations in order to have an advantage over their competitor in the present era of technological advancement and globalization. This is in line with Industry 5.0, which emphasizes a human-centric approach, employee well-being, and social goals beyond just employment and business prosperity (Saharan and Sharma, 2022).

19.3 LITERATURE REVIEW

19.3.1 WORKPLACE SPIRITUALITY

The purpose of WPS is to comprehend an individual's ultimate objectives in life, establish strong connections with colleagues and individuals in the workplace, and guarantee that one's beliefs and the company's objectives are harmonious or in alignment with each other (Milliman et al., 2003). According to Ashmos and Duchon (2000), WPS recognizes that individuals derive fulfillment and support from meaningful work within a broader community. WPS, or work passion syndrome, has been delineated by several scholars using different definitions. However, there is a general agreement that WPS is present when individuals have a strong dedication to their job and possess a clear goal. WPS is acknowledged as having significance at both the organizational level and at the individual level. At the individual level, employees experience a spiritual connection to their workplace and work. At the organizational level, the spiritual value of WPS becomes an essential component of the organizational culture, leading to improved behavior, decision-making, and resource allocation. Research on WPS has found that it has a significant impact on various aspects of organizational functioning. Specifically, it affects organization performance, stress levels, job satisfaction, work attitude, organizational commitment, job engagement, knowledge sharing behavior, innovative work behavior, organizational citizenship behavior (OCB), EE, employee productivity, employee sustainable behavior, job performance, and the factors of commitment, trust, and empowerment. Workplace satisfaction (WPS) is crucial since it reduces job-related stress among workers and enhances their dedication to the firm (Ashfaq and Irum, 2020). Furthermore, it is becoming increasingly significant due to its ability to address certain human resource difficulties (Garg, 2020). According to Yin and Mahrous (2022), the corporate world now recognizes that for a company to succeed in the digital age, it must demonstrate that its business practices are rooted in ethics, morals, and spirituality in the workplace.

19.3.2 ORGANIZATIONAL PERFORMANCE

OP refers to the position of an organization in the corporate world, which is determined by its accomplishments in relation to its objectives. This includes overall performance based on managerial skills, values, effectiveness, efficiency, leadership skills, recognition of desirable employer attributes, responsibility, competitiveness, and individual performance that can influence the performance of companies in the short, medium, and long term (Conţu, 2020). Performance of an organization is ultimately evaluated based on the tactics it employs to effectively utilize its operations, marketing, and human resources in order to satisfy stakeholders (Richard et al., 2009). The focus of OP is on the correlation between performance and organizational goals (effectiveness), stakeholder satisfaction (relevancy), and organizational resources (efficiency) (Jenatabadi, 2015; Peterson et al., 2003). Organizations can improve their performance by adopting a spiritual perspective, which can lead to increased efficiency, production, and profitability (Jurkiewicz and Giacalone, 2004). According to Lloyd (1990), firms that practice WPS showed superior performance compared to organizations that do not or have limited spiritual practices by 86%. Organizations can enhance their performance by implementing WPS (Neck and Milliman, 1994; Pandey et al., 2009). When individual knowledge is combined with organizational knowledge, it results in innovations that improve the performance of the organization (Matzler et al., 2008). There is a strong correlation between employee motivation and the efficacy and performance of an organization. The organization's effectiveness is enhanced by the performance of its employees (Dobre, 2013). The performance of an organization is influenced by various factors, including compensation and advancement opportunities, fairness, working conditions, employee benefits, job security, camaraderie, incentives, recognition, and a workplace that embraces spirituality. This integration of spirituality among employees has been found to enhance their performance (Bharadwaj and Jamal, 2020).

19.3.3 TRUST

According to Jurkiewicz and Giacalone (2004), trust is defined as the state of having the confidence to rely on the integrity and honesty of an organization and its representatives and according to Mayer and Gavin (2005) trust is defined as the readiness of one party to expose themselves to the acts of another party, with the expectation that the other party would carry out a specific activity that is significant to the trusting party, regardless of the capacity to supervise or control the other party.

19.4 WORKPLACE SPIRITUALITY, ORGANIZATIONAL PERFORMANCE, AND TRUST

The correlation between WPS and trust is complex, since WPS fosters an atmosphere of trust inside the firm. Spiritual organizations foster an environment of transparency and confidence, facilitating the dissemination of information, ideas, and social interaction between management and staff (Garg and Rastogi, 2006). In high-performance organizations, management and managers recognize that trust is the foundation for relationships at all levels of the organization. They establish strong ethical standards by treating people fairly and valuing employee loyalty. They build relationships with employees, which fosters trust and belief in the managers and coaches as honest and sincere role models. This support and protection from external work interference allows employees to work with integrity (de Waal, 2007). A higher level of WPS is associated with increased levels of employee flourishing and job engagement (Van der Walt, 2018) and is acknowledged for its ability to enhance employee trust (Van der Walt and de Klerk, 2014). The level of trust between workers and employers is a crucial parameter in determining the long-term viability of the business and EE level in their work (Breevaart et al., 2014). The organization's culture and interpersonal interactions, built on trust, foster individual performance and collaboration among employees, leading to harmony, motivation, and unanimity (Afsar and Rehman, 2015; Soliman et al., 2021). The organization's motivating culture, characterized by reciprocal connectivity, friendliness, and trustworthiness among workers in the workplace, leads to a rise in overall performance of the organization (Asgari and Heidariand, 2015; Hassan et al., 2016). Research shows that trust, in conjunction with spirituality and freedom, fosters a favorable work environment that positively impacts employee performance, learning, and teaching, and is attributed to improved coordination among individuals (Bird et al., 2012; Dirks and Ferrin, 2002; Hashim et al., 2020; Nedkovski et al., 2017). Trust in the workplace plays a crucial role in improving OP by acting as a mediator between human resource practices and performance. It also enhances social interactions among employees and has the potential to increase effectiveness and efficiency. Additionally, when an organization fosters trust, it encourages employees to take risks without fear of harassment. An organizational system based on trust among employees has a significant impact on organizational commitment, employee satisfaction, and organizational satisfaction. This system involves a bottom-up approach led by employees and a top-down approach driven by management and policies (Diffie-Couch, 1984; Zand, 1972).

19.5 THEORETICAL FRAMEWORK

WPS foster trust among employees, which contributes to a pleasant working environment, that is accomplished by meeting employees' desires and meeting their internal needs, which in turn makes work more meaningful and compassionate, and it also generates organizational values and a sense of community, which includes ethics, honesty, and integrity for team satisfaction (Daniel, 2010; Hassan et al., 2016; Marques et al., 2005). The same is being explained through various models that affect employees' performance and, in turn performance of the organization.

19.6 VARIOUS MODELS MEADIATING EFFECT OF OT IN THE RELATIONSHIP OF WPS AND OP

19.6.1 TRUST MEDIATES AMONG WPS AND JOB SATISFACTION (HASSAN ET AL., 2016)

The data was collected from universities, colleges, and the education sector of Pakistan by the qualitative method; and for measuring trust, the scale developed by Ayoko and Pekerti (2008) was used. The Minnesota Satisfaction Questionnaire (Weiss et al., 1967) was employed to measure job satisfaction, and the scale developed by Petchsawang and Duchon (2009) was utilized to assess work–life balance. The data was analyzed using statistical analysis software (SPSS), and demographic characteristics were examined along with reliability analysis and regression analysis to interpret the findings. The analysis determined that there is a strong positive relationship between WPS and trust, and trust plays a substantial role in mediating the influence of WPS on job satisfaction (Figure 19.1).

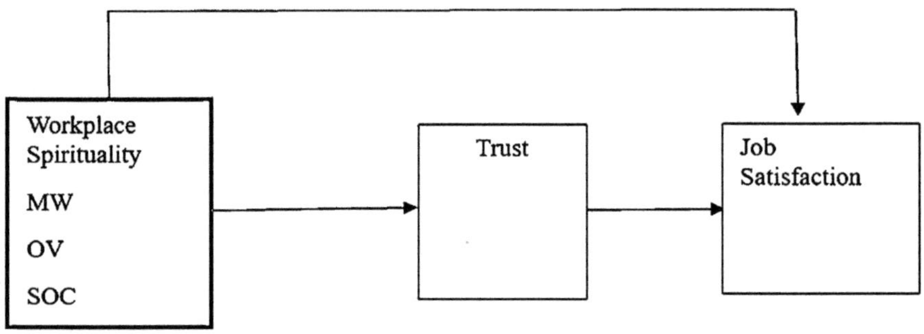

FIGURE 19.1 Model proposed by Hassan et al. (2016).

19.6.2 EFFECT OF ORGANIZATIONAL SPIRITUALITY WITH ORGANIZATIONAL TRUST (OT) AS MEDIATOR ON KNOWLEDGE SHARING ATTITUDE (KHARI AND SINHA, 2018)

Three items of Kolodinsky et al. (2008) organizational spirituality scale was used for measuring organizational spirituality, for measuring psychological flourishing eight items from scale of Diener et al. (2010), for measuring knowledge sharing attitude based on four items of Bock et al. (2005) scale and for measuring trust in organization four items from Nyhan and Marlowe (1997) scale were used. The findings demonstrate trust toward the organization, among coworkers, and between supervisors and workers, and concluded that OT and psychological flourishing fully mediated between organizational spirituality and knowledge sharing attitude (Figure 19.2).

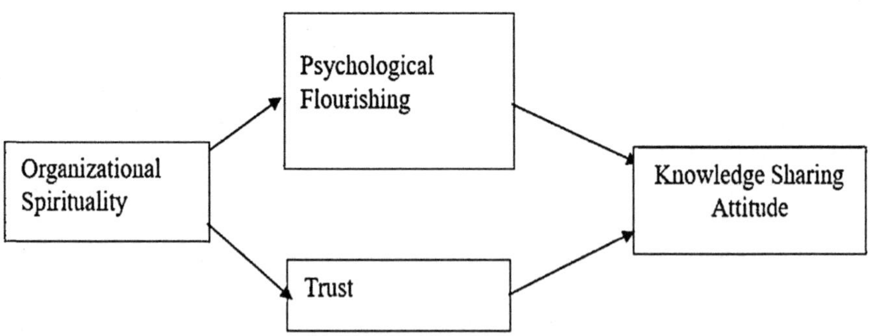

FIGURE 19.2 Model proposed by Khari and Sinha (2018).

19.6.3 WPS AND JOB PERFORMANCE, TRUST AS MEDIATOR (DANIEL, 2019)

Six items from Cook and Wall (1980) scale were used to measure trust. For measuring WPS, five items on inner life, seven items on meaningful work, and nine items on sense of community from Ashmos and Duchon's (2000) scale were used, and for job performance used nine items from the Mott group (1972) scale. Concluded after analyzing the data by using structural equation modeling that by incorporating trust as a mediator, it may assist by improving job performance, and there is no mediation of trust in the relation among sense of community and inner life with job performance (Figure 19.3).

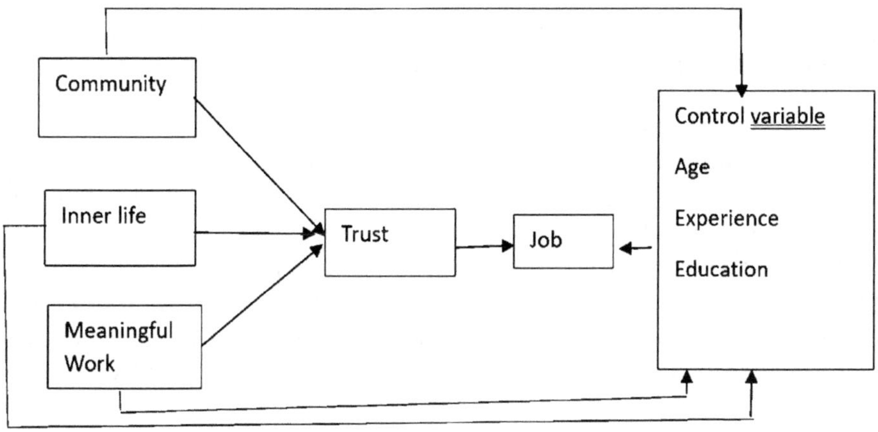

FIGURE 19.3 Model proposed by Daniel (2019).

19.6.4 WPS AND MEDIATING EFFECT OF TRUST ON ORGANIZATIONAL CITIZENSHIP BEHAVIOR (OCB) AND EMPLOYEE ENGAGEMENT (EE) FOR ENHANCEMENT OF PERFORMANCE (RANASINGHE, 2021)

Conceptual model based on existing literature established this hypothesis theoretically and concluded that WPS has impacts on employee OCB and EE, employee trust increases with the increase of WPS, and the increased employee trust enhances employee OCB and EE (Figure 19.4).

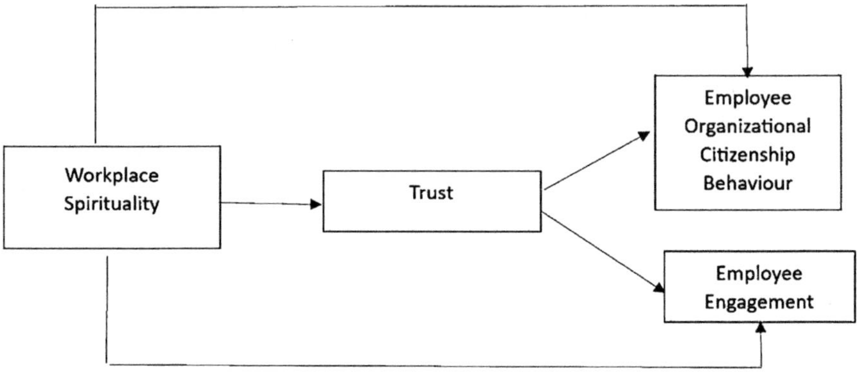

FIGURE 19.4 Model proposed by Ranasinghe (2021).

19.6.5 WPS AND MEDIATING EFFECT OF TRUST ON EMPLOYEE BURNOUT (KHARI AND SINHA, 2021)

Model based on literature, on the relationship among WPS, OT, and employee burnout, this was established that employee burnout has an adverse effect on employee commitment, EE, employee retention, personal accomplishment, emotional exhaustion, depersonalization, and their performance at work. Concluded that WPS positively affects organizational trust (OT) as this strengthens employee propensity to trust the organization, and employee's perception toward organizational integrity and OT mediates the negative relationship among WPS and employee burnout (Figure 19.5).

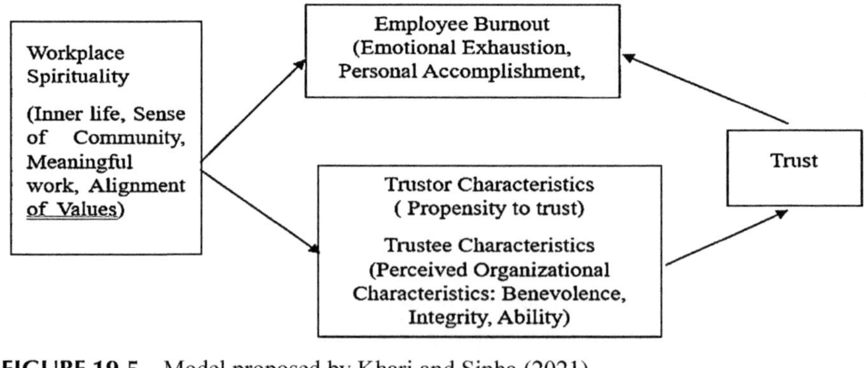

FIGURE 19.5 Model proposed by Khari and Sinha (2021).

19.6.6 TRUST AS MEDIATOR FOR WPS INFLUENCE ON KNOWLEDGE SHARING BEHAVIOR AND WORK ENGAGEMENT IN WORK (KHAN ET AL., 2022)

The WPS scale of Milliman et al. (2003) was used for measuring WPS. Twenty-one items of WPS consisting of six items for meaningful work, seven items for sense of community, and eight items for alignment with organization. For measuring trust, seven items from the trust scale of Robinson and Rousseau (1994), used the nine-items scale developed by Schaufeli et al. (2006) for measuring work engagement and for knowledge sharing behavior, five items scale of Bock et al. (2005) was used. For analyzing the data, SPSS was used. Concluded that WPS has a positive linkage with knowledge sharing behavior and work engagement, and WPS has a positive correlation with mediating variable trust with knowledge sharing behavior and trust with work engagement (Figure 19.6).

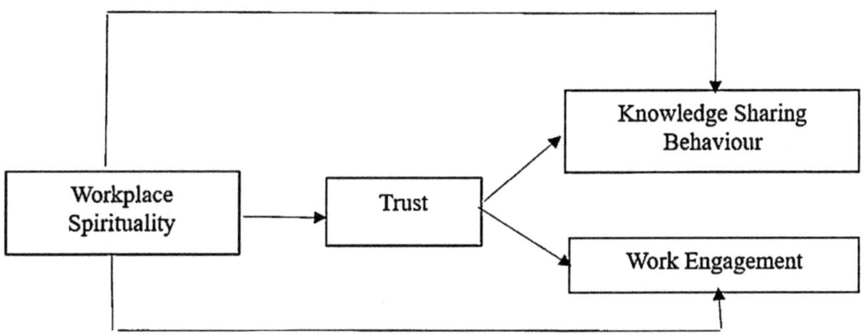

FIGURE 19.6 Model proposed by Khan et al. (2022).

19.6.7 ORGANIZATIONAL TRUST (OT) MEDIATING EFFECT AMONG WPS AND ORGANIZATIONAL CITIZENSHIP BEHAVIOR (OCB) (SÖKMEN AND YASREBDOOST, 2022)

Based on samples from two countries: Turkey and Iran. The questionnaire developed by Kinjerski and Skrypnek (2006) was used for measuring WPS, the Nyhan and Marlowe (1997) scale was used for measuring OT, and the Podsakoff et al. (1990) scale was used for measuring OCB. Concluded that WPS had a direct and positive action on OCB and OT, and OT has direct significant and positive effect on OCB in both the countries, but due to cultural differences between Turkey and Iran, in Turkey, the relationship between WPS and OT is positive, and OT mediates the relationship between WPS and OCB but in Iran there was no significant effect of WPS on OT and there was no OT mediation between WPS and OCB (Figure 19.7).

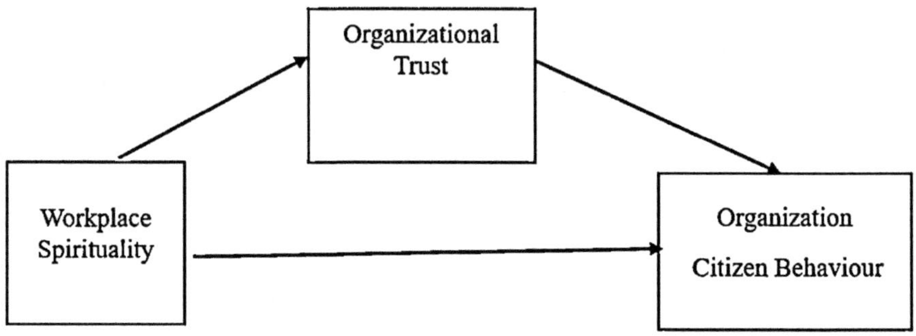

FIGURE 19.7 Model proposed by Sökmen and Yasrebdoost (2022).

19.7 DISCUSSION AND CONCLUSION

Hassan et al. (2016)'s study, sample consisted of teachers from the education sector in Pakistan, found that WPS is positively related to trust and trust mediates between WPS and job satisfaction. Khari and Sinha (2018) observed that employee's trust and psychological flourishing toward the organization increases when they perceive that the organization ensures its commitment to higher values, as OT and psychological flourishing fully mediate the result between organizational spirituality and knowledge sharing attitude. Daniel (2019) study was based on employees from education, health, manufacturing, and software sectors from north and central Mexico inferred that inner life and meaningful work has positive linkage with job performance, has an important role in enhancing employee performance by including trust as mediator and found that meaningful work has partial mediating effect with job performance and that, there is no mediation linkages among inner life, sense of community with job performance. Ranasinghe's (2021) work, based on the existing literature on the limited work in the Asian countries and in Sri Lanka, concluded that the employees trust increases with the increase in WPS, and the increased employee trust results in improved employee OCB. Khari and Sinha (2021) there is mediating effect of OT between WPS and employee burnout and suggested that the challenges for the organization for tackling employee burnout in the present situation, is to improve spirituality in the organization so that OT between the organization, employees and interpersonal trust improves, required for tackling employee burnout and that ultimately lead to enhancement of performance. Khan et al. (2022) established that for increased employee learning and information sharing behavior, a spiritual environment is necessary as WPS is related to work engagement through trust in the organizations. Organizations which promote trust establish closer relationships between employees, resulting in sharing their knowledge and experiences with other coworkers. Sökmen and Yasrebdoost (2022) study from Turkey's Gazi University and Iran's Azad University of Tabriz are different, found that in Turkey, WPS directly and positively affect OCB and OT directly and positively affect OCB and WPS has positive effect on OT and OT mediates the linkage between WPS and OCB, the finding of Azad University in Iran is different where WPS directly and positively affects OCB and OT positively and directly affects OCB however OT do not mediates between the relation of WPS and OCB as there is no significant effect of WPS on OT, the difference may be due to cultural differences in these two countries. OCB improves effectiveness of work groups and organization performance (Podsakoff and MacKenzie, 2014). OCB helps to improve OP (Jahangir et al., 2004). The conceptual model based on analysis of the literature that WPS enhances trust and OP by improving employees job satisfaction, employee and work engagement, knowledge sharing attitude and behavior, OCB.

The model is developed for the enhancement of OP. In this model, OT mediates between WPS and OP at Figure 19.8.

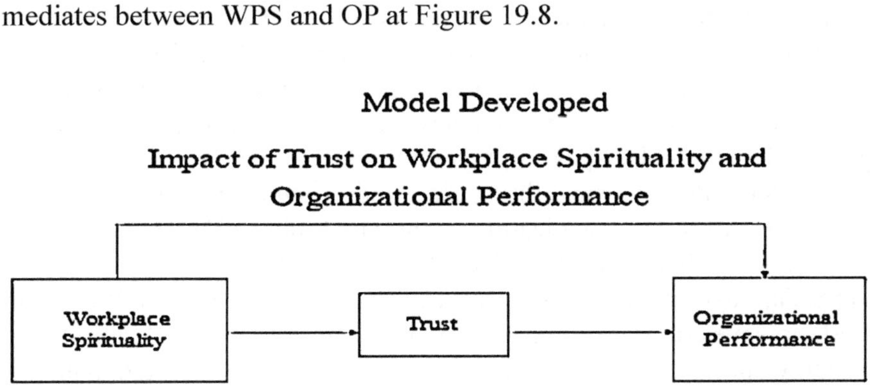

FIGURE 19.8 Proposed by authors.

19.8 IMPLICATIONS

The organizations are technologically driven with a focus on achieving higher efficiency and productivity, are shifting to a hybrid workforce model, and by creating digital workplaces with the provision for employees to work from remote and flexible locations. Industry 5.0 due to human-centric approach prioritizes the fundamental needs and interests of humans in the manufacturing process, moving away from development driven only by technology, and adopting a more focused approach on human and societal well-being that acknowledges the capacity of the industrial sector to contribute to social objectives that go beyond employment and economic expansion and aims to establish a resilient system that adheres to environmental limits and prioritizes the well-being of workers (Xu et al., 2021).

The future with technological advancement will reduce dependency on number of manpower while being more dependent on the capabilities of employed individuals, which leads to stress, burnout, and frustration can look toward WPS as the research on WPS established among managers and academicians that many of the human resource problems can be taken care which is critical in the present globally competitive environment. Organizations to succeed need to encourage spirituality at the workplace for enhancing trust among its associates at the workplace, which is important for employee's performance and job performance, as this ultimately improves organization performance (Daniel, 2019). WPS helps in building mutual trust between organizations and their workers, that increase in trust leads to greater satisfaction in work and OP, and is equally applicable to academics in the educational sector (Hassan et al., 2016). In order to remain competitive in the global market, the organization with spirituality and exhibits trust helps to develop a sense of connectedness of employees with each other and shapes employee attitude that increases their dynamic and strategic flexible capabilities and builds a collective knowledge base for the benefit of the organization (Khari and Sinha, 2018). Organizations may consider this for competitive advantage as there is a linkage among WPS, employee organization, citizen behavior, and EE being influenced by trust; however, due to limited work in the Asian countries and in Sri Lanka on these subjects, more work is required by the academics and managers (Ranasinghe, 2021). For the enhancement of organization performance in the long run, WPS aids in fostering trust in the organization by the employees, as this strengthens the positive attitudes and beliefs toward the organization (Khari and Sinha, 2021). A trusted atmosphere in a spiritual workplace encourages employees to improve knowledge sharing behavior among organizational members and ultimately helps the organization to perform better (Khan et al., 2022). The components that affect the relation among WPS, OT, and OCB at the regional, organizational, and national culture may be examined as the impacts are different (Daniel, 2019; Sökmen and Yasrebdoost, 2022). The clarity will help the managers and academics to acknowledge the significance of WPS and the mediating effect of trust for enhancing OP to be competitive in the present era of digitalization and globalization and Industry 4.0 and Industry 5.0 objectives.

19.9 LIMITATIONS

Hassan et al.'s (2016) study focused on the educational sector in Pakistan and recommended studies in the other sectors along with other variables, and others as moderators and mediators to check the impact on trust. Sinha and Khare's (2018) samples were limited to only three private telecom companies in India. Daniel's (2019) study, limited to the data from central and northern Mexico, which have different subcultures, and results may be different in other countries and different regions of the world. The work of Ranasinghe (2021) was limited to Sri Lanka. Future research to test the relationship among WPS, employee burnout, and trust may be explored (Khari and Sinha, 2021). Khan et al. (2022) work was limited to different organizations, but limited to one parent company. Sökmen and Yasrebdoost's (2022) research was limited to the University of Gazi in Turkey and Iran's Islamic Azad University, reflecting to the scope that results due to cultural differences between subcultures, cultures, and countries. It is significant to define the relationship among organizations may be influenced due to subcultures and cultures at the national level (Daniel, 2019; Ranasinghe, 2021; Sökmen and Yasrebdoost, 2022).

19.10 FUTURE STUDIES

Future research may be conducted in third world countries, including Sri Lanka and Asian countries, to find the degree of linkages among WPS, OP and trust in other sector with participants from private and public organizations and along with other variables and others moderators and mediators by qualitative and quantitative methods and to check the impact of overall organization performance and for generalization, More studies are required to access the impact of human–robot cooperative industrial systems on human psychological implications which is unclear, since new technologies have psychological effects on their users in line with global environment and Industry 5.0, human centric approach objective, with increased sample sizes impact is different and may be dependent on national culture, subculture, and organizational culture.

KEYWORDS

- **workplace spirituality**
- **trust**
- **organizational performance**
- **human resource management**
- **Industry 5.0**

REFERENCES

Afsar, B., Rehman, M. (2015). The relationship between workplace spirituality and innovative work behavior: The mediating role of perceived person–organization fit. *Journal of Management Spirituality & Religion*, *12*(4), 329–353. https://doi.org/10.1080/14766086.2015.1060515

Asgari, N., Heidari, H. (2015). Introducing a model of influencing factors of customer's trust and satisfaction in E-commerce area (Case study: Group discount sites in Iran). *Journal of Information Technology Management*, *7*(3), 655–674.

Ashfaq, M., Irum, S. (2020). Workplace spirituality paradigm and job stress. *PalArch's Journal of Archaeology of Egypt/Egyptology*, *17*(9), 2667–2673.

Ashmos, D. P., Duchon, D. (2000). Spirituality at work: A conceptualization and measure. *Journal of Management Inquiry*, *9*(2), 134–145.

Ayoko, O. B., Pekerti, A. A. (2008). The mediating and moderating effects of conflict and communication openness on workplace trust. *International Journal of Conflict Management*, *19*(4), 297–318. https://doi.org/10.1108/10444060810909275

Bharadwaj, S. H. U. B. H. A. N. G. I., Jamal, T. (2020). In search of spiritual workplaces: An empirical evidence of workplace spirituality and employee performance in the Indian IT industry. *International Journal of Scientific & Technology Research*, *9*(3), 1116–1124.

Bird, J. J., Wang, C., Watson, J., Murray, L. (2012). Teacher and principal perceptions of authentic leadership: Implications for trust, engagement, and intention to return. *Journal of School Leadership, 22*(3), 425–461. https://doi.org/10.1177/105268461202200302

Bock, G.-W., Zmud, R. W., Kim, Y.-G., Lee, J.-N. (2005). Behavioral intention formation in knowledge sharing: Examining the roles of extrinsic motivators, social-psychological forces, and organizational climate. *MIS Quarterly: Management Information Systems, 29*(1), 87. https://doi.org/10.2307/25148669

Breevaart, K., Bakker, A., Hetland, J., Demerouti, E., Olsen, O. K., Espevik, R. (2014). Daily transactional and transformational leadership and daily employee engagement. *Journal of Occupational and Organizational Psychology, 87*(1), 138–157.

Conțu, E. G. (2020). Organizational performance-theoretical and practical approaches; study on students' perceptions. *Proceedings of the International Conference on Business Excellence, 14*, 398–406.

Cook, J., Wall, T. (1980). New work attitude measures of trust, organizational commitment and personal need non-fulfilment. *Journal of Occupational Psychology, 53*(1), 39–52. https://doi.org/10.1111/j.2044-8325.1980.tb00005.x

Daniel, J. L. (2019). Trust as a mediator of workplace spirituality and job performance. *Journal of Business and Behavioral Sciences, 31*(2), 80–103.

Daniel, L. (2010). The effect of workplace spirituality on team effectiveness. *Journal of Management Development, 29*(5), 442–456.

de Waal, A. A. (2007). The characteristics of a high performance organization. *Business Strategy Series, 8*(3), 179–185. https://doi.org/10.1108/17515630710684178

Diener, E., Wirtz, D., Tov, W., Kim-Prieto, C., Choi, D.-W., Oishi, S., Biswas-Diener, R. (2010). New well-being measures: Short scales to assess flourishing and positive and negative feelings. *Social Indicators Research, 97*(2), 143–156. https://doi.org/10.1007/s11205-009-9493-y

Diffie-Couch, P. (1984). Building a feeling of trust in the company. *Supervisory Management, 29*(4), 31–36.

Dirks, K. T., Ferrin, D. L. (2002). Trust in leadership: Meta-analytic findings and implications for research and practice. *The Journal of Applied Psychology, 87*(4), 611–628. https://doi.org/10.1037/0021-9010.87.4.611

Dobre, O. I. (2013). Employee motivation and organizational performance. *Review of Applied Socio-Economic Research, 5*(1).

Garcia-Zamor, J.-C. (2003). Workplace spirituality and organizational performance. *Public Administration Review, 63*(3), 355–363. https://doi.org/10.1111/1540-6210.00295

Garg, N. (2020). Promoting organizational performance in Indian insurance industry: The roles of workplace spirituality and organizational citizenship behaviour. *Global Business Review, 21*(3), 834–849. https://doi.org/10.1177/0972150918778983

Garg, P., Rastogi, R. (2006). New model of job design: Motivating employees' performance. *Journal of Management Development, 25*(6), 572–587. https://doi.org/10.1108/02621710610670137

Hakan, K. U. R. U. (2023). Understanding employee wellness in industry 5.0: A systematic review. *Journal of Ekonomi, 5*(1), 32–35.

Hashim, R., Bakar, A., Noh, I., Mahyudin, H. A. (2020). Employees' job satisfaction and performance through working from home during the pandemic lockdown. *Environment-Behaviour Proceedings Journal, 5*(15), 461–467.

Hassan, M., Bin Nadeem, A., Akhter, A. (2016). Impact of workplace spirituality on job satisfaction: Mediating effect of trust. *Cogent Business & Management*, *3*(1), 1189808. https://doi.org/10.1080/23311975.2016.1189808

How the Coronavirus Outbreak Has—and Hasn't—Changed the Way Americans Work. (2020). Pew Research Report.

Ikumapayi, O. M., Afolalu, S. A., Ogedengbe, T. S., Kazeem, R. A., Akinlabi, E. T. (2023). Human-robot co-working improvement via revolutionary automation and robotic technologies–An overview. *Procedia Computer Science*, *217*, 1345–1353.

Jahangir, N., Akbar, M. M., Haq, M. (2004). Organizational citizenship behavior: Its nature and antecedents. *BRAC University Journal*, *1*(2), 75–85.

Jenatabadi, H. S. (2015). An overview of organizational performance index: Definitions and measurements. Available at SSRN 2599439.

Jurkiewicz, C. L., Giacalone, R. A. (2004). A values framework for measuring the impact of workplace spirituality on organizational performance. *Journal of Business Ethics*, *49*, 129–142.

Khan, J., Usman, M., Saeed, I., Ali, A., Nisar, H. G. (2022). Does workplace spirituality influence knowledge-sharing behavior and work engagement in work? Trust as a mediator. *Management Science Letters*, *12*(1), 51–66. https://doi.org/10.5267/j.msl.2021.8.001

Khari, C., Sinha, S. (2018). Organizational spirituality and knowledge sharing: A model of multiple mediation. *Global Journal of Flexible Systems Management*, *19*(4), 337–348. https://doi.org/10.1007/s40171-018-0197-5

Khari, C., Sinha, S. (2021). Workplace spirituality: An antidote to employee burnout. In: *The Palgrave Handbook of Workplace Well-Being* (pp. 671–693). Springer International Publishing.

Kinjerski, V., Skrypnek, B. J. (2006). Measuring the intangible: Development of the spirit at work scale. In: *Academy of management proceedings* (Vol. 2006, No. 1, pp. A1–A6). Academy of Management: Briarcliff Manor, NY, USA.

Kolodinsky, R. W., Giacalone, R. A., Jurkiewicz, C. L. (2008). Workplace values and outcomes: Exploring personal, organizational, and interactive workplace spirituality. *Journal of Business Ethics*, *81*(2), 465–480. https://doi.org/10.1007/s10551-007-9507-0

Leng, J., Sha, W., Wang, B., Zheng, P., Zhuang, C., Liu, Q., Wuest, T., Mourtzis, D., Wang, L. (2022). Industry 5.0: Prospect and retrospect. *Journal of Manufacturing Systems*, *65*, 279–295.

Lloyd, T. (1990). *The Nice Company*. Bloomsbury: London.

Marques, J., Dhiman, S., King, R. (2005). Spirituality in the workplace: Developing an integral model and a comprehensive definition. *Journal of American Academy of Business*, *7*(1), 81–91.

Matzler, K., Renzl, B., Müller, J., Herting, S., Mooradian, T. A. (2008). Personality traits and knowledge sharing. *Journal Of Economic Psychology*, *29*(3), 301–313. https://doi.org/10.1016/j.joep.2007.06.004

Mayer, R. C., Gavin, M. B. (2005). Trust in management and performance: Who minds the shop while the employees watch the boss. *Academy of Management Journal*, *48*(5), 874–888.

Milliman, J., Czaplewski, A. J., Ferguson, J. (2003). Workplace spirituality and employee work attitudes: An exploratory empirical assessment. *Journal of Organizational Change Management*, *16*(4), 426–447.

Mott, P. E. (1972). *The Characteristics of Effective Organizations*. Harper & Row: New York.

Neck, C. P., Milliman, J. F. (1994). Thought self-leadership: Finding spiritual fulfilment in organizational life. *Journal of Managerial Psychology*, *9*(6), 9–16.

Nedkovski, V., Guerci, M., De Battisti, F., Siletti, E. (2017). Organizational ethical climates and employee's trust in colleagues, the supervisor, and the organization. *Journal of Business Research*, *71*, 19–26. https://doi.org/10.1016/j.jbusres.2016.11.004

Nyhan, R. C., Marlowe, H. A. (1997). Development and psychometric properties of the organizational trust inventory. *Evaluation Review*, *21*(5), 614–635.

Pandey, A., Gupta, R. K., Arora, A. P. (2009). Spiritual climate of business organizations and its impact on customers' experience. *Journal of Business Ethics*, *88*(2), 313–332. https://doi.org/10.1007/s10551-008-9965-z

Petchsawang, P., Duchon, D. (2009). Measuring workplace spirituality in an Asian context. *Human Resource Development International*, *12*(4), 459–468. https://doi.org/10.1080/13678860903135912

Peterson, W., Gijsbers, G., Wilks, M. (2003). An organizational performance assessment system for agricultural research organizations: Concepts, methods, and procedures. *ISNAR Research Management Guidelines*.

Podsakoff, P. M., MacKenzie, S. B. (2014). Impact of organizational citizenship behavior on organizational performance: A review and suggestions for future research. *Organizational Citizenship Behavior and Contextual Performance*, *10*(2), 133–151.

Ranasinghe, V. R. (2021). Impact of workplace spirituality on employee organizational citizenship behaviour and employee engagement: Mediating effect of trust. *International Journal on Global Business Management & Research*, *10*(2), 9–20.

Richard, P. J., Devinney, T. M., Yip, G. S., Johnson, G. (2009). Measuring organizational performance: Towards methodological best practice. *Journal of Management*, *35*(3), 718–804.

Robinson, S. L., Rousseau, D. M. (1994). Violating the psychological contract: Not the exception but the norm. *Journal of Organizational Behavior*, *15*(3), 245–259. https://doi.org/10.1002/job.4030150306

Saharan, T., Sharma, H. (2022). Intellectual impact of spiritual wellbeing, self-determination, and employee outlook on industry 5.0. In: *Handbook of Research on Innovative Management Using AI in Industry 5.0* (pp. 118–135). IGI Global.

Schaufeli, W. B., Bakker, A. B., Salanova, M. (2006). The measurement of work engagement with a short questionnaire: A cross-national study. *Educational and Psychological Measurement*, *66*(4), 701–716.

Sökmen, A., Yasrebdoost, H. (2022). Comparing the mediation role of organizational trust on the effect of workplace spirituality on organizational citizenship behavior. *European Online Journal of Natural and Social Sciences*, *11*(1).

Soliman, M., Di Virgilio, F., Figueiredo, R., Sousa, M. J. (2021). The impact of workplace spirituality on lecturers' attitudes in tourism and hospitality higher education institutions. *Tourism Management Perspectives*, *38*(100826), 100826. https://doi.org/10.1016/j.tmp.2021.100826

Van der Walt, F. (2018). Workplace spirituality, work engagement and thriving at work. *SA Journal of Industrial Psychology*, *44*(1), 1–10.

van der Walt, F., de Klerk, J. J. (2014). Workplace spirituality and job satisfaction. *International Review of Psychiatry (Abingdon, England)*, *26*(3), 379–389. https://doi.org/10.3109/09540261.2014.908826

Vasconcelos, A. F. (2015). The spiritually-based organization: A theoretical review and its potential role in the third millennium. *Cadernos EBAPE BR*, *13*(1), 183–205. https://doi.org/10.1590/1679-395110386

Weiss, D. J., Dawis, R. V., England, G. W. (1967). Manual for the Minnesota satisfaction questionnaire. *Minnesota Studies in Vocational Rehabilitation*, *22*, 120.

Xu, X., Lu, Y., Vogel-Heuser, B., Wang, L. (2021). Industry 4.0 and industry 5.0—Inception, conception and perception. *Journal of Manufacturing Systems*, *61*, 530–535. https://doi.org/10.1016/j.jmsy.2021.10.006

Yin, E., Mahrous, A. (2022). Covid-19 global pandemic, workplace spirituality and the rise of spirituality-driven organisations in the post-digital era. *Journal of Humanities and Applied Social Sciences*, *4*(2), 79–93. https://doi.org/10.1108/jhass-11-2021-0177

Zand, D. E. (1972). Trust and managerial problem solving. *Administrative Science Quarterly*, *17*(2), 229. https://doi.org/10.2307/2393957

Index

K

L

M

For Product Safety Concerns and Information please contact our EU representative GPSR@taylorandfrancis.com Taylor & Francis Verlag GmbH, Kaufingerstraße 24, 80331 München, Germany

Batch number: 10725935

Printed by Printforce, the Netherlands